Wireless Sensor Networks

For a listing of recent titles in the *Artech House Microelectromechanical Systems (MEMS) Series*, turn to the back of this book

Wireless Sensor Networks

Nirupama Bulusu
Sanjay Jha

Editors

ARTECH
HOUSE

BOSTON | LONDON
artechhouse.com

Library of Congress Cataloging-in-Publication Data
Bulusu, Nirupama.
    Wireless sensor networks/Nirupama Bulusu, Sanjay Jha.
        p. cm. —(Artech House MEMS series)
    Includes bibliographical references and index.
    ISBN 1-58053-867-3 (alk. paper)
    1. Sensor networks. 2. Wireless LANs. I. Jha, Sanjay. II. Title. III. Series.
    TK 7872.D48B85 2005
    681′2.—dc22

                                                            2005043648

British Library Cataloguing in Publication Data
Bulusu, Nirupama
    Wireless sensor networks. —(Artech House MEMS series)
    1. Sensor networks 2. Wireless communication systems.
    I. Title III. Jha, Sanjay
    681.2

    ISBN-10: 1-58053-867-3

Cover design by Igor Valdman

International Standard Book Number: 1-58053-867-3

10 9 8 7 6 5 4 3 2 1

Contents

CHAPTER 6
Energy Conservation in Sensor Networks at the Link and Network Layers   75
John Heidemann and Wei Ye *(USC Information Sciences Institute)*

CHAPTER 7
Multihop Routing   87
Nirupama Bulusu *(Portland State University)*

CHAPTER 8
Reliable Transmission and Congestion Control   99
Özgür B. Akan *(Middle East Technical University, Turkey)* and Mehmet C. Vuran
(Georgia Institute of Technology)

CHAPTER 15

Sensor-Network Security, Privacy, and Fault Tolerance   215

Jing Deng, Richard Han, Shivkant Mishra (University of Colorado at Boulder)

CHAPTER 16

Habitat Monitoring with ZebraNet: Design and Experiences   235

P. Zhang, C. M. Sadler, T. Liu, I. Fischoff, M. Martonosi, S. A. Lyon,
D. I. Rubenstein (Princeton University)

CHAPTER 17

Sensor Webs in the Wild   259
Kevin A. Delin *(NASA Jet Populsion Laboratories)*

CHAPTER 18

Defense Systems: Self Healing Land Mines   273
William. M. Merrill, Lewis Girod, Brian Schiffer, Dustin McIntire, Guillaume Rava, Katayoun
Sohrabi, Fredric Newberg, Jeremy Elson, and William Kaiser *(Sensoria Corporation)*

CHAPTER 19

Workplace Applications for Sensor Networks   289
W. Steven Conner *(Intel Research and Development)*, John Heidemann
(USC Information Sciences Institute), Lakshman Krishnamurthy *(Intel Research
and Development)*, Xi Wang *(USC Information Sciences Institute)*, and
Mark Yarvis *(Intel Research and Development)*

Preface

Wireless sensor networks are one of the first real world examples of pervasive computing, the notion that small, smart, and cheap sensing and computing devices will eventually permeate the environment. Though the technology is still in its early days, the range of potential applications is mind-boggling. Tracking bushfires, microclimates, monitoring zebras, defense systems, letting businesses monitor and control their work spaces, and industries their manufacturing processes.

They present a range of computer systems challenges because they are closely coupled to the physical world with all its unpredictable variation, noise, and asynchrony; they involve many energy-constrained, resource-limited devices operating in concert; they must be largely self-organizing, self-maintaining and robust despite significant noise, loss, and failure.

The field of sensor networks contains a wealth of interesting and inter-related problems and clever solutions founded on solid underlying principles and paradigms. The motivation for this edited book is to identify and discuss these problems as well as principles. Each problem topic covered in the book contains contributions from leading international researchers, several of who initiated research on the problem they discuss.

Our intent is that the book should serve as the text for an introductory sensor networking class, at either the graduate or upper-division undergraduate level. We also believe that the book's focus on core concepts should be appealing to researchers and industry professionals who are interested in the opportunities and challenges arising from this emerging technology regime.

Approach

The design of sensor networks is an unavoidably multidisciplinary challenge. For example, there is a rich body of collaborative signal processing literature, which addresses the challenges of data fusion in a sensor network. Our focus however is on the building blocks required for the bootstrapping and autonomous operation of a sensor network. This first-of-its-kind book covers several fundamental building blocks for sensor networks in substantial depth from a systems perspective. Accordingly, our contributing authors cover each building block from first principles, clearly articulating the problem motivation, the design tradeoffs and rationale for solutions, and their workings and evaluation.

To make the underlying concepts more tangible, the book also contains invited contributions on four application case studies, which cover some of the earliest wireless sensor network deployments. Each case study focuses on a particular application domain, its implications for the sensor network architecture, and the lessons learnt from the design and implementation of the sensor network application. The basic strategy is to not only provide you with an understanding of the components of a sensor network, but also a feel of how all the pieces are selected and put together, depending on the application requirements.

Organization and Use

This book is organized into four parts.

- Part I introduces the set of core ideas that are used throughout the rest of the text. It overviews and categorizes the technological trends, leading applications, state-of-the art platform developments, future trends for sensor networks and their implications. In particular, it discusses what goes into a sensor network and the building blocks required.

- Part II surveys a wide range of building blocks for low-level sensor network self-organization and operation. It covers mechanisms for calibration, localization, and time-synchronization of sensor devices, energy conservation and harvesting to improve the lifetime of a sensor network, reliable multi-hop routing to gather sensor data, sensor coordinated actuation and deployment and tomography of large-scale sensor networks.

- Part III addresses middleware, i.e., high-level building blocks to support sensor network applications such as data-centric communications, storage, whole-network programming and security, privacy and fault tolerance.

- Part IV focuses on application case studies. Using examples from the application domains of habitat monitoring, defense and office environments, it shows how specific applications influence the architectural design of embedded-networked-systems.

This book can be used as a textbook or supplementary reading for upper-level undergraduate or graduate level courses on sensor networks, or for self study by researchers and engineering professionals. For an undergraduate course, the introductory material should be paced over several lectures to help students digest the material, probably at the expense of more advanced case studies.

For those of you using the book in self-study, we believe that the topics we have selected cover the core of networked sensor systems, and so we recommend that the book chapters be read sequentially. In addition, each chapter includes several additional references to help you locate supplementary material that is relevant to your specific area of interest.

There are several other things to note about the topic selection and organization. First, the book takes a unique approach to the topics of *energy conservation* and *security, privacy and fault-tolerance* by pulling all topics related to energy conservation or security together in one place – Chapters 6 and Chapter 15 respectively. They do this because the problems of energy conservation or security cannot be solved at any one level, and we want you to consider the various design options at the same time.

Second, even though Chapter 12 is the only chapter explicitly labeled *Enabling Data and Event-Centric Communications*, the topic is covered in substantial depth throughout the book. For example, Chapter 7 discusses the implications of data-centric communications on network routing, and Chapters 13 and 14 show how data-centric communications can be leveraged for storage and programming of the sensor network, respectively.

Acknowledgments

This book would not have been possible without the expertise and commitment of our contributing authors Özgür B. Akan, Tatiana Bokareva, W. Steven Conner, Kevin A. Delin, Jing Deng, Jeremy Elson, Saurabh Ganeriwal, Deepak Ganesan, Lewis Girod, Ramesh Govindan, Benjamin Greenstein, Richard Han, John Heidemann, Wendi Heinzelman, Andrew Howard, Aman Kansal, William Kaiser, Lakshman Krishnamurthy, Sam Madden, Dustin McIntire, William M. Merrill, Shivkant Mishra, Amy Murphy, Fredric Newberg, Mark Perillo, Sameera Poduri, Guillaume Rava, Brian Schiffer, Katayoun Sohrabi, Mani Srivastava, Gaurav Sukhatme, Mehmet C. Vuran, Xi Wang, and Matt Welsh, Mark Yarvis, Wei Ye, and Jerry Zhao. We are immensely indebted to them.

Several people reviewed drafts or otherwise provided input to various chapters. In particular, we would like to thank Deborah Estrin of the University of California at Los Angeles, Shana Jacob of Crossbow, Kay Romer of ETH Zurich, Joe Polastre of the University of California at Berkeley, Lewis Girod of the University of California at Los Angeles, Benjamin Greenstein of the University of California at Los Angeles, Jim Snow of Portland State University, Tatiana Bokareva of the University of New South Wales, Adrian Perrig of Carnegie Mellon University, Umesh Shankar of the University of California at Berkeley, Brian Anderson of the Australian National University, Todd Leen of the Oregon Health and Sciences University, Wu-chi Feng of Portland State University, Antonio Baptista of the Oregon Health and Sciences University, Pubudu Pathirana of Deakin University, Chun-tung Chou of the University of New South Wales, Alec Woo of Microsoft Research, Bhaskar Krishnamachari of the University of Southern California, Wen Hu of the University of New South Wales and Vladimir Bychkovsky of the Massachusetts Institute of Technology. Also, a special thanks to the graduate students at Portland State University for their feedback based on class-testing a draft version of the book.

We want to thank the National Science Foundation and the Defense Advanced Research Projects Agency in the USA, and the Australian Research Council for supporting our research and writing over the past several years.

We were fortunate to have several excellent people at Artech House carefully guide us through the myriad details of the book-writing process, especially Barbara Lovenvirth and Kate Remelt. We thank them for their hard work and patience.

Finally, we wish to thank the members of our families for their unconditional love and unwavering support.

Nirupama Bulusu
Sanjay Jha

CHAPTER 1

Introduction to Wireless Sensor Networks

Nirupama Bulusu

Wireless sensor networks (WSNs) are one of the first real-world examples of pervasive computing, the notion that small, smart, and cheap sensing and computing devices will eventually permeate the environment. Sensor networks combine distributed sensing, computation, and wireless communication. Sensor networks are touted as being as disruptive [1] and enabling a technology such as the Internet, with broad applications such as tracking bushfires and microclimates, monitoring zebras, conducting military surveillance, letting businesses monitor and control their work spaces, monitoring public exposure to contaminants, managing land use, and supporting safer structures. Why are WSNs generating so much attention now? How soon might we be able to use these applications? What are the systems challenges to widespread deployment of sensor networks? This chapter explores some of these issues and provides a foundation for the rest of the book.

1.1 The Push: Technological Advances

WSNs have garnered significant media attention in recent years. In 2003, MIT's *Technology Review* magazine [2] called sensor networks "One of the ten technologies that will change the world." According to a report released by ON World, a wireless research firm, more than half a billion nodes will ship for wireless sensor applications in 2010 for an end-user market worth at least $7 billion [3].

Wireless sensors have been available for decades. Until recently, however, the development and adoption of the technology has been hampered due to the cost and expense of the sensors. What are the recent technological advances driving WSNs to fruition? The first driving factor is the prolonged exponential growth in the underlying semiconductor technology. The number of transistors on a cost-effective chip doubles every year or two, following Moore's law [4]. Consequently, a given computing capacity becomes exponentially smaller physically and cheaper with each passing year. This scaling is predicted to continue for at least another 10 to 20 years before it eventually reaches some fundamental technical and economic limit [5]. The same semiconductor manufacturing techniques driving this miniaturization can also be used to build extremely small radios, as well as mechanical structures that sense fields and forces in the physical world.

The second driving factor is the miniaturization of energy capacity. Over the past 20 years, a AA nickel alkaline battery's capacity has risen from 0.4 to 1.2 Ah with fast recharging [6]. Moreover, the power consumption of a circuit is very strongly related to its performance capacity. A circuit can almost always be designed to require less energy to complete a task if given more time to complete it. Recently, a set of techniques has emerged to dynamically control the performance to minimize the power used [7].

The final driving factor is system-on-a-chip (SoC) integration technology. Microsensors, onboard processing, and wireless interfaces can be integrated at a very small scale and at relatively low power. In 5 to 10 years, "complete systems with computing, storage, communication, sensing, and energy storage could be as small as a cubic millimeter leading to very cheap, small form factor sensor devices"[8–10].

There lies the potential for sensor networks. Sensor networks link the information technologies, the wireless technologies that have revolutionized communications, and the sensor technologies that have revolutionized medicine and industrial process control. Sensor devices can be deeply embedded and densely deployed to enable up-close monitoring of a wide range of physical phenomena. This enables spatially and temporally dense environmental monitoring. The amplitude of most physical signals attenuates sharply with distance. Deeply embedded sensor networks can therefore "reveal phenomena that were previously unobservable" [11]. In the 1990s, the Internet forever transformed the way in which individuals and organizations interact with each other. Sensor networks promise to forever transform the way in which we interact with our physical world.

1.2   The Pull: Applications

It is envisioned that large-scale, distributed sensor networks will eventually cover and instrument the entire planet. They will continuously monitor and collect information on diverse phenomena, including endangered species, soil and air contaminants, patients, and man-made environments. Across this wide range of applications, sensor networks can help us understand and manage an increasingly interconnected physical world. According to D. Culler, D. Estrin, and M. Srivastava [12], sensor applications can be classified by the following functions:

- Monitoring space;
- Monitoring objects;
- Monitoring interactions of objects with each other and a space.

The first category includes environmental and habitat monitoring, precision agriculture, indoor climate control, military surveillance, treaty verification, and intelligent alarms. The second includes structural monitoring, ecophysiology, condition-based equipment maintenance, medical diagnostics, and urban terrain mapping. The most dramatic applications involve monitoring complex interactions (including wildlife habitats), disaster management, emergency response, ubiquitous computing environments, asset tracking, health care, and manufacturing process flow.

1.2.1 Monitoring Space

Ecological phenomena such as biocomplexity, the carbon cycle, climate change, and harmful algae cannot currently be observed at adequate spatial and temporal resolution. WSN technology can be used for improved modeling and forecasting of public health risks, as well as for real-time monitoring of existing exposure. For example, the University of California, Los Angeles (UCLA), Center for Embedded Network Sensing (CENS) [13] has an ongoing research project that involves monitoring the use of nitrate-laden, treated water for secondary agriculture. According to the prevailing wisdom, the impact of crop growth and the microbes in soil will together so diminish nitrate concentrations that by the time they reach the ground water, the nitrates will be at safe levels. However, the public health concern is that the nitrate level might actually contaminate the ground water. This is a very difficult process to verify because the soil is so heterogeneous that sampling at any one position is not necessarily representative. CENS is deploying a dense three-dimensional grid of small sensor devices that enables in situ observations of such important processes.

1.2.2 Monitoring Objects

Much monitoring of objects falls under the umbrella of condition-based maintenance. A physical structure, such as a machine, motor, or a building, has typical modes of vibration, acoustic emissions, and response to stimuli. Variations in these behaviors indicate wear, fatigue, or other mechanical changes; thus, monitoring them provides a useful indicator of equipment health. For example, a modern semiconductor fabrication plant can have several thousand vibration sensors attached to various pieces of routine machinery [14]. A team of electricians tours the plant every few months with a computing device that attaches to a sensor and logs a sample for a short period. The team then carries these logs back to a central computer, which analyzes them for signs of wear. Researchers at Intel's Heterogeneous Sensor Networks Group are investigating an alternative approach to maintenance using sensor networks [14], which reduces the infrastructure cost for each plant by a factor of eight or more. Each sensor device processes vibration data locally and transports it continuously to operations staff, improving the reliability of detecting structural faults.

1.3 Systems Challenges

The biggest impediment to widespread deployment and adoption of sensor networks comes from the systems challenges ahead of us [11, 15], including the following:

- *Immense scale.* To achieve dense instrumentation of complex physical systems with extremely small sensor devices, applications will deploy them in large numbers. Because of the limited capacity of individual sensor devices, the sensor-network application will derive its fidelity and availability from the quantity of partially redundant measurements and their correlation, rather than from the quality and precision of individual components.

- *Limited access.* Sensor devices will be embedded in environments that are either difficult or expensive to connect with wires. For example, wireless sensors are especially beneficial in industrial and commercial building settings where they can reduce the cost of wiring by up to 80%. As a consequence, they will be largely untethered, unattended, and resource constrained. Nodes will have to leverage wireless communication and rely on onboard and harvested energy sources such as batteries and solar cells. Inaccessibility, as well as the sheer number of devices per human, implies that they must operate without human attendance. At sufficient levels of efficiency, energy harvested from the environment can potentially allow arbitrary lifetimes, but the available energy bounds the amount of activity permitted per unit time.

- *Extreme dynamics.* Because sensor networks are closely coupled to the physical world with all its unpredictable variation, noise, and asynchrony, they will experience extreme dynamics. Environmental changes can often directly affect their performance. For example, environmental factors dramatically influence propagation characteristics of low-power radio frequency (RF) and can effectively cause time-varying wireless connectivity in the network. Sensor devices also experience extreme variation in demand. Most of the time, they observe no relevant events but must maintain passive vigilance, while consuming almost no power. However, when important events do occur, several sensor devices observe them concurrently. Flows of high- and low-level data among sensors and actuators must be efficiently interleaved, while meeting real-time demand and avoiding the transmission of redundant information. To cope with resource limitations in the presence of such dynamics, the sensor devices must continuously adapt their individual and joint behavior to resource and stimulus availability.

In order to operate autonomously for long periods of time in dynamic, heterogeneous environments, sensor networks must be adaptive and self-configuring. In order to scale to large numbers of devices, sensor networks must process the sensor data close to where the data are observed and thereby filter out the interesting events from the uninteresting ones. To meet these challenges, we require new frameworks for system design that are informed by direct, hands-on experience with the emerging technological regime.

1.4 Systems Taxonomy

As we discussed in Section 1.2, sensor-network applications are as varied as the physical environments in which we live and work. Yet, even with this heterogeneity, many opportunities and resources for exploiting commonality across them exist. One step toward identifying common building blocks is to define the systems taxonomy and applications so that we can identify and develop abstractions and features that can be parameterized and reused. D. Estrin et al. [15] divide the space of physically embedded systems according to their scale, variability, and autonomy along the critical dimensions of space and time. These dimensions, which we discuss below, are characterized by both the environmental stimuli being captured and the system elements.

1.4.1 Scale

Spatial and temporal scale concerns the sampling interval, the extent of overall system coverage, and the relative number of sensor nodes to input stimuli, and it is an important determinant of system design.

- *Sampling.* The physical phenomenon measured dictates the spatial and temporal sampling scale of the sensor network. High-frequency waves such as vibrations and acoustics require higher temporal and spatial sampling than phenomena (100 hertz or higher) such as temperature or light in a room, where spatial and temporal fluctuations are coarser grained. The sampling scale is also determined by the application. For example, if a structure is being monitored to detect structural faults, signatures of seismic response might be compared to "healthy" signatures at a relatively coarse grain. However, if data is being collected to generate profiles of structure response, fine-grained data is needed. Finer-grained sampling often requires innovative, collaborative signal-processing techniques [16–19].

- *Extent.* The spatial and temporal extent of systems also varies widely. At the high end of this spectrum are environment-monitoring systems such as a tsunami warning system, which can span on the order of tens of thousands of meters. Most existing sensor networks are several orders of magnitude smaller, such as those needed to cover a building or room. An example at the smaller end of the spectrum is the monitoring of a machinery surface. Systems intended to operate over extended periods of time and regions of space must be self-organizing because, as the system extent grows, it is impossible to manually configure and control the environment.

- *Density.* System density is a measure of sensor nodes per footprint of input stimuli. Higher-density systems provide greater opportunities for exploiting redundancy to eliminate noise and extend system lifetime. A high density of nodes to stimuli enables a large number of independent measurements and provides opportunities to combine these measurements to eliminate channel or coupling noise. Similarly, where density is high enough to allow oversampling, nodes can go to sleep for long periods and extend coverage over time.

1.4.2 Variability

Variability is a second differentiating characteristic of many systems. Relatively static systems can be optimized at design time. More variable systems must use runtime self-organization and might be fundamentally limited in the extent to which they are both variable and long-lived. Variability includes variability in system structure, task, and space (mobility).

- *Structure.* An engineered system that monitors a structure such as a building, bridge, or even an individual airplane has a highly static structure. At the other end are ad hoc sensor networks where sensor devices are randomly scattered in remote regions to study biocomplexity.

- *Task.* Variability in system task determines the extent to which we can optimize the system for a single mode of operation. Even a structurally static system might perform different tasks over time—for example, a structural system that periodically generates a structure profile could act as an event-monitoring system.

- *Space*. In many sensor networks, most or all of the nodes remain fixed in space once placed. However, many interesting, if longer-term, systems include mobile or portable sensing devices, such as sensors attached to vehicles or people. Moreover, the phenomena these systems monitor differ in the extent of mobility. A sensor network designed to manage the physical environment for a stationary human user faces different challenges than one designed to track humans moving quickly through that same space.

1.4.3 Autonomy

The degree of autonomy has some of the most significant and varied long-term consequences for system design. Greater autonomy increases the need for (1) multiple sensor modalities, and (2) translation between external requests and internal processing and the internal computation complexity.

- *Modalities*. Truly autonomous systems depend on multiple sensor modalities for robustness. Different modalities provide noise resilience to one another and can combine to eliminate noise and identify anomalous measurements.

- *Complexity*. Greater system autonomy also requires greater computational complexity. Detailed characterization systems are relatively low-autonomy systems because their intent is simply to deliver sensory information to a human user or external program. Event detection requires far more autonomy because the definition of interesting events must be programmed into the system, and the system must execute more complex queries and computations internally to process detailed measurements and identify events.

1.5 State of the Art

As we have discussed so far, WSNs merge a wide range of information and communication technology research spanning hardware, systems software, sources of energy, networking, and programming methodologies. This section describes state-of-the-art developments and trends that are significant in realizing the vision of pervasive sensor networks.

1.5.1 Sensing and Actuation: Microelectro Mechanical Systems

Interfacing with the physical world takes two forms: *sensing* and *actuation*. Whatever the sensed phenomena (temperature, light intensity), the sensor transduces a particular form of energy (heat, light) into information. Actuation lets nodes convert information into action. Actuation can also be used to improve sensing. An actuator moves part of itself, relocates spatially, or moves other items in the environment. Sensing and actuation are together the means of physical interaction between the nodes and the world around them.

How do sensors work?   When subjected to a variation in environmental conditions, many materials change their electrical characteristics. Sensors are manufactured so that these changes are predictable over a certain range. For example, a thermistor is a variable resistor that changes smoothly with temperature. An analog-to-digital converter (ADC) converts the voltage drop into a binary number that a microcontroller can store or process.

Sensors and actuators have undergone a revolution with the emergence of micro-electro-mechanical systems (MEMS) technology, in which mechanical devices such as accelerometers, barometers, and movable mirrors are constructed at minute size. Researchers can use the lithographic processes for etching transistors on silicon to carve out tiny mechanical structures, such as a microscopic springboard within an open cavity [15]. Gravitational forces or acceleration can deflect this cantilevered mass, causing powerful internal forces that cause changes in material properties or delicate alignments, which can be amplified and digitized. Improved equipment and technology result in smaller MEMS devices, leading to lower power consumption and improved device performance.

During the 1990s, MEMS technology moved from the laboratories into commercial products. Manufacturers used the first major commercial MEMS sensor, the accelerometer, to trigger automotive airbag release. Whereas high-precision piezoelectric accelerometers cost hundreds of dollars, MEMS provided sufficient precision for a few dollars. Once the devices entered mass production, they could ride the CMOS technology growth of modern chips to become increasingly accurate, while remaining inexpensive. Internet packets bounce off submillimeter mirrors that switch photons between different optical fibers. MEMS technologies are successful in these applications because they provide far superior performance or the same performance at lower prices.

MEMS sensors can cheaply and efficiently sense a wide variety of physical phenomena, such as physical forces, chemical concentrations, and environmental factors. These structures consume a few milliwatts and only need to be turned on a fraction of the time. With extremely efficient ADCs, the sensor subsystem has an energy profile similar to the processor. MEMS sensors constitute a critical building block for embedded devices.

A common problem in both sensing and actuation is uncertainty. The physical world is a partially observable, dynamic system. Sensors and actuators are physical devices with inherent accuracy and precision limitations. Thus, sensor-measured data is necessarily an approximation to actual values. In a large system of distributed nodes, this implies that we need some form of filtering at each node before we can meaningfully use the data. We can also achieve increased accuracy and fault tolerance by redundancy, using sensors with overlapping fields of view. This raises interesting challenges for sensor placement and fusion, especially in the context of very large networks. In addition to uncertainty, there is the further problem of latency in actuation. For closed-loop control, stochastic latency can cause instability and unreliable behavior. The field of robotics provides the foundation for real-time embedded decision making in physical environments under uncertainty [20, 21].

1.5.2   Energy

Why is energy important?   Energy constraints dominate algorithm and system design trade-offs for small embedded sensor devices. The lifetime of a WSN depends on the energy that can be stored and harvested by individual sensor devices. Energy storage has advanced substantially but not at the same pace as silicon-based processing, storage, and sensing. The most widely used energy-storage devices are batteries. Over the past 20 years, a AA nickel alkaline (NiCd and NiMH) battery's capacity has risen from 0.4 to 1.2 Ah with fast recharging [22]. Most batteries store approximately 1 J/mm^3 of energy. Several factors influence the choice of technology for a particular application. Fuel-based alternatives potentially have a much higher energy density but are currently under active development.

The choice of an energy-storage solution depends on the application, with the trade-offs being energy density, longevity, and recharge times. Zinc-based batteries used in hearing aids have high energy density but high leakage, so they are best for high usage over short duration. Lithium batteries offer higher energy density with fewer memory effects but longer recharge times. Recent polymer-based batteries have excellent energy density, can be manufactured in a range of form factors, and are flexible, but they are also expensive. Researchers have fabricated tiny, 1-mm^3, lead-acid batteries, which can be packaged directly with logic. Fuel cells potentially have 10 times the energy density of batteries, if only the fuel is considered. However, the additional volume of the membrane, storage, and housing lowers this by a factor between nearly two to five.

The most common form of energy harvesting is solar panels. Recently, researchers have begun exploring avenues for harvesting the mechanical energy associated with specific applications, such as shoe flexing, button pushing, window vibration, or airflow in ducts [23].

Tables 1.1 and 1.2 show sample battery and scavenging energy ratings according to a 2001 study conducted by L. Doherty et al. [24].

According to K. Pister [15], with existing technology, a cubic millimeter of battery space has enough energy to perform roughly 1 billion 32-bit computations, take 100 million sensor samples, or send and receive 10 million bits of data. As all the system's layers become optimized for energy consumed per operation, these numbers will increase by at least an order of magnitude and some by several orders.

Table 1.1   Battery Energy Ratings

Battery Type	Energy Ratings (joules/cm^3)
Nonrechargeable lithium	2,880
Zinc-air	3,780
Alkaline	1,190
Rechargeable lithium	1,080
Nickel metal hydride (NiMHd)	864
Fuel cells (based on methanol)	8,900
Hydrocarbon fuels (for use in microheat engines)	10,500

Table 1.2   Scavenging Energy Ratings

Type	Energy Rating
Solar (outdoors midday)	15 mW/cm^2
Solar (indoor office lighting)	10 μW/cm^2
Vibrations (from microwave oven casing)	200 μW/cm^3
Temperature gradient: (from a 10° C temperature gradient)	15 μW/cm^3

1.5.3   Wireless Communication

For the past several years, sensors have been integrated with many appliances, vehicles, and gadgets. The key distinction in WSNs is the addition of wireless communication—enabling devices to not only sense but also to process and communicate sensor readings to other devices, thus translating the physical world into information.

Conventional CMOS technology can now be used to manufacture radio components. This has led to a proliferation of pagers, walkie-talkies, cell phones, and wireless local area networks for mobile laptops. However, the amount of energy required to communicate wirelessly increases rapidly with distance. Moreover, obstruction, such as people, walls, and other sources of interference, further attenuate the signal.

Wireless local area networks and cell phones consume hundreds of milliwatts of power and rely on a powerful infrastructure. The wireless radios used for sensor networks typically consume about 20 mW, and their range typically is measured in tens of meters.

For small devices to cover long distances, the network must route the information hop by hop through nodes, much as routers move information across the Internet. Communication is the dominant energy consumer, with each bit transmitted costing as much energy as about 1,000 instructions. Whenever possible, sensor networks process data within the network.

1.5.4   Embedded Devices

The unique aspect of the devices used in WSNs is that they integrate wireless communication, sensing, and computation. A sensor-network node's hardware consists of a microprocessor, data storage, sensors, ADCs, a data transceiver, microcontrollers, and an energy source. G. J. Pottie and W. Kaiser at the UCLA developed the first examples of such devices and coined the term *wireless integrated networked sensors* to refer to them [25].

Decreasing lithographic feature size offers two options for wireless integrated networked sensors: (1) packing more computing and storage capacity on to a chip, or (2) packing the same capacity into a progressively smaller area with an even greater reduction in power consumption. At the latter extreme, researchers have conducted design studies on the feasibility of building an entire system, including power storage, processing, sensing, and communication, within a cubic millimeter [8–10]. This is because size and performance improvements in the node's processing, storage, communications, and sensing subsystems reduce power consumpion and allow a corresponding decrease in the size and cost of the power supply as well.

Miniaturization allows simple microcontrollers to operate near 1 mW, while running at about 10 MHz. Most of the circuits can be powered off, so the standby power can be about 1 μW. If such a device is active 1% of the time, its average power consumption is just a few microwatts. This scale of energy can be harvested using solar cells, or mechanical vibrations, as discussed in Section 1.6.2. A typical cubic-centimeter battery stores about 1,000 mAh, so centimeter-scale devices can run almost indefinitely in many environments.

Research conducted at the University of California, Berkeley, produced one of the first widely used microsensor nodes, called the *mote* [26]. The design goals for the mote according to D. Culler were,

- "It must be extremely energy efficient, especially in the low duty-cycle vigilance mode, and it must be extremely facile with event bursts."
- "It must meet hard real-time constraints, such as sampling the radio signal within bit windows, while also handling asynchronous sensor events and supporting localized data processing algorithms."
- "It also must be robust and reprogrammable in the field."

Crossbow [27] currently produces the mote in volume as shown in Figure 1.1, and it is used widely for research and educational purposes. The core building block of the mote is a 1"×1.5" motherboard consisting of a low-power microcontroller, a low-power 900-MHz radio, nonvolatile memory, network programming support by Light Emitting Diodes, (LEDs), and a vertical expansion bus connector. The microcontroller contains the Flash program and SRAM data storage, an ADC, and external input/output (I/O) (standard and direct ports). A second small microcontroller lets the node reprogram itself from network data. The motherboard has sensors and actuators to monitor its own operation, such as battery-voltge sensor, radio-signal-strength sensing and control, and LED display. The microcontroller's external interface is exposed in a standardized form on the expansion connector, providing analog, digital, direct I/O, and serial bus interconnections. Sensor packs for the specific applications, including thermistors, photo detectors, accelerometers, magnetometers, and humidity, pressure, or actuator connections, are stacked like tiny PC104 boards. The processor dissipates several nanojoules per 8-bit instruction.

The Stargate Gateway, an XScale platform, and the Mote Interface Boards (MIBs), developed by Crossbow, interface motes to personal computers, the Internet, and existing wired or wireless networks and protocols.

Figure 1.1 The Crossbow MICA mote next to an MICA programming board. (*Source:* Crossbow, Inc.)

Along with the Mica mote, a number of hardware platforms have emerged in the past few years. Many of these platforms are designed for advanced sensing applications and trade off the cost and form factor of the MICA for increased sensor sampling, computation, communication, or storage capabilities. These include the MIT μAMPS node, Sensoria WINSng nodes, Ember nodes, Intel Imotes, and the TelOS motes from Moteiv. A detailed on-line hardware survey can be found in [28].

1.5.5 Operating Systems

Several distinct operating systems approaches have emerged to simplify the programming of embedded devices in the first few years.

Initially, real-time operating systems, such as Vxworks [29], GeoWorks [30], and Chorus [31], compact Linux variants, and Windows CE, scaled down their footprints and added Transport Control Protocol/Internet Protocol (TCP/IP) capabilities. PalmOS provided data exchange and synchronization with infrastructure machines but little support for the concurrency associated with interactive communication.

In 2001, J. Hill et al. [26] argued that to make the networked, embedded node an effective vehicle for developing algorithms and applications, modularity, in addition to efficiency, would be critical. A modular, structured runtime environment should provide the scheduling, device interface, networking, and resource-management primitives for the network. Moreover, it must support several concurrent flows of data from sensors to the network to controllers.

Their approach was to provide fine-grained multithreading for concurrency via a component-based tiny operating system environment called *TinyOS*. Designed initially to be an operating system for the mote, TinyOS provides a framework for dealing with extensive concurrency and fine-grained power management, while providing substantial modularity for robustness and application-specific optimization. The TinyOS framework establishes the rules for constructing reusable components that can support extensive concurrency on limited processing resources.

TinyOS is also built around a lightweight event scheduler, where all program execution is performed in tasks that run to completion. An application in TinyOS may be viewed as a graph of components (see Figure 1.2). TinyOS uses a special description language for composing a system of smaller components [32] that are statically linked with the kernel to a complete image of the system. After linking, modifying the system is not possible [33].

In order to provide runtime reprogramming for TinyOS, P. Levis and D. Culler have developed Maté [33], a virtual machine for TinyOS devices. Code for the virtual machine can be downloaded into the system at runtime. The virtual machine is specifically designed for the needs of typical sensor-network applications. Similarly, the MagnetOS [34] system uses a virtual Java machine to distribute applications across the sensor network. The advantages of using a virtual machine instead of native machine code is that the virtual machine code can be made smaller, thus reducing the energy consumption of transporting the code over the network. The drawback is the increased energy spent in interpreting the code for long running programs.

The Mantis system [35] uses a traditional preemptive multithreaded model of operation. Mantis enables reprogramming of both the entire operating system and parts of the program memory by downloading a program image onto electronically

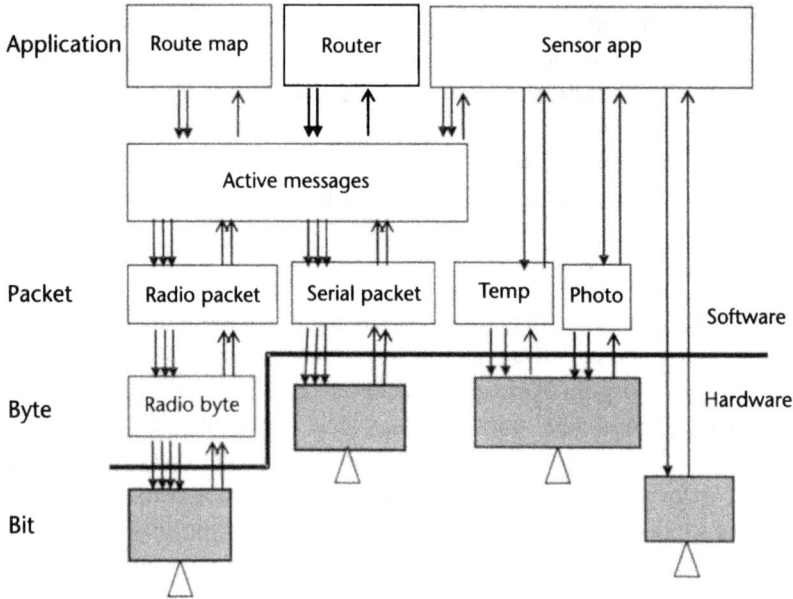

Figure 1.2 TinyOS application: A graph of components.

erasable programmable read-only memory (EEPROM), from where it can be burned into flash read only memory (ROM). Due to the multithreaded semantics, every Mantis program must have stack space allocated from the system heap, and locking mechanisms must be used to achieve mutual exclusion of shared variables.

Contiki [36] provides a dynamic structure that allows programs and drivers to be replaced during runtime and without relinking. Contiki programs use native code and can therefore be used for all types of programs, including low-level device drivers without loss of execution efficiency. In contrast, Contiki uses an event-based scheduler without preemption, thus avoiding allocation of multiple stacks and locking mechanisms. Preemptive multithreading is provided by a library that can be linked with programs that explicitly require it. The code size for the event kernel in Contiki is larger than TinyOS. Because of the flexibility it provides, compile-time code optimization cannot be performed to the extent possible with TinyOS.

SensorWare [37] provides an abstract scripting language for programming sensors, but their target platforms are not as resource constrained as the motes. Similarly, the EmStar environment [38] is designed for less resource-constrained systems.

1.5.6 Distributed System Architecture

The long-term success of the WSN technology will be heavily influence by the distributed system architecture. Miniaturization and wireless communication impose constraints that require significant changes in the overall architecture to achieve the desired functionalities.

Wireless communication is the primary consumer of finite energy [25]. Consequently, we cannot realize long-lived autonomous systems by simply streaming all the sensor data out of the nodes for processing by traditional computing elements. Instead of building a network of sensors that all output high-bandwidth bit streams, we must construct distributed systems whose outputs are at a higher semantic level—compact detection, identification, tracking, pattern matching, and so forth. To support this architecture, important systems-oriented trends are emerging. Two important pieces of this architecture are *self-configuring networks* and *data-centric systems*.

1.5.6.1   Self-Configuring Networks

Consider the task of tracking a moving object using a wireless, distributed sensor network. Before nodes can accomplish this task,

* The sensor nodes must be deployed so they sufficiently cover the area of interest.
* They must coordinate access to the wireless media to start bootstrapping.
* They must all know their positions in a consistent coordinate system. In order to track the object collaboratively, they must be synchronized in time with their neighbors.
* Any faults in sensor readings must be detected and corrected.
* Sensor nodes must form a connected network topology and establish routes to collaborate with each other and report events. Network mechanisms are required to ensure reliable reporting of important events.

The primary objective of self-configuration is to let sensor devices autonomously coordinate among themselves to ensure adequate deployment coverage, form a connected network, establish a spatial coordinate system, synchronize themselves in time, and calibrate their sensor readings. An extremely important secondary objective is to exploit the node redundancy for energy conservation and management to achieve longer, unattended system lifetimes. Techniques for coordinating and adapting node sleep schedules to support a range of trade-offs between fidelity, latency, and efficiency are also emerging [39–42]. The idea of self-configuration in the presence of wireless links and node mobility emerged from the earlier work on ad hoc routing [43], but ad hoc routing protocols generally support the traditional Internet model of shipping data from one edge of the network to another. Chapters 2 to 11 in this book explore the various issues in low-level network self-configuration.

1.5.6.2   Data-Centric Systems

Small form factor wireless sensor nodes cannot afford to ship all data to the edges for outside processing and, fortunately, do not need to operate according to the same layering restrictions as Internet networks when it comes to application-layer data processing at intermediate hops.

Directed diffusion, designed and implemented at the University of Southern California's Information Sciences Institute promotes in-network processing by building on a data-centric instead of an address-centric architecture for the distributed system or network. Using data naming as the lowest level of system organization supports flexible and efficient in-network processing [44–46]. In Directed Diffusion, sensors publish and clients subscribe to data. Both identify data by attributes such as "the southwest" or "acoustic sensors."

Directed Diffusion decouples data identity from node identity. To do this, it uses a simple typing mechanism and encoding of attribute-based naming with simple matching rules. Names are sets of attributes. Each attribute is a tuple, including keys, values, and operations. User-provided code or filters can be distributed to the sensor network to perform application-specific, in-network processing tasks such as data aggregation, caching, and collaborative signal processing [11].

Once in-network data processing is in place, we must develop programming abstractions that allow a user to program an entire sensor network instead of individual sensor nodes.

Chapters 12 to 15 will investigate the issues involved in building, using, and programming data-centric sensor networks.

Figure 1.3 shows the layers of distributed system architecture for sensor networks. This layering is not as strict as in the Internet regime. Indeed, cross-layer optimizations and abstractions will be needed to optimize the performance of the entire sensor network. Chapters 16 to 19 provide case studies of some of the earliest sensor-network deployments and show how specific applications influence the distributed system architecture.

1.6   Future Trends

In the coming few years, significant information and communication technology research and development will no doubt be needed to realize the potential of sensor networks. In particular, research must address the concerns of robustness, scaling, data integrity, and data fusion. Building on the successes of the past few years, early versions of sensor networks are being applied to several environmental, engineering,

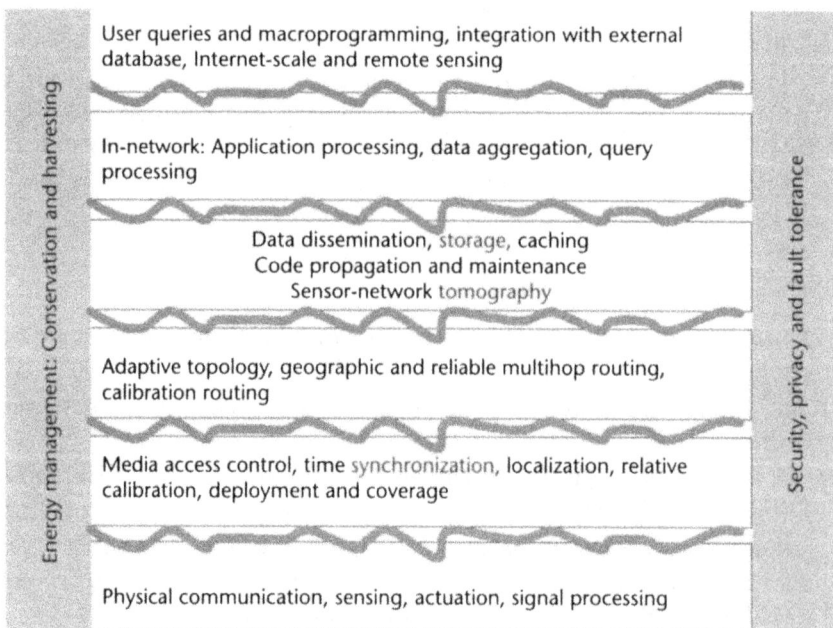

Figure 1.3   A distributed system architecture for sensor networks.

and commercial applications [47]. Let us examine some of the emerging research trends in the sensor networks regime that are informed by experience with these deployments.

Heterogeneity.   An important trend is the increased reliance on *heterogeneous, tiered architectures* in which fewer higher-end elements complement the more limited capabilities of widely and densely dispersed nodes. Very small devices will inevitably possess limited storage and computing resources as well as limited bandwidth with which to interact with the outside world. Introducing heterogeneous systems, where some system elements have greater capacity, is desirable.

In a heterogeneous architecture, the smallest system elements help achieve spatial diversity and short-range sensing, whereas the resource-rich elements implement more sophisticated and performance-intensive processing functions, such as digital signal processing, localization, and long-term storage. Moreover, these resource-rich devices can include robotic elements that traverse the sensor field, delivering energy to depleted batteries or computing localization coordinates for ad hoc collections of smaller nodes.

Another advantage of resource-rich elements is that they provide more complex sensing capabilities. For example, vision provides an important, orthogonal sensing modality to traditional sensing applications but requires greater systems complexity. Scalar sensor data can be validated via video and vice versa. Cyclops, shown in Figure 1.4, is an embedded image-capturing and interpretation module [48], which can be integrated with the Crossbow MICA motes. Panoptes, shown in Figure 1.5, is a Stargate-based video sensing platform that can interact wirelessly with networks of motes [49].

Figure 1.4 The Cyclops image capture and interpretation module. (*Source:* UCLA CENS.)

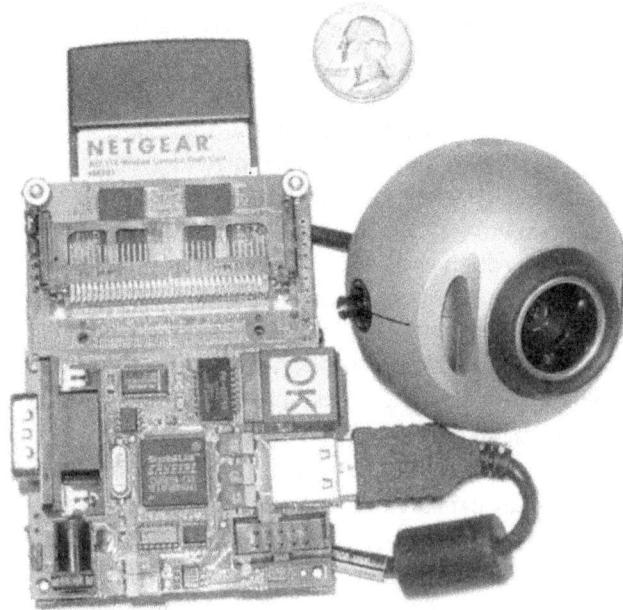

Figure 1.5   The Panoptes video sensing platform. (*Source:* Wuchi Feng.)

A final advantage of heterogeneity is mobility. WSNs will increasingly exploit motion in addition to sensing. The injection of small amounts of mobility into these observing systems can tremendously improve their sensing coverage and effectiveness. As an example, consider how the reception of a cell phone is improved with very small movements of orientation or position. Several research projects exploit robotic technology to support high-density sensing and sampling in both air and water. For example, to determine the preconditions for the development of harmful algae blooms, robotic nodes move around on the water surface and autonomously collect samples of microorganisms at both the water surface and below, which can then be correlated with sensed environmental microcondition [13].

The Robomote [50] is an example of a robotic microsensor platform with the same network interface as the microsensor nodes. It consists of a motor control board stacked on a microsensor node, and the resulting assembly is mounted on a motorized chassis with two wheels. The motor control board regulates wheel speeds and provides range information from two forward- and one rear-looking infrared emitters. While the Robomote is a resource-constrained platform for autonomous mobility, the Networked Infomechanical Systems (NIMS) platform [51] is a resource-rich, controlled-mobility platform, wherein the motion of mobile devices is controlled along cable wires, allowing rich three-dimensional sensing.

Real-time event detection.   Ad hoc networks of wireless seismometers could be deployed to hop data out in the presence of potentially massive disruptions and at challenging data rates. The technology must enable scientists to deploy large wireless grids quickly for data collection in response to important seismic or volcanic events. Ultimately, systems must be programmed with event detection inside the network so that these systems can provide real-time monitoring and alerts, in addition to their essential data-gathering function.

Dense WSNs will also present tremendous opportunities for integration with the Internet and with remote-sensing capabilities. Establishing "ground truth" for interpretation of remote-sensing images is an obvious application. However, the possibilities also include real-time adjustment of in situ assets, position, focus, attention and so on in response to more global phenomena observed via remote sensing.

Legal and social implications.   The deployment of WSNs in our environments brings forth several pressing social issues. For example, if a sensor network has been deployed to warn the public of exposure to harmful chemicals, how do we verify and validate the integrity of such a system? If a sensor network is monitoring people continuously as they work and play, how can we ensure their privacy? Finally, will sensor devices be biodegradable? What will be the impact from leaving millions of energy-depleted sensor nodes in the environment? For these problems, techniques for verifying system integrity, preserving anonymity, and harvesting energy from the environment, respectively, are expected to be enabling technologies. Nevertheless, social and legal policy will ultimately be inevitable.

The ability to form networks of devices interacting with the physical world opens broad avenues for information technology beyond the highly connected, responsive home or workspace. Sensor devices are being embedded in the civil infrastructure such as buildings, bridges, waterways, highways, and protected regions to monitor structural health and detect crucial events. Researchers are also investigating whether sensor networks can fundamentally alter the practice of numerous scientific endeavors, such as studies of complex ecosystems, by providing in situ monitoring and measurement at unprecedented levels of temporal and spatial density without disturbing the complex systems under study [47]. Projects such as the National Ecological Monitoring Network [52] aim to create a global sensor-network grid for habitat monitoring. Ultimately, high-resolution WSNs could dramatically alter our work environments, enable more efficient energy management in buildings and resource usage in agriculture, reduce public exposure to polluted air and water, provide access to safer food and shelter, allow first responders to magnify their effectiveness in reacting to natural and human disasters, and be used in many other applications whose limits lie only in our imagination.

References

[1]   Christensen, C. M., *The Innovator's Dilemma: When New Technologies Cause Great Firms to Fail*, Boston, MA: Harvard Business School Press, 1997.

[2]   MIT Technology Review at www.techreview.com.

[3]   On World at http://www.onworld.com/html/wirelesssensorsrprts.htm.

[4]   Moore, G., "Progress in Digital Integrated Electronics," *IEDM Tech. Digest*, 1975, pp. 11–13.

[5]   Borkar, S., "Design Challenges of Technology Scaling," *IEEE Micro*, Vol. 19, No. 4, 1999, pp. 23–29.

[6]   Computer Science and Telecommunication Board, *Embedded Everywhere: A Research Agenda for Networked Systems of Embedded Computers*, Washington, D.C.: National Research Council, 2001.

[7]   DARPA Power Aware Computing/Communication Program, at http://www.darpa.mil/ito/research/pacc/.

[8]   Kahn, J. M., R. H. Katz, and K. S. J. Pister, "Next Century Challenges: Mobile Networking for Smart Dust," *Proc. ACM conf. mobile and computing networking (MobiCom) 2000*, Boston, MA: August 2000 ACM Press.

[9]   Pister, K. S. J., and B. E. Boser, "Smart Dust: Wireless Networks of Millimeter-Scale Sensor Nodes," Electronics Research Laboratory Research Summary, Electronic Research Lab, University of California, Berkeley, 1999.

[10] Warneke, B., et al., "Smart Dust: Communicating with a Cubic-Millimeter Computer," *IEEE Computer*, Vol. 34, No. 1, January 2001, pp. 44–51.

[11] Estrin, D., et al., "Next Century Challenges: Scalable Coordination in Sensor Networks," *Proc. ACM Conf. Mobile and Computing Networking (MobiCom)*, New York: ACM Press, 1999.

[12] Culler, D., D. Estrin, and M. Srivastava, "Overview of Sensor Networks," *IEEE Computer*, August 2004, Vol. 37, No.8, pp. 41–49.

[13] UCLA Center for Embedded Networked Sensing at http://cens.ucla.edu.

[14] Krishnamurthy, L., et al., "Wireless Sensor Networks in Intel Fabrication Plants," May 2004, at http://www.intel.com/research/vert_manuf_condmaint.htm.

[15] Estrin, D., et al., "Connecting the Physical World with Pervasive Networks," *IEEE Pervasive Computing*, Vol. 1, No. 1, January–March 2002, pp. 59–69.

[16] Chu, M., H. Haussecker, and F. Zhao, "Scalable Information-driven Sensor Querying and Routing for Ad Hoc Heterogeneous Sensor Networks," *International Journal of High Performance Computing Applications*, 2002.

[17] Zhao, F., J. Shin, and J. Reich, "Information-driven Dynamic Sensor Collaboration for Target Tracking," *IEEE Signal Processing*, 2002, Vol. 19, No. 2, pp. 61–72.

[18] Pradhan, S., J. Kusuma, and K. Ramchandran, "Distributed Compression in a Dense Sensor Network," *IEEE Signal Processing*, 2002, Vol. 19, No. 2, pp. 51–60.

[19] Li, D., et al., "Detection, Classification and Tracking of Targets in Distributed Sensor Networks," *IEEE Signal Processing*, 2002, Vol. 19, No. 2, pp. 17–29.

[20] Mataric, M. J., "Behavior-based Control: Examples from Navigation, Learning, and Group Behavior," *Journal of Experimental and Theoretical Artificial Intelligence*, Vol. 9, No. 2, 3, 1997, pp. 323–336.

[21] Arkin, R., "Toward the Unification of Navigational Planning and Reactive Control," *Proc. AAAI Spring Symposium, AAAI*, Menlo Park, CA, 1989.

[22] The National Academies Press, at http://books.nap.edu/books/0309059348/html/index.html.

[23] Context Aware Computing, MIT Media Lab, at http://www.media.mit.edu/context.

[24] Doherty, L., et al., "Energy and Performance Considerations for Smart Dust," *Int'l J. Parallel Distributed Systems and Networks*, Vol. 4, No. 3, 2001, pp. 121–133.

[25] Pottie, G. J., and W. Kaiser, "Wireless Integrated Networked Sensors," *Communications of the ACM*, Vol. 43, No. 5, May 2000.

[26] Hill, J., et al., "System Architecture Directions for Networked Sensors," *Proc. ACM Conference on Architectural Support for Programming Languages and Operating Systems (ASPLOS)*, New York, 2000.

[27] Crossbow Technology Incorporated, at http://www.xbow.com.

[28] An on-line sensor hardware survey, at http://www.cse.unsw.edu.au/~sensar/hardware/hardware_survey.html.

[29] Wind River, at http://www.windriver.com.

[30] Geoworks, at http://www.geoworks.com.

[31] Sun Microsystems, at http://www.sun.com/chorusos.

[32] Gay, D., et al., "The *nesC* Language: A Holistic Approach to Network Embedded Systems," *Proc. ACM SIGPLAN 2003 Conference on Programming Language Design and Implementation (PLDI)*, San Diego, CA, 2003.

[33] Levis, P., and D. Culler, "Maté: A Tiny Virtual Machine for Sensor Networks." *Proc. 10th International Conference on Architectural Support for Programming Languages and Operating Systems (ASPLOS X)*, San Jose, CA, 2002.

[34] Barr, R., et al., "On the Need for System-Level Support for Ad Hoc and Sensor Networks," *SIGOPS Oper. Syst. Rev.*, Vol. 36, No. 2, 2002.

[35] Bhatti, S., et al., "MANTIS OS: An Embedded Multithreaded Operating System for Wireless Micro Sensor Platforms," *ACM/Kluwer Mobile Networks & Applications (MONET)*, to appear 2005.

[36] Dunkels, A., B. Grönvall, and T. Voigt, "Contiki—a Lightweight and Flexible Operating System for Tiny Networked Sensors," *Proc. 1st IEEE Workshop on Embedded Networked Sensors 2004 (IEEE EmNetS-I)*, Tampa, FL, November 2004.

[37] Boulis, A., and M. B. Srivastava, "A Framework for Efficient and Programmable Sensor Networks," *Proc. OPENARCH 2002*, New York, N.Y., June 2002.

[38] Girod, L., et al., "EmStar: A Software Environment for Developing and Deploying Wireless Sensor Networks," *Proc. USENIX General Track*, Boston, MA, 2004 pp. 283–296.

[39] Chen, B., et al., "Span: An Energy-Efficient Coordination Algorithm for Topology Maintenance in Ad Hoc Wireless Networks," *Proc. 7th ACM Conf. Mobile and Computing and Networking (MobiCom)*, New York: ACM Press, 2001.

[40] Cerpa, A., and D. Estrin, "ASCENT: Adaptive Self-configuring Sensor Network Topologies," *Proc. IEEE Infocom*, Piscataway, NJ, 2002.

[41] Schurgers, C., V. Tsiatsis, and M. Srivastava, "STEM Topology Management for Efficient Sensor Networks," *Proc. IEEE Aerospace Conf.*, Los Alamitos, CA: IEEE CS Press, 2002.

[42] Xu, Y., J. Heidemann, and D. Estrin, "Geography-Informed Energy Conservation for Ad Hoc Routing," *Proc. 7th Ann. ACM/IEEE Int'l Conf. Mobile Computing and Networking (MobiCom)*, New York: ACM Press, 2001.

[43] Johnson, D., and D. Maltz, "Protocols for Adaptive Wireless and Mobile Networking," *IEEE Personal Communications*, Vol. 3, No. 1, February 1996.

[44] Heidemann, J., "Building Efficient Wireless Sensor Networks with Low-Level Naming," *ACM Symp. Operating Systems Principles*, New York: ACM Press, 2001.

[45] Adjie-Winoto, W., et al., "The Design and Implementation of an Intentional Naming System," *ACM Symposium on Operating System Principles*, New York: ACM Press, 1999.

[46] Bonnet, P., J. E. Gehrke, and P. Seshadri, "Querying the Physical World," *IEEE Personal Comunications*, Vol. 7, No. 5, October 2000, pp. 10–15.

[47] Cerpa, A., et al., "Habitat Monitoring: Application Driver for Wireless Communications Technology," *Proc. ACM SIGCOMM Workshop on Data Communications in Latin America and the Caribbean*, New York: ACM Press, 2001.

[48] Cyclops image sensor, at http://www.cens.ucla.edu/~mhr/cyclops/Cyclops.htm.

[49] Feng W., et al., "Panoptes: Scalable Low-Power Video Sensor Networking Technologies," *Proc. ACM Multimedia*, Berkeley, CA, 2003, pp. 562–571.

[50] Sibley, G. T., M. H. Rahimi, and G. S. Sukhatme, "Robomote: A Tiny Mobile Robot Platform for Large-Scale Sensor Networks," *Proc. IEEE Int'l Conf. Robotics and Automation*, Washington, D.C., 2002, pp. 1119–1124.

[51] Kaiser, W. J., et al., "Networked Infomechanical Systems (NIMS) for Ambient Intelligence," *UCLA CENS Technical Report #31*, December 5, 2003.

[52] National Ecological Observatory Network, at http://www.neoninc.org.

CHAPTER 2

Potential Field Methods for Mobile-Sensor-Network Deployment

Andrew Howard and Sameera Poduri

This chapter considers the problem of deploying a mobile sensor network in an unknown environment using a virtual potential field. A mobile sensor network is composed of a collection of nodes, each of which has sensing, computation, communication, and locomotion capabilities. Such networks are capable of self-deployment; that is, starting from some compact, initial configuration, the nodes may spread out such that the area covered by the network is maximized. This chapter describes a distributed potential field method for network deployment and shows how this method may be applied to maximize sensor coverage, while maintaining full network connectivity.

2.1 Introduction

It is locomotion capability that distinguishes a mobile sensor network from its static cousins; locomotion facilitates a number of useful network capabilities, including the ability to self-deploy (in which nodes autonomously position themselves in the environment) and self-repair (should some nodes fail, other nodes can reposition themselves to compensate). Mobile sensor networks have a range of potential applications, including search-and-rescue operations and emergency environment monitoring. Consider, for example, a toxic chemical spill in a residential neighborhood: a mobile sensor network, whose nodes are equipped with chemical detectors, can rapidly (and autonomously) deploy into the neighborhood, make real-time observations of the location and concentration of hazardous material, and relay this information to crews located at a safe distance. From this example, we consider the problem of deployment for both *coverage* [1] and *constrained coverage* [2]. This chapter describes a deployment algorithm that seeks to maximize the net area covered by the nodes' sensors, while simultaneously maintaining full network connectivity. Emergency monitoring tasks also present two additional challenges: (1) prior models of the environment will be either incomplete, inaccurate, or unavailable, and (2) processing and communication requirements for any centralized deployment algorithm are likely to scale as $O(n)$ in network size (and networks may contain hundreds or thousands of nodes). Consequently, we consider only deployment algorithms that are fully distributed (with either local communication or no communication at all) and use only local sensed information (no maps or a priori knowledge).

The specific approach described here is based on the notion of virtual potential fields; that is, nodes are treated as virtual particles and are subject to virtual forces that either attract or repel them from obstacles and each other. Thus, for example, coverage can be achieved using purely repulsive forces between pairs of nodes and between nodes and obstacles (the network spreads through the environment in a manner analogous to a gas diffusing through a solid). Note that there is no explicit reasoning about coverage in this approach: coverage is a global property of the network that emerges from local interactions between nodes.

Maintaining network connectivity (a constrained-coverage problem) is only slightly more challenging. Once again, connectivity is a global property of the network, but evidence from the theory of random graphs indicates that network connectivity is strongly correlated with the average node degree (i.e., number of neighbors with which each node can communicate) [3]. Node degree is a local constraint and can be used to modulate the forces between nodes. Thus, for example, any node whose degree falls below a certain threshold may replace repulsive internode forces with an attractive force; in Section 2.6 we will show that this simple mechanism is extremely effective at maintaining full connectivity.

The remainder of this chapter is organized as follows: we enumerate some of the basic properties of the mobile sensor network, develop the basic theory of potential fields, and apply this theory to both coverage and constrained-coverage problems. To illustrate the approach and explore some of its properties, we present results from a simulated network with 100 nodes.

2.2   Related Work

The concept of *coverage* as a paradigm for evaluating many-robot systems was introduced by D. W. Gage [1]. Gage defines three basic types of coverage: blanket coverage, where the objective is to achieve a static arrangement of nodes that maximizes the total detection area; barrier coverage, where the objective is to minimize the probability of undetected penetration through the barrier; and sweep coverage, which is more-or-less equivalent to a moving barrier. According to this taxonomy, the deployment problem described in this chapter is a blanket-coverage problem.

Potential field techniques for robotic applications were first described by O. Khatib [4] and have since been widely used in the mobile robotics community for tasks such as local navigation and obstacle avoidance. The related concept of *motor schemas*, which utilizes the superposition of spatial vector fields to generate behavior, was introduced by R. C. Arkin [5]. Both techniques have since been applied to the problem of formation control for groups of mobile robots [6, 7]. The formation problem is similar, in some respects, to the deployment problem described in this chapter, in that the robots will attempt to maintain a formation based on local sensing and computation. The deployment problem is also similar, in some respects, to the multirobot exploration-and-mapping problem. Here, the aim is to build a global map of the environment by sequentially visiting each location with one or more robots. This problem has been considered by a number of authors [8–11], who use a variety of techniques ranging from topological matching [8] to fuzzy inference [12] and particle filters [13]. Two good examples are provided by R.Simmons et al. [10]

and W. Burgard et al. [11], both of whom build global maps, apply heuristics to select goal locations for exploration, and use explicit communication to prevent more than one robot from heading for the same goal. This approach to exploration contrasts markedly with the approach to deployment described in this chapter; potential field methods are able to achieve good coverage without global maps, without communication, and without explicit reasoning.

The deployment problem described here is similar to that described by N. Bulusu, J. Heidemann, and D. Estrin [14], who consider the problem of adaptive beacon placement for localization in large-scale WSNs. These networks rely on RF-intensity information to determine the location of nodes; appropriate placement of RF-beacons is therefore critical. The authors describe an empirical algorithm that adaptively determines the optimal beacon locations.

Finally, we note that the problem of deployment is related to the traditional art gallery problem in computational geometry [15]. The art gallery problem seeks to determine, for some polygonal environment, the minimum number of cameras that can be placed such that the entire environment is observed. While there exists a number of algorithms designed to solve the art gallery problem, all of these assume that we possess good prior models of the environment.

2.3   Sensor-Node Capabilities

For the discussion that follows, we make a number of assumptions regarding the capabilities of sensor nodes and the environment they inhabit. Specifically, the environment is planar; nodes have omnidirectional sensors (for measuring the phenomena of interest, the location of obstacles, and the range and bearing of nearby nodes); and nodes have holonomic locomotion (such that they can move freely in the plane). We assume that sensors can distinguish between obstacles and nodes but cannot determine individual node identities (i.e., nodes are anonymous). In practice, such sensors can be constructed using scanning laser range finders (with retroreflectors to identify nodes), Infrared (IR) transceivers, or omnidirectional cameras. For holonomic locomotion, one can employ either a genuine omnidirectional mechanism or a more common differential drive (where the rotational degree of freedom can be employed to make such robots holonomic with respect to linear motion in the plane).

Note that we also make a number of important *nonassumptions*: the environment need not be static, and localization and communication are not required (although they may be available).

2.4   Potential Fields: Theory and Implementation

Potential fields are a commonly used and well-understood method in mobile robotics, where they are typically applied to tasks such as local navigation and obstacle avoidance [4, 5]. Here, we adapt the method to create a fully distributed deployment algorithm for mobile sensor networks.

Consider a single network node. This node is subject to a virtual force F that is defined as the gradient of a scalar potential field U:

$$F = -\nabla U \tag{2.1}$$

We generally divide the potential field into two components: the field U_o due to obstacles, and the field U_n due to other nodes; these fields give rise to forces F_o and F_n, respectively. These potentials can be superimposed, such that

$$U = U_o + U_n \text{ and } F = F_o + F_n \tag{2.2}$$

Consider the potential field due to obstacles. If we imagine that each node and each obstacle carries an electric charge, we can write down an expression for the resultant "electrostatic" potential:

$$U_o = k_o = \sum_i \frac{1}{r_i} \tag{2.3}$$

The summation is over all obstacles that can be seen by the node; k_o is a constant describing the strength of the field; and r_i is the Euclidean distance between the node and obstacle i. Let x denote the position of the node, and let x_i denote the position of obstacle i. Using these definitions, the total force F_o due to obstacles can be computed using (2.1), as follows:

$$F_o = -\frac{dU_o}{dx} = -\sum_i \frac{dU_o}{dr_i} \supseteq \frac{dr_i}{dx} \tag{2.4}$$

Inserting the appropriate derivatives yields

$$F_o = -k_o \sum_i \frac{1}{r_i^2} \supseteq \frac{r_i}{r_i} \tag{2.5}$$

where $r_i = x_i - x$ and $r_i = |r_i|$. Note that the force is expressed entirely in terms of the relative positions r_i of obstacles rather than their absolute positions x_i. This allows us to compute the force directly from sensor data without the need for global localization. Figure 2.1 shows the potential field U_o and force field F_o generated by this potential in a simple environment. This particular environment contains two minima at the bottom left and top right; in the absence of any other forces, the node will eventually settle at one of these two locations.

Consider now the potential field U_o due to other nodes. By analogy with the obstacle field, we can derive expressions for the potential U_n and force F_n by replacing a summation over visible obstacles with a summation over visible nodes; thus:

$$U_n = k_n \sum_i \frac{1}{r_i} \text{ and } F_n = -k_n \sum_i \frac{1}{r_i^2} \supseteq \frac{r_i}{r_i} \tag{2.6}$$

where r_i is the relative position of node i. In Sections 2.5 and 2.6, we will show how variants of this potential can be used to achieve both coverage and constrained coverage.

The idealized trajectory of a node subject to force F can be computed using an appropriate equation of motion. We use an equation of the following form:

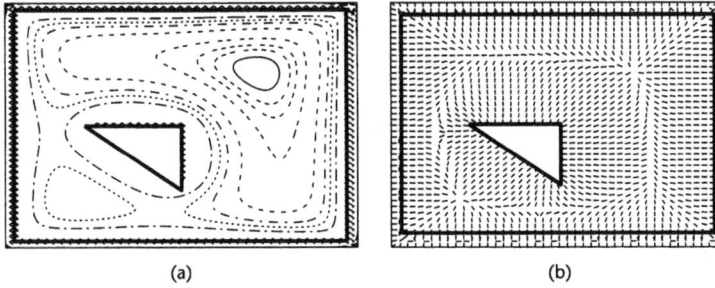

(a)   (b)

Figure 2.1   (a) Potential field generated by a simple environment; the contours show the lines of equal potential. (b) Force fields generated by this potential; the arrows indicate the direction (but not magnitude) of the force.

$$ma = F - \mu v \qquad\qquad (2.7)$$

where a denotes the acceleration of the node and m denotes its mass (since the mass can be set to one without loss of generality, we generally omit this parameter). The second term on the right-hand side of this equation is a *viscous friction* term in which μ is the viscosity coefficient and v is the node velocity. This term is used to ensure that, in the absence of external forces, the node will eventually come to a standstill (the node's velocity will approach zero asymptotically). Given a fixed set of potentials and a static environment, the viscous friction term also guarantees that the network *as a whole* will ultimately reach a state of static equilibrium (i.e., a state in which all nodes have stopped moving). Note that this is not the case in dynamic environments, in which energy is effectively added to the system whenever an object is moved by some agency other than the network itself (this is akin to stirring a container full of fluid). The same is true if one is using switching potentials of the kind described in Section 2.6.

Our discussion up to this point has focused entirely on a virtual physical system (i.e., one in which the forces, accelerations, masses, and the like are entirely imaginary). This virtual physical system must, however, be mapped onto a real physical system made up of real nodes. On the sensing side, we must be aware that real sensors have noise, limited range, and blind spots; the virtual forces computed using real sensor data may thus differ markedly from those predicted by theory. Moreover, in contrast with our theoretical analysis, there is no guarantee that internode forces will be symmetric: node A may be repelled by node B, but node B may not have observed node A. The impact of such effects must be determined empirically.

On the actuation side, real nodes generally have some form of velocity controller, and one must therefore define a *control law* that translates a virtual force into a commanded velocity. In deriving this control law, we must be cognizant of the fact that real nodes are not "free particles": they have both kinematic and dynamic constraints. The kinematic constraints can be largely ignored if we make the assumption that the nodes have holonomic drive mechanisms (as we did in Section 2.3). Dynamic constraints, however, cannot be ignored: the node will have both a maximum velocity and a maximum acceleration, and these must be captured by the control law. One such law is expressed algorithmically as follows. Let v denote the commanded velocity at some time t, and let Δv denote the change in the commanded velocity between times t and $\tau + \Delta\tau$ (the controller has a fixed cycle time of $\Delta\tau$

seconds). The change in commanded velocity is determined using a piecewise-constant approximation to the equation of motion (2.7):

$$\Delta v \leftarrow (F - \mu v)\Delta t \qquad (2.8)$$

The x and y components of Δv are subsequently clipped to the domain $[-a_{max}, +a_{max}]$, where a_{max} denotes the largest allowable change in velocity. The commanded velocity v is determined using:

$$v \leftarrow v + \Delta v \qquad (2.9)$$

and is then clipped to the domain $[-v_{max}, +v_{max}]$, where v_{max} is the maximum allowed velocity.

Using this control law, the real node dynamics will closely approximate those described by the equation of motion. There are, however, two regimes in which the correspondence will fail. First, for small v, the viscous friction term will tend to produce oscillation rather than asymptotic convergence to zero velocity; this kind of behavior is typical of discrete control systems and can be eliminated by introducing a velocity dead-band. Second, large accelerations and velocities will simply be clipped, in which case the deviation from the virtual dynamics may become arbitrarily large. This deviation is significant only if it prevents the network as a whole from reaching static equilibrium or if it significantly increases the time taken to reach this equilibrium. We assert (without proof) that the acceleration and velocity limits act like additional nonlinear friction terms and that, therefore, these limits will not prevent the system from reaching static equilibrium. The limits may, however, impact the time taken to reach equilibrium, and this impact must be determined empirically.

2.5 Coverage

The problem of deployment for coverage can be succinctly stated as follows: from some initial configuration, move the nodes such that the net area covered by the network is maximized. We make the following definitions. Let R_n denote the maximum range of the sensor used for detecting obstacles and other robots. We assume this sensor is omnidirectional, requires line of sight, and has binary sensing characteristics (i.e., nodes that are unoccluded and within range R_n are detected; nodes that are occluded or outside this range are undetected). Each node also has a sensor for monitoring the environment with maximum range R_s and identical characteristics to the node/obstacle sensor (omnidirectional, line of sight, and binary).

A point in the environment is said to be *covered* if it is unoccluded and lies within range R_s of at least one node; the *total network coverage* is the integral of all such points. To scale out the effects of network size, we define the *normalized per-node coverage* C_s:

$$C_s = \frac{(total\ network\ coverage)}{n\pi R_s^2} \qquad (2.10)$$

where n is the network size. This is our key performance metric: it has a value of one if, and only if, all nodes are deployed such that their sensor fields are nonoverlapping (in which case the total network coverage is maximized). Note that the total network coverage is difficult to compute analytically; in practice, we compute it by ray-tracing on a grid, using line-of-sight range data.

Given these definitions, we can write down an internode potential suitable for coverage:

$$U_n = k_n \sum_i \frac{1}{\min(r_i, R_s)} \tag{2.11}$$

where the summation is over all detected nodes (for which $r_i < R_n$) (this potential is illustrated in Figure 2.2). Note that there are two regimes for this potential: if $R_n > R_s$, nodes are repelled until their sensor fields no longer overlap, at which point the repulsion force drops to zero; if $R_n > R_s$, nodes are repelled until they can no longer detect one another. In the latter case, we do not expect the network to achieve optimal coverage (i.e., $C_s < 1$, since sensor fields may still overlap).

We have tested this deployment algorithm using a high-fidelity physics-based simulator that accurately models vehicle kinematics, dynamics, and sensor feedback [16]. We tested two conditions: an open space representing an ideal outdoor environment and a highly structured space representing a typical indoor environment (Figure 2.3). In both cases, the initial configuration of nodes was random and compact. The simulated network contains 100 identical nodes, each of which is equipped with a 360° scanning laser range finder mounted atop an omnidirectional platform. We use the laser for obstacle and node detection, as well as the primary

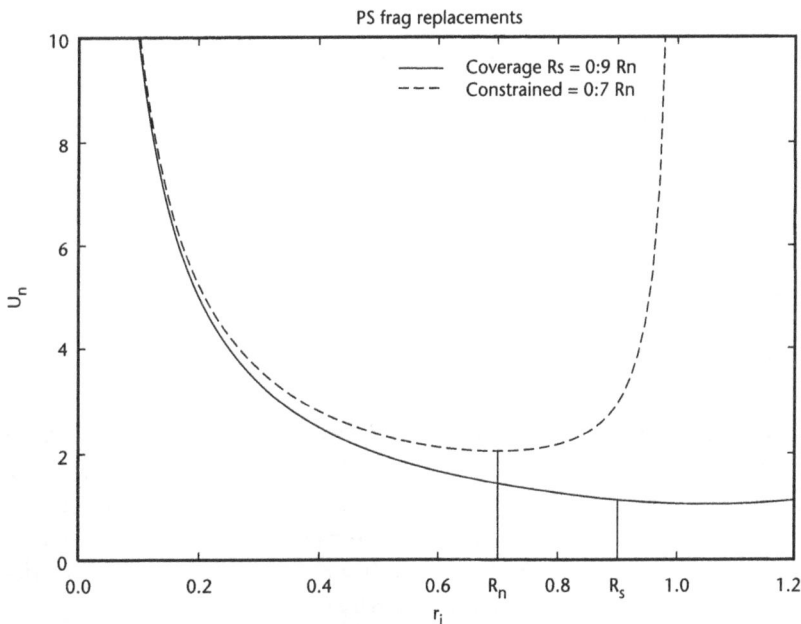

Figure 2.2 Individual internode potentials for coverage and constrained coverage. The coverage potential repels nodes until their sensor fields no longer interact ($R_s < r_i$). The constrained-coverage potential adds an attractive component when the node degree falls below a set threshold, with an equilibrium point at range $r_i = \eta R_n$.

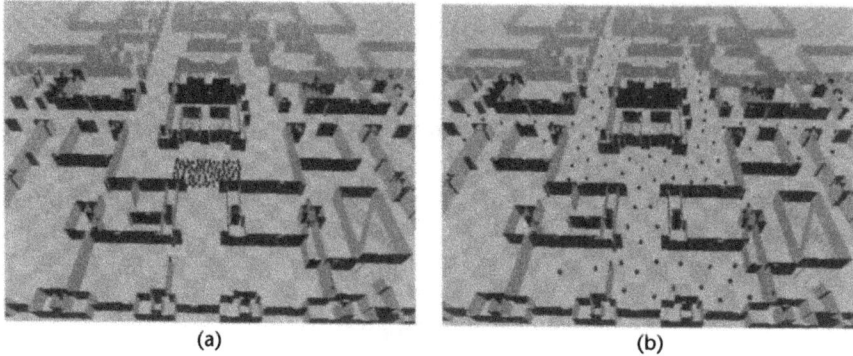

Figure 2.3   (a) Initial and (b) final states of the network in a structured environment.

coverage sensor. For these experiments, the maximum range of the sensor was set to $R_n = R_s = 8.0m$.

Figure 2.4 shows the time evolution of the network in both open and structured spaces. In open space, the mean node separation rises monotonically to reach the limit of the sensor range (i.e., nodes push themselves apart until they can no longer detect each other). Since $R_s = R_n$, this naturally leads to optimal coverage ($C_s = 1$). The results for a structured space are much more interesting: mean node separation rises monotonically but plateaus well short of the maximum sensor range. In this environment, line-of-sight effects dominate: nodes cannot be repelled by nodes they cannot see, even if those nodes are within the nominal sensor range. Thus, the steady-state value for mean separation is determined primarily by the structure of the environment rather than the structure of the network. One can confirm this by applying an arbitrary scaling factor to the environment; as the scaling factor goes to infinity, one recovers the open-space result.

Unsurprisingly, the coverage value for structured space is significantly lower than that for open space (around $C_s = 0.2$). This low value partly reflects the lower mean node separation, and partly reflects the line-of-sight characteristics of the coverage sensor; that is, for a structured environment, the maximum theoretical coverage may be significantly less than one.

2.6   Constrained Coverage

For monitoring tasks, coverage alone is an insufficient criterion for deployment; we also require a network that is fully connected in the sense that any node can communicate with any other node (thus, data from any node can be accessed from anywhere in the network). This leads to the following problem statement for constrained coverage: from some initial configuration, deploy nodes such that the coverage area is maximized, subject to the constraint that the network is fully connected. As noted in Section 2.1, connectivity is a global property of the network but is known to be strongly correlated with average node degree [3]. We therefore reformulate the constrained-coverage problem with a local constraint as follows: from some initial configuration, deploy nodes such the coverage area is maximized, subject to the constraint that each node has a degree of at least D. The value of D must

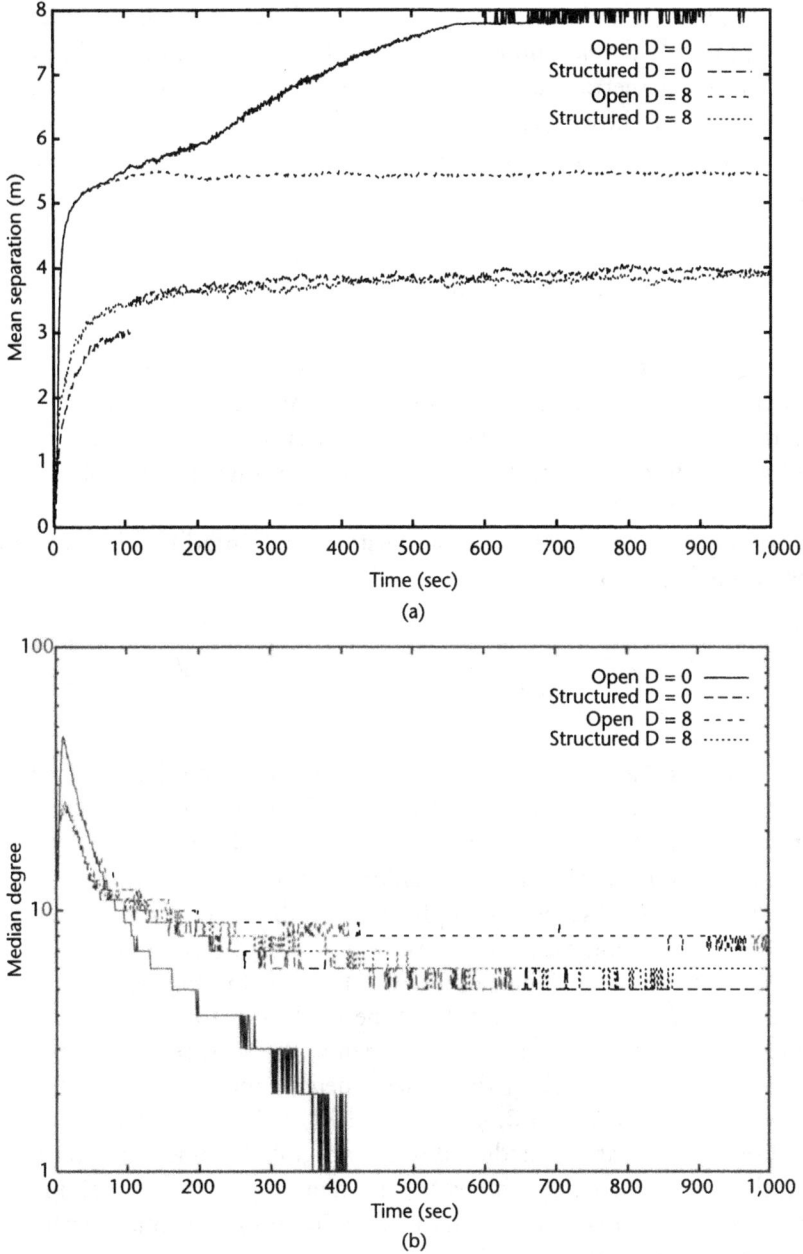

Figure 2.4 Time evolution plots for the network, showing the mean node separation and the normalized network coverage C_s. Each graph shows results from four experiments: (a) unconstrained coverage ($D = 0$) in open and structured environments and (b) constrained coverage ($D = 8$) in open and structured environments.

be set a priori and determined empirically. We make the follow definitions. Let R_c denote the maximum communication range. Communication is assumed to be omnidirectional, symmetric, and binary (nodes can transmit and receive from any node within range R_c). To simplify the discussion that follows, we also make the assumption that communication requires line of sight between nodes and that

$R_c > R_n$; that is, nodes can communicate with any node they can detect. Using this assumption, we define a node's degree d to be the total number of observable nodes.

Numerous metrics have been designed to measured global network connectivity; in our case, however, we are particularly interested in the ability of nodes to communicate with one another and, therefore, define the *normalized global connectivity* C_c to be

$$C_c = \frac{2}{n(n-1)} \sum_{i,j>i} Path(i,j) \tag{2.12}$$

where $Path(i,j)$ returns a value of one if there exists a path between nodes i and j (or, equivalently, if nodes i and j belong to the same subnet) and zero otherwise. This metric effectively measures the probability that any two nodes, selected at random, will be able to communicate with one another; it has a value of one if, and only if, the network is fully connected.

From these definitions, we can write down a modified internode potential for constrained coverage:

$$U_n = k_n \sum_i \left[\frac{1}{\min(r_i, R_s)} + \frac{(1-\eta)^2}{\eta^2} \frac{1}{R_n - r_i} H(d - D - 1) \right] \tag{2.13}$$

where $H(z)$ is the Heaviside step function [defined such that $H(z)$ returns a value of one if z is nonnegative and zero otherwise]. This function is used to "switch on" an attractive potential whenever the node degree d falls below some threshold D. The potential is tuned such that the equilibrium point between attractive and repulsive potentials lies at a distance of ηR_n from the node. η is thus interpreted as a *safety factor*, with $\eta = 1$ being maximally unsafe, as the equilibrium point will lie at the limit of the detection range. This potential is illustrated in Figure 2.2.

Figures 2.4 and 2.5 show the time evolution for a network using this constrained-coverage algorithm in both open and structured environments (the initial conditions are identical to those described in the previous section). The results shown are for $D = 8, D = 8, \eta = 0.75$ and $R_c = R_s = R_n = 8m$.

Looking first at the median degree and global connectivity plots, it is apparent that the constrained algorithm is able to maintain the desired minimum degree, and as a result, the global connectivity C_p remains at or near one. Contrast this with the result for the unconstrained algorithm in open space: here, the node degree and connectivity both drop to zero, yielding a completely disconnected network.

Looking at the separation and coverage plots, one observes an interesting phenomenon: while coverage in open space is significantly degraded ($C_s = 0.4$) compared with the constrained algorithm, the performance in the structured space is almost identical. In open space, the constrained algorithm implicitly limits the node separation: comparing the separation and degree plots, it is apparent that node separation plateaus when the median degree approaches D. In the structured space, however, the line-of-sight constraints are such that median node degree never drops below five or six, even when the unconstrained algorithm is used. Thus, for this environment, the two algorithms are basically identical in function and performance.

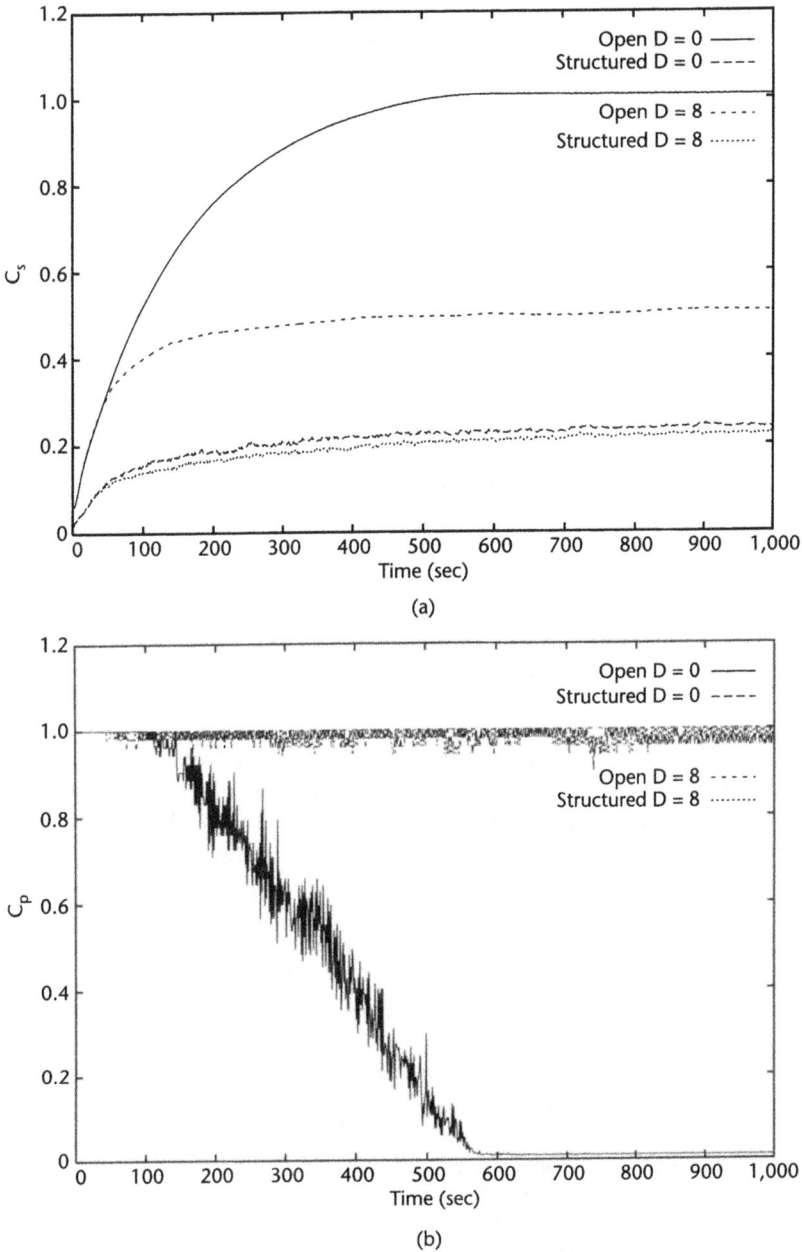

Figure 2.5 Time evolution plots for the network, showing the median node degree and the global network connectivity C_p. Each plot shows results from four experiments: (a) unconstrained coverage ($D = 0$) in open and structured environments, and (b) constrained coverage ($D = 8$) in open and structured environments.

One can conduct similar experiments for other values of D: smaller values generate better coverage in open space but are less likely to yield fully connected networks. At $D = 0$, one recovers the unconstrained-coverage algorithm from the previous section. Encouragingly, the results from structured environments indicate a weak correlation between D, coverage, and connectivity. This suggests that having selected a value of D suitable for open spaces, one may use this same value in any structured environment, regardless of its composition. Put another way, this parameter does not require tuning.

2.7 Conclusion

As a method for deploying mobile sensor networks, potential fields have a number of attractive features, including local sensing, distributed control, and no requirement for localization or communication. Through simulation, we have demonstrated that potential fields can be used to deploy fully connected networks with up to 100 nodes, and we have every reason to believe that the method will scale to much larger networks. It should also be noted that while we have restricted the discussion to static environments, potential fields are intrinsically adaptive and are therefore well suited to dynamic problems. Consider, for example, a network deployed across a small, closed room: if the door to the room is opened, the network will immediately spread beyond its original confines and seek a new equilibrium state in the larger environment.

References

[1] Gage, D. W., "Command Control for Many-Robot Systems," *Proc. AUVS-92, the Nineteenth Annual AUVS Technical Symposium*, Hunstville, AL, June 1992, pp. 22–24. Reprinted in *Unmanned Systems Magazine*, Vol. 10, No. 4, fall 1992, pp. 28–34.

[2] Poduri, S., and G. S. Sukhatme, "Constrained Coverage for Mobile Sensor Networks," *Proc. IEEE International Conference on Robotics and Automation*, New Orleans, LA, May 2004, pp. 165–179.

[3] Xue, F., and P. R. Kumar, "The Number of Neighbors Needed for Connectivity of Wireless Networks," *Wireless Networks*, Vol. 10, No. 2, March 2004, pp. 169–181.

[4] Khatib, O., "Real-Time Obstacle Avoidance for Manipulators and Mobile Robots," *International Journal of Robotics Research*, Vol. 5, No. 1, 1986, pp. 90–98.

[5] Arkin, R. C., Motor Schema Based Mobile Robot Navigation, *International Journal of Robotics Research*, Vol. 8, No. 4, 1989, pp. 92–112.

[6] Scheider, F. E., D. Wildermuth, and H. L. Wolf, "Motion Coordination in Formations of Multiple Mobile Robots Using a Potential Field Approach," In L. E. Parker, G. W. Bekey, and J. Barhen, (eds.), *Distributed Autonomous Robotics Systems*, Vol. 4, Springer, Knoxville, TN, 2000, pp. 305–314.

[7] Balch, T., and M. Hybinette, "Behavior-based Coordination of Large-Scale Robot Formations," *Proc. Fourth International Conference on Multiagent Systems (ICMAS '00)*, Boston, MA, July 2000, pp. 363–364.

[8] Dedeoglu, G., and G. S. Sukhatme, "Landmark-based Matching Algorithms for Cooperative Mapping by Autonomous Robots," In L. E. Parker, G. W. Bekey, and J. Barhen, (eds.), *Distributed Autonomous Robotics Systems*, Vol. 4, Springer, Knoxville, TN, 2000, pp. 251–260.

[9] Thrun, S., W. Burgard, and D. Fox, "A Real-Time Algorithm for Mobile Robot Mapping with Applications to Multi-robot and 3D Mapping," *Proc. IEEE International Conference on Robotics and Automation (ICRA2000)*, Vol. 1, San Francisco, CA, 2000, pp. 321–328.

[10] Simmons, R., et al., "Coordination for Multi-robot Exploration and Mapping," *Proc. Seventeenth National Conference on Artificial Intelligence (AAAI-2000)*, Austin, TX, 2000, pp. 852–858.

[11] Burgard, W., et al., "Collaborative Multi-robot Exploration," *Proc. IEEE International Conference on Robotics and Automation (ICRA)*, Vol. 1, San Francisco, CA, 2000, pp. 476–481.

[12] López-Sánchez, M., et al., "Map Generation by Cooperative Low-Cost Robots in Structured Unknown Environments," *Autonomous Robots*, Vol. 5, No. 1, 1998, pp. 53–61.

[13] Thrun, S., et al., "Robust Monte Carlo Localization for Mobile Robots," *Artificial Intelligence Journal*, Vol. 128, No. 1–2, 2001, pp. 99–141.

[14] Bulusu, N., J. Heidemann, and D. Estrin, "Adaptive Beacon Placement," *Proc. 21st International Conference on Distributed Computing Systems (ICDCS-21)*, Phoenix, AZ, April 2001.

[15] O'Rourke, J. *Art Gallery Theorems and Algorithms*, New York: Oxford University Press, 1987.

[16] Koenig, N., and A. Howard, "Design and Use Paradigms for Gazebo, an Open-Source Multi-robot Simulator," *Proc. IEEE/RSJ International Conference on Intelligent Robots and Systems*, Sendai, Japan, September 2004.

CHAPTER 3

Sensor Fault Detection and Calibration

Nirupama Bulusu and Tatiana Bokareva

Once deployed in the physical world, even factory-calibrated sensors are prone to numerous faults that corrupt sensor data. For example, as part of an environmental observation and forecasting system, sensors deployed in the Columbia River Estuary (CORIE) gather information on physical dynamics and changes in estuary habitat [1]. Of these, salinity sensors are particularly susceptible to *biofouling* (the growth of biological material on the sensor), which gradually degrades sensor response and corrupts critical data. In a large sensor network, it is cumbersome to detect and fix these errors manually and in a timely fashion. Moreover, the sensors are deployed in a physically inaccessible environment.

3.1 Introduction

The measured value of a sensor often has a deviation from the true physical state. This deviation can often be statistically characterized and is known as *bias*, as shown in Figure 3.1. The bias of a sensor is typically determined in the factory through *calibration*. Calibration is the process of mapping raw sensor readings into corrected values by identifying and correcting systematic bias.

This motivates automatic, in situ fault detection and sensor calibration, enabling the sensor network to identify phenomena such as biofouling early and minimize data loss. Sensor calibration can either be *active* or *passive*. Active calibration techniques

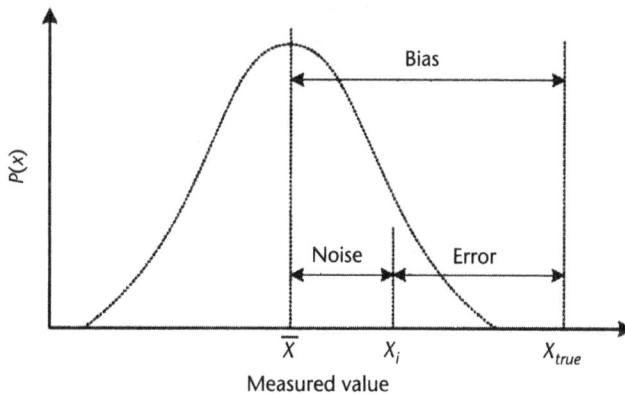

Figure 3.1 Properties of a sensor—bias and error. Note that even after the sensor bias is accounted for, there may still be a residual error due to random noise.

rely on an active excitation source or stimuli and are typically implemented via actuator elements. For example, C. J. Taylor [2] calibrates a network of cameras using active lights. Active calibration techniques can compensate for errors caused by physical coupling, but they may not always be viable, as they are expensive, not scalable, and applicable only in the context of a limited set of sensing modalities. In contrast, passive calibration techniques exploit spatiotemporal correlations among sensor data and require no additional actuation infrastructure.

What makes passive sensor calibration challenging? Let us revisit the example of the CORIE system. Early biofouling detection is made difficult by the normal variability of salinity sensors. Tides cause the salinity measurements to vary from near river salinity to near ocean salinity twice a day. The temporal pattern of salinity penetration varies spatially in the estuary and is further impacted by changes in weather such as winds and precipitation as well as ocean conditions. Moreover, there is a scarcity of biofouling onset examples and a further variability in the biofouling signature from episode to episode. For example, the time from onset to complete biofouling can take anywhere from 3 weeks to 5 months. Detectors that monitor salinity alone cannot distinguish normal decreases in salinity from early biofouling. The consequence is a high false-alarm rate, where biofouling would be reported even when it has not occurred.

The CORIE example illustrates the limitations of fault detectors that only accrue information from a single sensor over time. Approaches to automatic sensor calibration must therefore utilize alternate information sources, such as correlated sensor modalities as well as collocated sensors. At a single site, natural salinity decreases can be recognized by monitoring a correlated source of information such as temperature that is not corrupted by biofouling [1]. When several collocated sensor nodes measure the same physical phenomena, they can dynamically collaborate so as to improve sensor calibration in a large, unattended, densely deployed sensor network [3, 4].

We discuss these approaches in the rest of this chapter, which is organized as follows. Section 3.2 discusses a methodology for defining continuous valued sensors and their failures. Section 3.3 discusses a passive sensor calibration technique that leverages correlated sensor modalities to detect sensor faults. Section 3.4 covers a distributed in situ sensor calibration that leverages the redundancy of densely deployed sensor networks. Section 3.5 presents conclusions.

3.2 Continuous Valued Sensors

In this section, we review a methodology for characterizing continuous valued sensors and their failures. This methodology was developed by K. Marzullo originally for process control systems [3], but it can be applied broadly to WSNs.

3.2.1 Physical State Variables and Concrete and Abstract Sensors

A *physical state variable* can take on any real value at arbitrary times. A *concrete sensor* is a device that can be used to sample a physical state variable. An example is a thermometer, which records temperature. The concrete sensor has a limited accuracy, which is exacerbated by network and processor scheduling delays. Moreover, the control program may be interested in the sensor value (temperature) at a time when the thermometer was not sampled. This requires interpolation, which in turn requires knowledge of the physical process being monitored.

Abstract sensors address this issue. An abstract sensor is a piecewise continuous function from a physical state variable to a dense interval of real numbers. An abstract sensor is correct if (1) it is not too inaccurate, and (2) it always includes the value of the physical variable. Figure 3.2 illustrates the relationship between physical state variables, concrete sensors, and abstract sensors.

3.2.2 Fault-Tolerant Abstract Sensors

An abstract sensor will fail when the underlying concrete sensor fails. There are three failure classes:

- *Fail-stop failures*, in which a failed abstract sensor can be detected;
- *Arbitrary failures with bounded inaccuracy*, in which an abstract sensor that is too inaccurate (the numerical value of its accuracy is too large) can be detected;
- *Arbitrary failures*, in which an abstract sensor can fail arbitrarily.

3.2.3 Fault-Tolerant Sensor Averaging

How do we construct an abstract sensor that is tolerant of failures? The solution lies in *replication*. Suppose we are given n independent abstract sensors and an assumption that no more than f sensors can fail. We intuit that (1) intervals containing the correct value must intersect, and (2) any point not contained in at least $(n - f)$ intervals must not be correct.

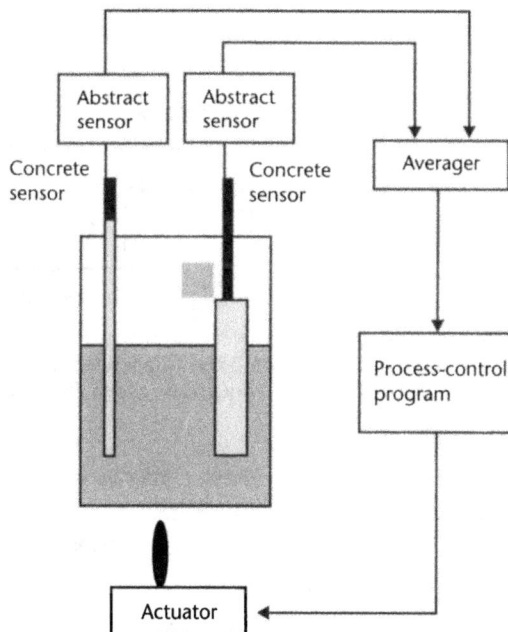

Figure 3.2 Physical state variables, concrete sensors, and abstract sensors. A physical state variable sampled by a concrete sensor is transformed into an abstract sensor.

Marzullo's algorithm for fault-tolerant sensor averaging finds a cover of $(n - f)$ cliques [i.e., finds the smallest value (*low*) and the largest value (*high*) that are contained in at least $(n - f)$ intervals]. The solution is an interval [*low ... high*], which is a fault-tolerant abstract sensor. This is illustrated in Figure 3.3.

The advantages of this algorithm are that it has an $O(n \log n)$ runtime, and the results may even be better than the best sensor. A drawback is that the final interval may still contain incorrect values. For example, in Figure 3.3, let us suppose that the faulty sensor is a. The fault-tolerant abstract sensor obtained using the averaging algorithm is larger than the interval common to b and c and contains some incorrect values common to a and c. This is an artifact of the interval continuity assumption.

3.2.4 Replicated Sensor Performance

The sensor's accuracy is determined by the width of the interval that is an abstract sensor value. If the ratio f/n of the number of faulty to nonfaulty abstract sensors is too large, then one cannot bound the inaccuracy of the resulting abstract sensor. Table 3.1 summarizes the theoretical upper bound f_{max} for the maximum number of faulty sensors for a given n for different failure models.

3.2.5 Limitations

Marzullo's work shows that local sensing can be bound through sensor replication, even in the presence of sensor failures. This methodology does not address discrete sensors, like one denoting whether or not a door is open, or multivalued sensors, like one that returns the altitude and azimuth of an airplane. Moreover, the approach is centralized and limited to a single sensor. Section 3.3 explores how correlated sensing modalities can be leveraged at a sensor node. Section 3.4 explores a distributed, in situ sensor calibration technique for a sensor network that leverages sensor redundancy.

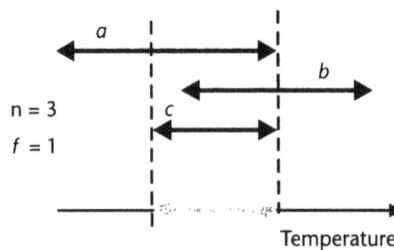

Figure 3.3 Fault-tolerant sensor averaging with $n = 3$ sensors and no more than $f = 1$ faulty sensor. The intervals a, b, c are the values of the three independent abstract sensors. The gray interval depicts the fault-tolerant abstract sensor computed by the fault-tolerant-averaging algorithm.

Table 3.1 Upper Bounds for Number of Faulty Sensors Relative to Given Number of Sensors n, for Fault-Tolerant Sensor Averaging under Various Failure Models

Failure Model	f_{max}	Min n: f = 1	Min n: f = 2
Arbitrary failures; unbounded inaccuracy	$\lfloor (n - 1)/3 \rfloor$	4	9
Arbitrary failures; bounded inaccuracy	$\lfloor (n - 2)/2 \rfloor$	4	6
Fail-stop failures	$n - 1$	2	3

3.3 Correlated Sensing Modalities

In this section, we show how correlated sensing modalities can be used for fault detection, when one sensor modality (e.g., temperature) is not susceptible to the same source of biofouling as the other sensing modality (e.g., salinity), using the example of the CORIE sensors developed by C. Archer, T. K. Leen, and A. Baptista [1].

Salinity and temperature at a CORIE station are products of the same mixing process of ocean and river waters, so we expect those values will be correlated. Assuming linear mixing of ocean and river waters, measured salinity S_m and temperature T_m are linear functions of ocean $\{S_o, T_o\}$ and river $\{S_r, T_r\}$ values

$$S_m = \alpha'(t)So + (1 - \alpha'(t))S_r \tag{3.1}$$

$$T_m = \alpha'(t)So + (1 + \alpha'(t))T_r \tag{3.2}$$

where $\alpha'(t)$ is the mixing coefficient at time t. River salinity S_r is close to zero.

Consequently, the estimated mixing coefficient $\alpha'(t)$ should be well correlated with salinity $S_m \approx \alpha'S_o$ and is modeled as a joint Gaussian distribution. When biofouling occurs, the salinity measurement is suppressed relative to the true value via a linear degradation function with unknown slope θ and onset time I'.

A natural framework for detecting degradation that grows with time is the standard sequential likelihood method from classical pattern recognition. It uses a parameterized biofouling model, which identifies the occurrence h and onset time I' of biofouling and works as follows.

Assume a sequence of measurements (salinity and temperature), $y_n, n = 1, ..., N$, where N is the current time. Construct probability densities for such sequences for both clean sensors and for biofouled sensors. The likelihood ratio test (LRT) is a statistical test of the goodness-of-fit between two models [5]. Construct an LRT h with these distributions. A threshold λ is set to provide a maximum false-alarm rate on historical data. The discriminant function h depends on (1) the parameters of the clean model, which are determined from historical data, and (2) the slope parameter θ and the onset time I', which are fit on-line so as to maximize h.

If the sequence is coming from a clean sensor, the fit should give $\theta \approx 0$ and, hence, $h \approx 0$, and we will detect no event (assuming $\lambda > 0$). When h is above the chosen threshold λ, the detector signals a biofouled sensor.

Real-time detectors installed during the summer of 2001 produced no false alarms yet detected all episodes of sensor degradation before the field staff scheduled these sensors for cleaning. The biofouling detectors essentially doubled the amount of useful data coming from the CORIE sensors.

3.4 Sensor Redundancy

We now discuss a two-phase, postdeployment calibration algorithm for large-scale, dense sensor networks that leverages sensor redundancy, developed by V. Bychkovskiy et al. [4].

In the first phase, the algorithm derives relative-calibration relationships between pairs of colocated sensors. The key idea in this phase is to use temporal correlation of signals received at neighboring sensors to derive the function relating their bias in amplitude. The second phase maximizes the consistency of pairwise calibration functions among groups of sensor nodes.

3.4.1 Pairwise Calibration

Pairwise calibration consists of (1) relative calibration of individual sensors, (2) calibration routing, and (3) distributed baseline sensors. Figure 3.4 illustrates pairwise calibration.

Relative calibration is calibration of individual sensors relative to each other instead of to some absolute reference. If temperature sensor A is biased by $+2°$, and temperature sensor B is biased by $-3°$, then the relative calibration $F(A, B)$ from A to B is $-5°$. Relative calibration consists of (1) identifying local events by comparing time-series data of two sensors and deriving local mappings via (2) filtering and (3) regression.

- *Identifying local events.* The assumption here is that if local changes in time-series data of the two sensors are highly correlated, the same phenomenon is being observed at both sensors. In the simplest case, linear correlation between the two sensors can be used to identify events.

- *Filtering.* Irrelevant data points can be filtered out by weighing each data point based on correlation and discarding points below a threshold.

- *Regression.* Assuming that the nature of the calibration relationship is known, the parameters for the calibration relationship are derived via regression.

Relative calibration is scalable and independent of known external stimuli but does not relate values of sensors that are far apart.

For sensors that are far apart, *calibration routing* translates a sensor value as it traverses the network. The translation error may grow with the number of

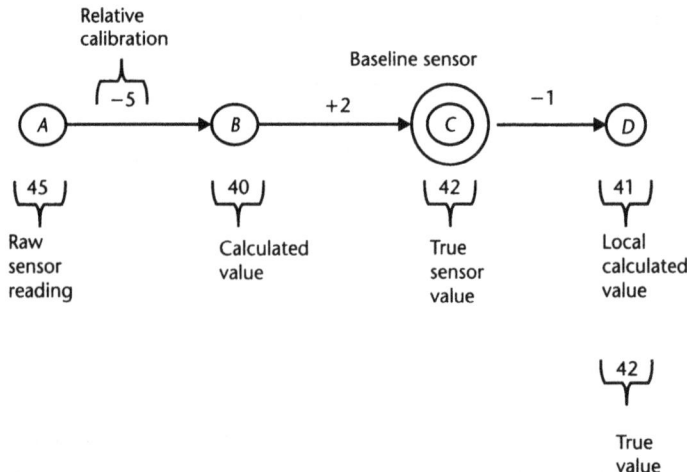

Figure 3.4 Pairwise calibration. The directed edges show the relative-calibration relationship between individual sensors. Calibration routing translates a sensor value (in this case, the raw sensor reading of A) as it traverses the network. The baseline sensor C, in conjunction with calibration routing, transforms the measurements from any sensor to the absolute scale.

hops between the source and the destination and may be further exacerbated by the choice of path. This is addressed by consistency maximization, described in Section 3.4.2.

A *distributed baseline* is a set of "better" sensors distributed among the ordinary sensors to provide an absolute reference. In practice, "better" may mean that these sensors are more expensive or that they have external means of recalibration. The relative-calibration procedure does not distinguish baseline sensors from ordinary sensors. In conjunction with calibration routing, this transforms measurements from any sensor in the network to the absolute scale.

3.4.2 Consistency Maximization

Because the pairwise calibration algorithm does not ensure global consistency, errors accumulated over different forwarding paths in calibration routing are likely to be different. This may lead to inconsistent behavior of a sensing application because the same condition may be observed differently by the distributed parts of the application.

After pairwise calibration relationships have been established, the network forms a calibration graph (CG). Each edge on this graph represents a relative-calibration relationship. If all pairwise calibration relationships (edges) in this graph represent the true relative-calibration relationships, the graph must be consistent. That is, the value delivered from some sensor A to sensor B would be the same independent of the path taken. More formally, *a CG is consistent if, and only if, a convolution of calibration relationships over any cycle in the graph is a null transformation.* For example, in Figure 3.5 the convolution of calibration relationships over cycles C_1, C_2, and C_3 must all be a null transformation.

The computation of an optimal solution to consistency maximization has exponential complexity. A simple algorithm optimizes local consistency of pairwise relationships by considering only short cycles. It works as follows. For each edge going from A to B, find a set of short independent paths from A to B. For each path,

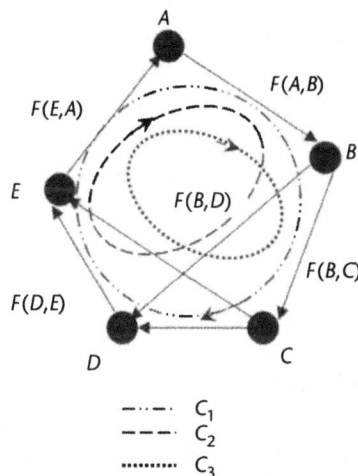

Figure 3.5 Consistency maximization. $F(I, J)$ represents the relative-calibration relationship from sensor I to sensor J. In order to maximize consistency, the summation of $F(I, J)$ values traversed on each cycle C_1, C_2, C_3 must be zero.

translate a set of sensor values x using calibration routing, which results in a set of values y. Determine the parameters of the calibration relationship between A and B by performing a weighted least-squares fit on x and y. The weights for the constraints are allocated based on the length of the translation path.

3.4.3 Discussion

Much of the difficulty in sensor error detection and calibration lies in a lack of reference common to all sensors. Bychkovskiy's technique uses local events as sources of calibration. This requires (1) physical collocation of the sensors being calibrated, and (2) the ability to model the sensed phenomena at the scale of intersensor separation. A small number of baseline sensors are used to relate the measurements to the absolute scale. With in-place sensor calibration, there is no access to ground truth (i.e., the actual value of the physical phenomenon). Thus, a major research challenge is to develop a methodology to evaluate in situ sensor calibration techniques given the lack of ground truth.

3.5 Conclusion

Once deployed in the physical world, factory-calibrated sensors are prone to numerous faults, such as biofouling, that corrupt sensors and sensor data. This motivates techniques for automatic sensor calibration. Sensor calibration is challenging because of the normal variability of sensor measurements, the scarcity of biofouling onset examples, and the variability of the biofouling signature.

Sensor calibration is not only required to ensure data integrity in the sensor network but also to perform fundamental network coordination tasks such as localization and time synchronization [6]. Distributed sensor calibration, in turn, requires localization and time synchronization of spatially distributed sensors (see Chapters 4 and 5).

Sensor calibration can be either active or passive. In this chapter, we discussed two approaches to passive calibration that exploit redundancy in sensing modalities and redundancy in sensor deployment, respectively. In the first approach, natural decreases in one sensor modality (e.g., salinity) can be recognized by monitoring a correlated source of information that is not corrupted by biofouling. In the second approach, densely deployed colocated sensors coordinate with each other to calibrate themselves. The second approach assumes a linear relative-calibration relationship among sensors, which may not be true in several applications. Nonparametric estimation techniques to dynamically derive sensor models and mapping functions for calibration relationships are an emerging area of research [6–11].

Each of the proposed techniques has limitations. In some applications, an orthogonal sensing modality alone may not suffice to detect sensor faults. Conversely, spatiotemporal correlation with collocated sensors may not work if many sensors in a region are corrupted. Sensor-network applications may therefore need to combine several of these approaches to meet their data integrity objectives.

A promising research direction is mobility-enhanced calibration, wherein a mobile actuator could be used to calibrate static sensor nodes [12, 13]. By offering multiple sensing perspectives, mobility can detect calibration errors that may not be detected through coordination among statically collocated sensor nodes.

Distributed calibration techniques for sensor networks and their evaluation methodologies are a relatively nascent area of research. Nevertheless, sensor calibration and fault tolerance will be critical building blocks to meet the system-integrity and quality-assurance requirements of sensor-network applications.

References

[1] Archer, C., T. K. Leen, and A. Baptista, "Parameterized Novelty Detection for Environmental Sensor Monitoring," In Thrun S., Saul L., and Obermayer, K. (eds.), *Advances in Neural Information Processing Systems 16*, Cambridge, MA: The MIT Press, 2000.

[2] Taylor, C. J., "A Scheme for Calibrating Smart Camera Networks Using Active Lights," *Proc. ACM SenSYS 2004*, Baltimore, MD, November 2004.

[3] Marzullo, K., "Tolerating Failures of Continuous Valued Sensors," *ACM Transactions on Computer Systems*, Vol. 8, No. 4, 1990, pp. 284–384.

[4] Bychkovskiy, V., et al., "A Collaborative Approach to In-Place Sensor Calibration," In Feng Zhao and Leonidas J. Guibas, (eds.), *Proc. Information Processing in Sensor Networks, Second International Workshop, IPSN 2003*, Palo Alto, CA, April 22–23, 2003.

[5] Myers, R. *Classical and Modern Regression with Applications*. Boston, MA: Duxbury Press, 1990.

[6] Whitehouse, K., and D. Culler, "Calibration as Parameter Estimation in Sensor Networks," *Proc. ACM WSNA*, Atlanta, GA, 2002, p. 67.

[7] Ihler, A., et al., "Nonparametric Belief Propagation for Sensor Network Self-calibration," *Proc. IEEE IPSN 2004*, Berkeley, CA, April 2004.

[8] Whitehouse, K., and D. Culler, "Macro-Calibration in Sensor/Actuator Networks," *Mobile Networks and Applications*, Vol. 8, No. 4, 2003, pp. 463–472.

[9] Zhang, Q., and R. Pless, "Extrinsic Calibration of a Camera and Laser Range Finder," *Proc. IEEE International Conference on Intelligent Robots and Systems (IROS)*, Sendai, Japan, 2004.

[10] Feng, J., S. Megerian, and M. Potkonjak, "Model-based Calibration for Sensor Networks," *Proc. IEEE International Conference on Sensors*, Toronto, Ontario, Canada, October 2003.

[11] Mukhopadhyay, S., D. Panigrahi, S. Dey, "Model Based Error Correction for Wireless Sensor Networks," *Proc. 1st IEEE Communications Society Conference on Sensor and Ad Hoc Communications and Networks (SECON '04)*, Santa Clara, CA, October 2004, pp. 575–584.

[12] Kansal, A., et al., "Sensing Uncertainty Reduction Using Low Complexity Actuation," *Proc. ACM Third International Symposium on Information Processing in Sensor Networks (IPSN) 2004*, Berkeley, CA, April 2004.

[13] Ganeriwal, S., A. Kansal, and M. B. Srivastava, "Self-aware Actuation for Fault Repair in Sensor Networks," *Proc. IEEE International Conference on Robotics and Automation (ICRA) 2004*.

[14] Ma, D., and J. M. Hollerbach, "Identifying Mass Parameters for Gravity Compensation and Automatic Torque Sensor Calibration," *Proc. IEEE International Conf. Robotics and Automation*, New Orleans, LA, April 1996, pp. 661–666.

CHAPTER 4

Localization

Nirupama Bulusu

4.1 Introduction

Localization is a mechanism for autonomously discovering and establishing spatial relationships among objects; autonomous localization of sensor devices is a crucial systems building block for embedded sensor-network applications [1]. Because embedded sensor-network applications are coupled to the physical world, localization measures and gives a context to the physical coupling. Many of these systems are embedded to monitor or control the behavior of physical systems as opposed to strictly virtual information systems; therefore, nodes often need to determine their action based on their physical location (am I the right sensor to monitor a particular object?).

Localization benefits span several layers [2] of the sensor-network application. At the application layer, localization is indispensable for sensor-network applications that select services based on location [3, 4] and for sensor networks that achieve power conservation by combining data from multiple sensors. At the network layer, location information on a scale with the transmission range can directly enable geographic routing algorithms that propagate information more efficiently through a multihop network [5].

One of the best-known and most widely used localization systems is the global positioning system (GPS). GPS [6] solves the problem of localization in outdoor environments for personal digital assistant (PDA)-class nodes. However, GPS does not always meet the operational (e.g., low power), cost, or environmental constraints (e.g., indoor operation) for sensor networks.

This has motivated significant research in sensor-network localization. Several issues render the localization problem more challenging for large-scale, densely distributed sensor networks than for many other domains. As we have argued in chapter one, sensor networks must satisfy several physical constraints. In order to be untethered and deeply embedded, sensor nodes must have a small form factor and provide their own energy. The system overall must tolerate ad hoc deployment and unattended operation without infrastructure support. Thus, any deployable localization solution must scale to large areas and to large numbers of devices and must accommodate the device constraints of small sensor nodes.

Approaches to sensor-network localization typically deploy a small number of known nodes, also known as *references*, *beacons*, or *anchors*. They estimate the coordinates of unknown nodes from geometric constraints to the known nodes (and sometimes to unknown nodes as well). Figure 4.1 provides an illustration of some of the various types of constraints.

In Section 4.2, we explore the application requirements that influence the design of a localization system. We study the various components that constitute a localization system and review the current state of the art in ranging and positioning in Section 4.3. In Section 4.4, we review recent developments in robust sensor-network localization, and we conclude in Section 4.5.

4.2 Application Requirements

As sensor-network applications are widely varied, so are their localization requirements. They include the following:

- *Granularity and scale.* What are the smallest and largest measurable distances? For example, local coordinate systems for a sensor network deployed in a building might scale from centimeters to hundreds of meters, whereas GPS coordinates for a tsunami warning system have a global scale and a granularity on the order of meters.

- *Accuracy and precision.* How close should the answer be to the ground truth (accuracy), and how consistent are the answers (precision)?

- *Relative or absolute positioning.* Do we need to determine the locations of sensor devices relative to each other, or do we need an absolute frame of reference?

- *Dynamics.* Are the environments static or dynamic? Do they involve mobile or fixed sensor devices? How frequently should sensor devices be localized?

- *Cost.* What is the desired cost of individual sensor nodes? What is the desired infrastructure and installation cost?

- *Form factor.* How large are sensor devices? How much power can they have?

- *Energy and communications constraints.* What are the energy and communication constraints in the system? How do sensor devices coordinate with each other? Who initiates the localization process?

- *Environment.* What is the environment in which the sensor device must be localized—indoors, outdoors, underwater, underground, or on Mars? Each environment poses its own unique challenges for localization. Thus, an indoor

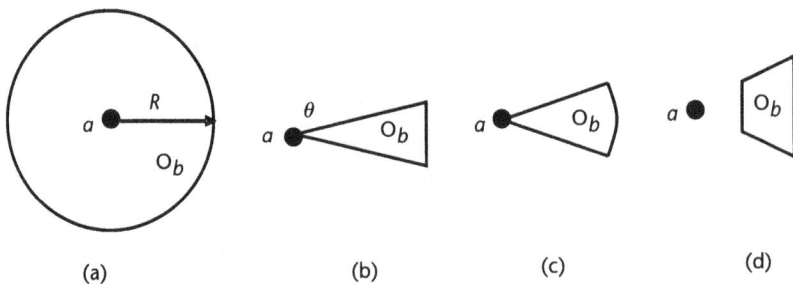

Figure 4.1 Examples of geometric constraints: (a) radial, (b) angular, (c) quadrant, and (d) trapezoid. Node *B* must lie in the region anchored by node *A*.

localization system should account for multipath effects, whereas an outdoor localization system should account for weather variations.

Because application requirements are complex and widely varied, no single localization technology can serve all applications. This motivates a spectrum of localization solutions, which we review in the next section.

4.3 Localization Systems Components

Generally, the process of node localization consists of three stages (see Figure 4.2). First, we estimate the constraints that characterize the absolute or relative spatial relations between pairs of sensor nodes. Second, we combine these constraints to establish coordinates of individual sensor nodes. Finally, we stitch and refine the location measurements across large groups of sensors.

The design of a localization system encompasses the selection of (1) a measurement technique to determine constraints in spatial relationships (such as distance and angles), (2) a system architecture, and (3) robust position-estimation algorithms.

4.3.1 Measurement Techniques

We classify the various measurement techniques into two broad categories based on the granularity of the constraints measured. Approaches that infer fine-grained information, such as the distance to a reference point based on signal strength or timing measurements, fall into the category of *fine-grained measurement techniques*; those that infer coarse-grained information, such as proximity to a given reference point, are categorized as *coarse-grained measurement techniques*.

4.3.1.1 Fine-Grained Measurement Techniques

Fine-grained measurement techniques can be classified further into range-finding and directionality-based methods, depending on whether ranges or angles relative

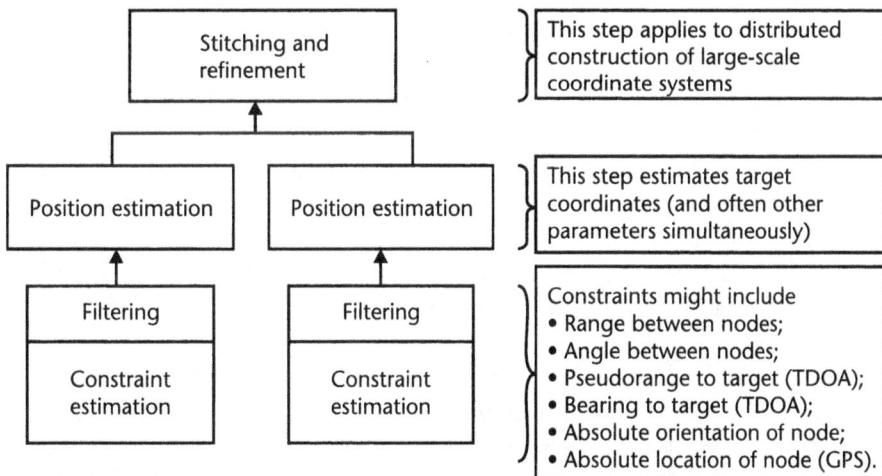

Figure 4.2 Stages of sensor-network localization.

to reference points are being inferred. Additionally, signal-pattern-matching methods are also included in fine-grained localization methods.

In range-finding methods, the ranges of the receiver node to several reference points are determined by one of several timing- or signal-strength-based techniques. The position of the node can then be computed using multilateration (see, e.g., [7]). We discuss timing- and signal-strength-based range-finding methods separately.

Timing.   The distance between the receiver node and a reference point can be inferred from the time of flight of the communication signal. The time of flight may be calculated using the timing-advance technique, which measures the how much the timing of the measuring unit has to be advanced in order for the received signal to fit into the correct time slot. This technique is used in GPS [1] and Pinpoint's Local Positioning System (LPS) [8]. GPS measures one-way flight time, whereas LPS measures round-trip time (thereby eliminating the need for time synchronization).

GPS [6] is a wide-area radio positioning system. In GPS, each satellite transmits a unique code, a copy of which is created in real time in the user set receiver by the internal electronics. The receiver then gradually time-shifts its internal clock until it corresponds with the received code, an event called *lock-on*. Once locked onto a satellite, the receiver can determine the exact timing of the received signal in reference to its own internal clock. If that clock were perfectly synchronized with the satellite's atomic clocks, the distance to each satellite could be determined by subtracting a known transmission time from the calculated receive time. In real GPS receivers, the internal clock is not quite accurate enough. An inaccuracy of a mere microsecond corresponds to a 300m error.

Pinpoint's 3D-iD system is an LPS that covers an entire three-dimensional (3-D) indoor space and is capable of determining the 3-D location of items within that space. The LPS subdivides the interior of the building into cell areas that vary in size with the desired level of coverage. The cells are each handled by a cell controller attached by a coaxial cable to up to 16 antennas. It provides an accuracy of 10m for most indoor applications, although some may require accuracy of 2m. The main drawback of this system is that it is centralized and requires significant infrastructural setup.

An innovative approach to calculate one-way time of flight is to make explicit time-of-arrival measurements based on two distinct modalities of communication, *ultrasound* and *radio*, as in the Active Bat system [7] and more recently in [9–12]. These two modalities travel at vastly different speeds ($350\ ms^{-1}$ and $3 \times 10^{-8}\ ms^{-1}$, respectively), enabling the radio signal to be used for synchronization between the transmitter and the receiver and the ultrasound signal to be used for ranging, as illustrated in Figure 4.3. The Active Bat system, however, relies on significant effort for deployment indoors. Ultrasound systems may not work very well outdoors because they all use a single transmission frequency (40 kHz); hence, there is a high probability of interference from other ultrasound sources.

Signal strength.   An important characteristic of radio propagation is that attenuation of the radio signal typically increases with the distance between the transmitter and receiver. Radio-propagation models in various environments have been well researched [13] and have traditionally focused on predicting the average received-signal strength at a given distance from the transmitter using

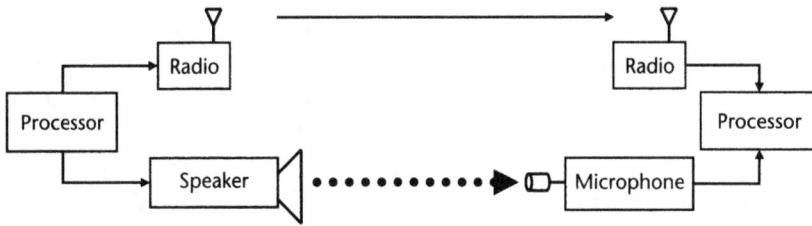

Figure 4.3   Ranging with radio and acoustics. The sender transmits a radio and an acoustic signal at the same time. The radio signal arrives first in a negligible amount of time. The receiver uses the arrival time of the radio to compute the time of flight of the acoustic signal, requiring no explicit synchronization with the sender.

large-scale propagation models, as well as the variability of the signal strength in close spatial proximity to a location using small-scale or fading models. In the RADAR system [14], P. Bahl and V. N. Padmanabhan propose estimating distance based on signal strength in indoor environments. They compute distance from measured signal strength by applying a wall attenuation factor (WAF)–based signal-propagation model. The distance information is then used to locate a user by triangulation. This approach, however, yields lower accuracies than RF mapping of signal strengths corresponding to various locations for their system. Their RF-mapping-based approach is quite effective indoors but requires extensive infrastructural effort, making it unsuitable for rapid or ad hoc deployment.

Signal-pattern matching.   Another fine-grained measurement technique is the proprietary Location Pattern Matching technology used in U.S. Wireless Corporation's RadioCamera system [15]. Instead of exploiting signal timing or signal strength, it relies on signal-structure characteristics. It turns the multipath phenomenon to surprisingly good use: by combining the multipath pattern with other signal characteristics, it creates a signature unique to a given location. The RadioCamera system includes a signal-signature database for a location grid of a specific service area. To generate this database, a vehicle drives through the coverage area transmitting signals to a monitoring site. The system analyzes the incoming signals, compiles a unique signature for each square in the location grid, and stores it in the database. Neighboring grid points are spaced about 30m apart. To determine the position of a mobile transmitter, the RadioCamera system matches the transmitter's signal signature to an entry in the database. The system can use data from only a single point to determine location. Moving traffic and changes in foliage or weather do not affect the system's capabilities. The major drawback of this technique, as with RADAR [14], is the substantial effort needed for generation of the signal-signature database. Consequently, it is difficult to use in ad hoc–deployment scenarios.

Directionality.   Another way of estimating location is to compute the angle of each reference point with respect to the mobile node in some reference frame. The position of the mobile node can then be computed using triangulation methods.

An important example of a directionality-based system is the VOR/VORTAC station [16], which was used for long-distance aviation navigation prior to GPS. The VOR station transmits a unique omnidirectional signal that allows an aircraft aloft to determine its bearing relative to the VOR station. The VOR signal is electrically phased so that the received signal is different in various parts of the 360° circle. By determining which of the 360 different radials it is receiving, the aircraft can determine the direction of each VOR station relative to its current position.

Small-aperture direction finding is yet another directionality-based technique used in cellular networks. It requires a complex antenna array at each cell-site location. The antenna arrays can, in principle, work together to determine the angle (relative to the cell site) from which a cellular signal originates. When several cell sites can determine their respective angles of arrival, the cell phone location can be estimated by triangulation. Two drawbacks to this approach make it inapplicable to our application domain. The cost of the complex antenna array implies that it can be placed only at the cell sites. Second, the cell sites are responsible for determining the location of the mobile node, which will not scale well when we have a large number of such nodes and desire a receiver-based approach.

Directionality-based methods are considered not to be very effective in indoor environments because of multipath effects prevalent in these environments.

4.3.1.2 Coarse-Grained Measurement Techniques

Coarse-grained techniques estimate proximity from connectivity measurements. The *Active Badge* [17] system was one of the earliest indoor localization systems. Each person or object is tagged with an Active Badge. Every 10 seconds, the badge transmits a unique IR signal that is received by sensors placed at fixed positions within a building and relayed to the location-manager software. The location-manager software is able to provide information about the person's location to the requesting services and applications.

Another system based on IR technology is described in [18] and is the precursor to the current HiBall tracking system [19] used for virtual reality applications. This system requires IR transmitters to be located at fixed positions inside the ceiling of the building. An optical sensor sitting on a head-mounted unit senses the IR beacons, and system software determines the position of the person.

Both of these IR-based solutions perform quite well in indoor environments because IR range is fairly small and can be limited to the logical boundaries of a region, such as a room (bounded by walls). The short range of IR, which facilitates location, is also a major drawback of these systems because the building has to be wired with a significant number of sensors. In the few places where such systems have been deployed, sensors have been physically wired in every room of the building. Such a system scales poorly and incurs significant installation, configuration, and maintenance costs. IR also tends to perform poorly in the presence of direct sunlight and, hence, cannot be used outdoors.

More recent work uses radio signals to infer proximity [1]. This technique works reasonably well in open, uncluttered environments. On the other hand, the same technique cannot be applied in indoor environments because RF propagation in indoor environments suffers from severe multipath effects that make it impossible to characterize radio proximity.

4.3.2 System Architecture

Localization systems using similar measurement techniques can differ considerably in their system architecture. For example, Active Bat [7], GALORE Panel [20],

Cricket [10], and AHLoS [12] all use the same ranging technology (radio plus ultra-sound), but their system architectures are centralized, hierarchical, decentralized, and iterative respectively. The choice of the system architecture is influenced by application requirements—such as the need for highly accurate or real-time position estimation.

The system could be either *tightly coupled,* using beacons that are wired to a centralized controller and placed at fixed positions, or *loosely coupled,* using beacons that are wireless. Loosely coupled systems could further be classified as hierarchical, decentralized, or iterative.

Tightly coupled systems.   Several traditional and mature localization technologies use a tightly coupled system architecture. These include the Active Bat system [7] developed for sentient computing applications and the HiBall tracker designed for virtual reality applications. These applications have high accuracy and real-time tracking requirements.

In the Active Bat system, the beacons are passive and record the time of flight of signals received from active devices. Problems of time synchronization and coordination among beacons are easily resolved because these systems are wired and have a centralized controller. These systems therefore achieve high accuracy, but the centralized position estimation limits the number of devices the systems can simultaneously track (HiBall). Moreover, wiring impedes deployment. How can we achieve the accuracy of these systems outdoors, where deployment cannot be controlled, and wiring is infeasible?

Loosely coupled systems.   Motivated by scaling and deployment concerns, sensor localization systems are not centralized and are typically wireless. They sacrifice the accuracy of tightly coupled systems for ease of deployment and scalability to large numbers of devices.

Cricket uses a completely decentralized approach. It deploys a system of active beacons, each of which periodically transmits an advertisement containing its position. Clients compute their position based on the advertisements that they receive.

GALORE uses a more pragmatic hierarchical approach to support small sensor devices such as the motes. A group of beacons use cooperative ranging to localize themselves. Each beacon then provides a location service to a herd of motes. After ranges to all the motes from a beacon are determined, the beacon computes the location for all the motes it serves.

Finally, AHLoS devices compute their location from a small number of beacons or anchors using an iterative process, which we describe in Section 4.3.3.

Because beacons are deployed in an ad hoc manner in all of these systems, coverage and coordination among beacons must be addressed for large-scale ad hoc deployment.

4.3.3   Robust Positioning Algorithms

Besides ranging technologies and system architecture, a third component of a localization system is a problem formulation and algorithm solution for robust position estimation and formation of a coordinate system. We comment on four popular techniques below.

- *Monte Carlo localization.* In the field of mobile robotics, localization has been referred to as the most fundamental problem with regard to providing a mobile robot with autonomous capabilities. Environmental obstructions such as walls, moving people, and objects can greatly interfere with the sensing capabilities of a mobile robot. Statistical techniques provide a means to represent uncertainty in sensor measurements. For example, robot localization is formulated as an instance of a general, statistical inference problem [21]. Monte Carlo localization algorithms represent a robot's measurement estimates probabilistically by a set of weighted hypotheses that approximate the posterior under a common Bayesian formulation of the localization problem. These algorithms are versatile, resource-adaptive, and robust under a range of circumstances and can be applied to sensor networks. However, they are not as computationally efficient as other localization algorithms.

- *Convex optimization.* One way to formalize the node-localization problem is to express relations between pairs of sensor nodes (proximity, angles) as a set of convex constraints. L. Doherty, L. El Ghaoui, and K. S. J. Pister's convex-constraint-satisfaction approach [22] formulates the localization problem as a feasibility problem with radial or angular constraints. This convex-constraint problem is in turn solved by semidefinite programming (an interior point method) to find a globally optimal solution. To work well, the technique requires anchor nodes to be placed at the corners. Only in this configuration are the constraints tight enough to yield a useful configuration. When all anchors are located in the interior of the network, the position estimation can easily collapse toward the center, which leads to large estimation errors.

- *Iterative multilateration.* Multilateration is the process of estimating a node's location from ranges to three or more beacons. If not all nodes have ranges to three beacons, then their positions must be computed through an iterative process, such as in [12]. In iterative multilateration, all sensor nodes are preplaced and listening. Nodes have two modes—*localized* and *unlocalized*. Nodes

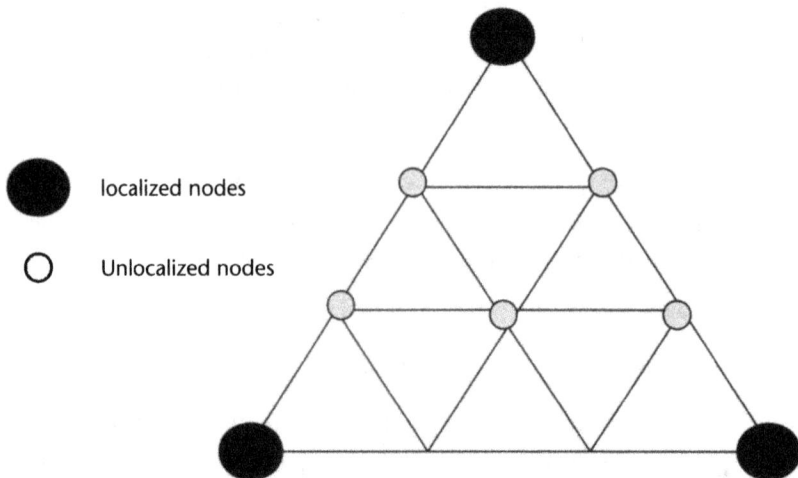

Figure 4.4 Iterative multilateration. Initially only the three beacon vertices of the triangle are localized. Eventually, all sensor nodes are localized as the triangle gets reflected inwards.

determine distance from transmissions heard. Iterative multilateration, illustrated in Figure 4.4, works as follows:

1. *Localized mode.* Nodes broadcast their position.
2. *Unlocalized mode.* Nodes listen for position broadcasts. If a broadcast from a node B_i at (x, y) is heard, determine distance to B_i. If a node has heard three or more broadcasts, it determines its position and switches to unlocalized mode.

Iterative multilateration is a completely decentralized algorithm. However, it incurs additional overhead in terms of energy and communication costs.

- *Multidimensional scaling (MDS).* MDS [23], a technique from mathematical psychology, is a set of data analysis techniques that displays distancelike data as a geometrical picture. Using a matrix of proximity (or distance) data among sensor nodes, MDS finds a configuration of points in multidimensional space such that the interpoint distances are related to the provided proximity data by some transformation (e.g., a linear transformation). The MDSMAP algorithm proposed by Shang, Y., et al. in [23] works as follows:

  1. Compute the shortest paths between pairs of all nodes in the region of consideration. The shortest path distances are used to construct the distance matrix for MDS.
  2. Apply classical MDS to the distance matrix, retaining the first two (or three) largest eigenvalues and eigenvectors to construct a two-dimensional (2-D), or 3-D, relative map.
  3. Given sufficient anchor nodes (2- or 3-D), transform the relative map to an absolute map based on the absolute positions of anchor nodes.

In this section, we have reviewed the components of a localization system, system-architecture design choices, measurement techniques, and algorithms for position estimation. Iterative position estimation algorithms, such as MDS and iterative multilateration, may not be robust when the multihop paths are exaggerated due to obstructions such as walls and obstructions. In the next section, we review recent developments to ensure the robustness of sensor-network localization.

4.4   Recent Developments

State-of-the-art localization systems such as Cricket, AHLoS, and the GALORE panel can estimate device locations to an accuracy of a few centimeters, given adequate availability of reference nodes or beacons. Recent research has begun to develop a theory of sensor-network localization to mathematically characterize these conditions and constraints.

When are sensor nodes uniquely localizable?   T. Eren et al. [24] address this problem using structural graph rigidity. They propose the construction of grounded graphs, wherein each vertex represents a network node, and two vertices in the graph are connected if the distance between the two is known. They show that a network has unique localization if, and only if, its corresponding grounding graph is generically globally rigid (i.e., the distance between each pair of nodes is unambiguous). Figure 4.5 shows an example of a generically globally rigid graph.

For networks based on random locations and a communication radius r, they provide some necessary and some sufficient conditions for the network to be uniquely localizable with high probability, as summarized in Table 4.1. In 2-D space, with the number of sensor nodes $n \geq 4$, they show that *3-connectivity* is a necessary condition, and *6-connectivity* is a sufficient condition for the sensor network to be uniquely localizable with high probability. For a system like AHLoS, which uses iterative multilateration, they show that a high beacon density is required for sensor nodes to be uniquely localizable in linear time. They calculate the expected time of computation for sensor localization as a function of the number of beacons and the sensing or ranging radius, both of which can be expressed in terms of the number of nodes in sensor network n.

Eren et al.'s work provides an important mathematical foundation for sensor-network localizability, but their analysis is limited to static sensor networks and distance constraints. Unique localizability in three dimensions with angular constraints and mobility are open problems for future research.

How can we ensure that sensor nodes are uniquely localizable?   This requires that we have a spatially dense deployment of beacon or anchor nodes, which might not always be feasible or cost-effective. A promising recent approach to address this problem is mobility-enhanced localization [25, 26], wherein a mobile beacon or user can be used to emulate spatially dense beacon deployment. In the Robust Extended

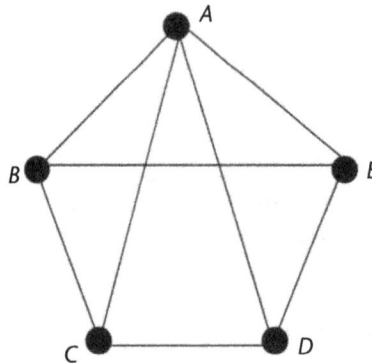

Figure 4.5 A generically globally rigid graph.

Table 4.1 Expected Time of Computation for Localization as a Function of Number of Beacons and Sensing Radius

Beacons	Sensing Radius	Expected Time of Computation
$O(1)$	$O\left(\sqrt{(\log n)/n}\right)$	$O\left(\sqrt{n/(\log n)}\right)$
$O(n/(\log n))$	$O\left(\sqrt{(\log n)/n}\right)$	$O\left(\sqrt{(\log n)}\right)$
$O(n)$	$O\left(\sqrt{(\log n)/n}\right)$	$O(1)$

Kalman Filter (REKF) localization system [26], a mobile robot estimates the location of static sensor devices in its coverage area using radio-signal strength. This has improved coverage properties over traditional localization approaches because fewer beacons are required. Moreover, because the robot receiver is mobile, over a period of time, the fading noise in radio-signal-strength measurements can be eliminated, which would not be possible with a static transmitter-receiver pair.

How do we make sensor localization secure? In many sensor applications involving humans, ensuring the security and privacy of location information may be vital. SeRLoC [27] is a scheme for secure localization that uses lightweight encryption of location beacons and authentication of locator identifiers. This allows sensor localization to be secure even when the sensor network is under attack.

Widespread deployment of sensor networks enables location-based access control, which requires secure location verification. N. Sastry, U. Shankar, and D. Wagner [28] developed the Echo protocol for secure verification of location claims. The protocol begins when the verifier node sends a packet containing a nonce to the prover using radio; the prover immediately echoes the packet back to the verifier using ultrasound. If the elapsed time from the initial transmission of the echo packet to its reception is more than expected, then the verifier node rejects the prover's claim; otherwise, it accepts.

To ensure privacy of location data, M. Gruteser et al. [29] use a privacy-aware location sensor network through a distributed anonymity algorithm that is applied before the location data is accessed.

4.5 Conclusion

In our previous work [1], we argued that autonomous localization of wireless sensor devices would be a crucial building block for sensor networks. Since then, a very substantial body of theory and systems research on sensor-network localization has emerged. The design of a localization system typically encompasses constraint estimation (e.g., a ranging technology and algorithm), system architecture, and a positioning algorithm. In this chapter, we have reviewed state-of-the-art ranging techniques and positioning algorithms and commented on the conditions to ensure unique localizability of sensor nodes.

State-of-the-art localization systems, now commercially available, can estimate device locations with an accuracy of few centimeters, given adequate constraints. Both theory [24, 25] and analysis [30, 31] show that it is important to ensure the availability of adequate constraints in ad hoc–deployed localization systems for sensor-network localizability. As we have explored, a promising recent approach to improve the localizability of a sensor network is mobility-enhanced localization. Nevertheless, significant challenges persist to ensure that localization is self-configuring and robust not only in laboratory settings but also in unknown environments where real-world applications will be deployed.

Both radio and acoustic ranging are vulnerable to obstructions in the environment. Ultrawideband radio ranging is being proposed as an alternative ranging technology, which can penetrate objects and walls [32], although it has challenging time-synchronization requirements.

Increasingly, sensor networks will be deployed in underwater and underground environments for applications ranging from habitat monitoring to oil-field exploration. New ranging technologies must be developed that operate robustly in such environments.

Both the localization needs of an application and the design space of localization solutions vary widely. Because of the wide range of application requirements, a localization solution for an application must be carefully chosen according to its needs.

References

[1] Bulusu, N., Heidemann J., and Estrin D., "GPS-less Low Cost Outdoor Localization for Very Small Devices," *IEEE Wireless Communications*, Vol. 7, No. 5, 2000, pp. 28–34.

[2] Zimmerman, H., "OSI Reference Model—The ISO Model of Architeciure for Open Systems Interconnection," *IEEE Transactions on Communications*, Vol. COM-28, No. 4, 1980, pp. 425–432.

[3] Harter, A., et al., "The Anatomy of a Context-Aware Application," *Proc. ACM MOBICOM 1999*, Seattle, WA, August 1999, pp. 59–68.

[4] Hightower, J., and Borriello G., "Location Systems for Ubiquitous Computing," *IEEE Computer*, Vol. 34, No. 8, 2001, pp. 57–66.

[5] Karp, B., and Kung H. T., "GPSR: Greedy Perimeter Stateless Routing for Wireless Networks," *Proc. ACM MOBICOM 2000*, Boston, MA, August 2000, pp. 243–254.

[6] Hofmann-Wellenhof, B., H. Lichtenegger, and Collins J., *Global Positioning System: Theory and Practice*, 4th ed., Springer Verlag, Berlin, Germany, 1997.

[7] Ward, A., Jones A., and Hopper A., "A New Location Technique for the Active Office," *IEEE Personal Communications*, Vol. 4, No. 5, October 1997, pp. 42–47.

[8] Werb, J., and Lanzl C., "Designing a Positioning System for Finding Things and People Indoors," *IEEE Spectrum*, Vol. 35, No. 9, September 1998, pp. 71–78.

[9] Girod, L., "Development and Characterization of an Acoustic Rangefinder," Tech. rep. 00-728, Computer Science Department, University of Southern California, April 2000.

[10] Priyantha, N., Chakraborty A., and Balakrishnan H., "The Cricket Location Support System," *Proc. ACM MOBICOM*, Boston, MA, August 2000.

[11] Priyantha, N., et al., "The Cricket Compass for Context-Aware Mobile Applications," *Proc. ACM MOBICOM*, Rome, Italy, August 2001, pp. 32–43.

[12] Savvides, A., Han C. C., and Srivastava M. B., "Dynamic Fine-Grained Localization in Wireless Ad Hoc Sensor Networks," *Proc. ACM MOBICOM 2001*, Rome, Italy, August 2001, pp. 166–179.

[13] Rappaport, T. S., *Wireless Communications—Principles and Practice*, Upper Saddle River, NJ: Prentice Hall, 1996.

[14] Bahl, P., and Padmanabhan V. N., "Radar: An In-Building RF-based User Location and Tracking System," *Proc. IEEE INFOCOM 2000*, Vol. 2, Tel Aviv, Israel, March 2000, pp. 775–784.

[15] http://www.sss-mag.com/eq11.html.

[16] http://www.navfitsm.addr.com/vor_Aau.htm.

[17] Want, R., et al., "The Active Badge Location System," *ACM Transactions on Information Systems*, Vol. 10, No. 1, January 1992, pp. 91–102.

[18] Azuma, R., "Tracking Requirements for Augmented Reality," *Communications of the ACM*, Vol. 36, No. 7, July 1993, pp. 50–55.

[19] Welch, G., et al., "The HiBall Tracker: High-Performance Wide-Area Tracking for Virtual and Augmented Environments," *Symposium on Virtual Reality Software and Technology*, December 1999, London, U.K., pp. 1–10.

[20] Girod, L., et al., "Locating Tiny Sensors in Time and Space: A Case Study," *Proc. ICCD 2002*, Freiburg, Germany, September 2002.

[21] Fox, D., et al., "Bayesian Filtering for Location Estimation," *IEEE Pervasive Computing*, IEEE Computer Society Press, 2003, Vol. 2, No. 3, pp. 24–33.

[22] Doherty, L., El Ghaoui L., and Pister K. S. J., "Convex Position Estimation in Wireless Sensor Networks," *Proc. IEEE Infocom 2001*, Anchorage, AK, April 2001.

[23] Shang, Y., et al., "Localization from Mere Connectivity," *Proc. ACM MOBIHOC 2003*, Annapolis, MD, June 2003.

[24] Eren, T., et al., "Rigidity, Computation, and Randomization in Network Localization," *Proc. International Annual Joint Conference of the IEEE Computer and Communications Societies (INFOCOM)*, Hong Kong, China, March 2004, pp. 2673–2684.

[25] Moore, D., et al., "Robust Distributed Network Localization with Noisy Range Measurements," *Proc. ACM SenSYS 2004*, Baltimore, MD, November 2004.

[26] Pathirana, P., et al., "Node Localization Using Mobile Robots in Delay-Tolerant Sensor Networks," *IEEE Transactions on Mobile Computing*, Vol. 4, No. 4, July 2005.

[27] Lazos, L., and Poovendran R., "SeRLoc: Secure Range-Independent Localization for Wireless Sensor Networks," *Proc. 2004 ACM Workshop on Wireless Security*, Philadelphia, PA., 2004, pp. 21–30.

[28] Sastry, N., Shankar U., and Wagner D., "Secure Verification of Location Claims," *Proc. 2003 ACM Workshop on Wireless Security*, San Diego, CA, 2003, pp. 1–10.

[29] Gruteser, M., et al., "Privacy-Aware Location Sensor Networks," *Proc. Workshop on Hot Topics in Operating Systems (HotOS)*, Lihue, HI, 2003.

[30] He, T., et al., "Range Free Localization Schemes for Large Scale Sensor Networks," *Proc. ACM Mobicom 2003*, San Diego, CA, 2003, pp. 81–95.

[31] Bulusu, N., "Self-configuring Localization Systems," Ph.D thesis, University of California, Los Angeles, CA, October 2002.

[32] Aetherwire corporation, at http://www.aetherwire.com.

CHAPTER 5

Time Synchronization

Saurabh Ganeriwal, Jeremy Elson, and Mani B. Srivastava

5.1 Introduction

Time synchronization is critical to a sensor network at many layers of its design. It enables better duty cycling of the radio, accurate localization, beamforming, and other collaborative signal processing. Examples of existing sensor-network applications that need precise time include measuring the time of flight of sound [1]; distributing an acoustic beam-forming array [2]; forming a low-power time division multiple access (TDMA) radio schedule [3]; integrating a time series of proximity detections into a velocity estimate [4]; suppressing redundant messages by recognizing duplicate detections of the same event by different sensors [5]; ordering logging of events during system debugging; integrating multisensor data; or coordinating on future action [6].

The time-synchronization problem has been investigated thoroughly on the Internet and in local-area networks (LANs). Many existing algorithms rely on time information from GPS. However, GPS has drawbacks: it is not available ubiquitously (e.g., under foliage, indoors, underwater) and requires a relatively high-power receiver, which is not feasible on small and cheap sensor nodes. This motivates the need for developing pure software-based approaches to achieve in-network time synchronization. Over the years, many protocols have been developed for maintaining synchronization in computer networks [7–10]. Among these, Network Time Protocol (NTP) [10] stands out by virtue of its scalability, self-configuration in large multihop networks, robustness regarding failures, and ubiquitous deployment. Furthermore, the combination of NTP and GPS has proven very successful, achieving accuracy on the order of a few microseconds. Many in the sensor-network research community have asked why GPS and NTP cannot be used in the context of sensor networks. The essence of the answer is that sensor networks are, in some ways, a radical departure from traditional networking; they break many of the assumptions on which prior work is based. While existing solutions are critical building blocks, we argue that the time-synchronization requirements are drastically different in the context of WSNs.

A plethora of applications envisioned for sensor networks impose a diverse set of requirements on timing-synchronization accuracy. For example, acoustic applications require precision of several microseconds, while sensor tasking works on the timescale of hours or days. Local collaborations often require only a pair of neighbors to be synchronized, while global queries require a unique global timescale.

Event triggers may only require momentary synchronization, while data logging or debugging often require an eternally persistent timescale. Communication with a user requires an external, human timescale such as Coordinated Universal Time (UTC), whereas only relative time is important for purely in-network comparisons. Thereby, a one-size-fits-all solution such as NTP or GPS is not suitable for these networks. Any time-synchronization approach for WSNs should be *multimodal*. The scheme should not only be able to adapt itself to the runtime system dynamics but should also be able to optimize on different axes (e.g., scope, availability, precision, and persistence of the timescale). For this, it is paramount that the scheme have easily changeable parameters that can be customized for specific applications. Without *tunability*, the synchronization scheme might be wasting precious resources such as energy or bandwidth to provide an error bound that is not even needed by the application. For example, the same periodicity of synchronization messages should not be used for acoustic target detection (accuracy of 50–100 μs) and synchronized sampling in seismic sensor networks (accuracy of 1 ms).

Furthermore, this new class of networks has a large density of nodes and very limited energy resources at every node; this leads to scalability requirements while limiting the resources that can be used to achieve them. The need for energy efficiency violates a number of assumptions routinely made by classical synchronization algorithms (such as NTP)—that using the CPU in moderation is free, listening to the network is free, and occasional packet transmissions have a negligible impact on the node lifetime.

Having realized the inadequacy of existing synchronization approaches, there is a need to develop a simple, scalable, and energy-efficient solution to the problem of timing synchronization in sensor networks that is also flexible enough to meet the desired levels of accuracy and algorithmic overhead.

5.2 Models for Time Synchronization

After observing the diversity in the need for time-synchronization requirements by different applications, researchers in sensor networks have studied the problem in the context of different models:

- *Virtual clocks.* The first and perhaps the simplest type of model concentrates on just maintaining the relative notion of time between nodes. This approach is motivated by the applications that need to establish a temporal order of events without caring about the absolute time of their occurrence. Schemes proposed in [11] and [12] take this idea to its extreme by doing away with physical clocks entirely in favor of virtual clocks that only tick when an event occurs. In [12], K. Romer achieves 1-ms precision in establishing a correct chronology of events in sensor networks.

- *Internal synchronization.* This model is not extendible to scenarios where the nodes need to maintain synchronized clocks, such as distributed signal processing, synchronized sampling, planned coordinated action, and so forth. In many such applications, the exact time at which the plan is executed is far less important than ensuring that all nodes act simultaneously. Examples include

TDMA frame scheduling, distributed robotics, and signal processing. This class of applications requires internal or relative synchronization: the network must be internally consistent, but its relationship to outside time standards is not needed or known. Reference Broadcast Synchronization (RBS) [13] aims to provide relative synchronization in sensor networks.

- *External synchronization.* Perhaps the most complex ("always-on") model is the one where every node maintains a clock that is synchronized with respect to a reference node in the network. The aim here is to maintain a global and unique timescale throughout the network. Although this model consumes maximum energy, it is a superset of all the models. Therefore, if the clocks are absolutely synchronized, they are relatively synchronized, too, and a correct chronology of events can also be detected. This model is best suited to the class of applications that require external synchronization (i.e., each node is synchronized to an outside timescale such as UTC). For example, to understand a request such as, Show me all targets detected between 2 p.m. and 8 p.m., the network and the requestor must share a common time frame. Similarly, external synchronization is also needed to archive data as the notion of time, while data storage must be consistent with the notion of time that an external user will have while retrieving it.

- *Hybrid synchronization.* Applications that are a hybrid of internal and external synchronization can also be imagined. For example, in some cases, nodes need a frequency standard but do not need a standard time. For instance, some applications may need to know the length of a second but are not concerned with which second is currently elapsing according to UTC. Applications that measure phenomena in the physical world often fall into this category. For example, acoustic-ranging and beam-forming applications assume that they know the speed of sound: approximately 345 m/s. To convert propagation delay to distance, the delay must be measured using the same definition of a second as that which is implicitly part of the speed-of-sound constant. Similarly, motion-tracking applications need the correct definition of a second to report the speed of target in units (meters per second) that are meaningful. Timing-sync Protocol for Sensor Networks (TPSN) [14] and Flooding Time Synchronization Protocol (FTSP) [15] aim to provide external synchronization as well as the hybrid of external/internal synchronization in the network.

5.3   Sources of Delay

The greatest hurdle to precise network time synchronization is nondeterminism. Latency estimates are confounded by random events that lead to asymmetric round-trip message-delivery delays; this contributes directly to synchronization error. Figure 5.1 shows the decomposition of packet delay when the packet traverses over a wireless link. We designate the node that initiates the packet exchange as the sender and the node that responds to this message as the receiver. In this discussion, we will borrow terms from a typical layered architecture used in traditional computer networks. We briefly analyze each component shown in Figure 5.1.

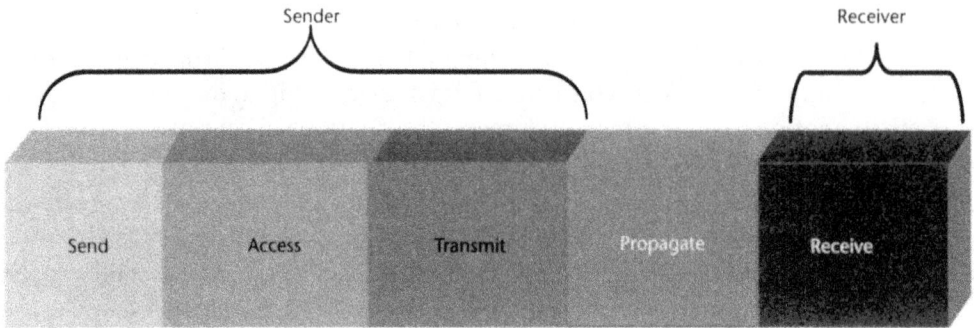

Figure 5.1   Decomposition of packet delay over a wireless link.

- *Send time.* When a node decides to transmit a packet, it is scheduled as an operating system task. There is time spent in actually constructing the packet at the application layer, after which it is passed to the lower layers for transmission. This includes kernel protocol processing and variable delays introduced by the operating system (e.g., context switches and system call overhead incurred by the synchronization application).

- *Access time.* After reaching the Medium Access Control (MAC) layer, the packet waits until it can access the channel. Access time refers to this waiting time. This delay is specific to wireless networks resulting from the property of multiple senders sharing the medium for packet transmission. This is the most critical factor contributing to packet delay. Moreover, it is highly variable in nature and is specific to the MAC protocol employed by the sensor node.

- *Transmission time.* This refers to the time spent at the physical layer when a packet is transmitted bit by bit. This delay is deterministic in nature and can be estimated using the packet size and the radio speed. The software implementation of the transmitter will have minor variations due to the response time for interrupts.

- *Propagation time.* This is the actual time taken by the packet to traverse the wireless link from the sender to the receiver. The absolute value of this delay is negligible as compared to other sources of packet latency.

- *Receive time.* This is the sum of the time taken in receiving the bits at the network interface and passing them to the MAC layer, the time taken in constructing a packet from bits and passing it to the application layer, and the time taken to notify the host of the arrival of the packet. The value of receive time changes due to the variable delays introduced by the operating system, which could be in the interrupt handler of the network driver, system calls, context switches, and so forth.

We note that communication takes place in bits, and a node optimizes by performing events in parallel. Thus, when a bit is being coded for transmission, another bit could be in the air or being received at the other end simultaneously. Thus, this decomposition is just an approximation when done at the packet level instead of the bit level.

5.4   Approaches to Synchronization Algorithms

Consider two nodes, A and B, which maintain a 32-bit register as a clock that is triggered by a crystal oscillator. This is the only notion of time that a node has. We assume that these two nodes might have been started at completely random times; as a result, there exists a clock offset, represented by $\Delta^{A\text{->}B}$ between them. The objective of any synchronization algorithm is to estimate this offset, represented by $\hat{\Delta}^{A\text{->}B}$, so that one of the nodes, say A, can correct its clock to get synchronized to B. The performance of different synchronization algorithms is typically gauged by the metric of error, defined as the difference between the estimated offset, $\hat{\Delta}^{A\text{->}B}$, and the actual offset, $\Delta^{A\text{->}B}$, between the nodes. In this chapter, we focus on synchronizing a pair of nodes, as this is the fundamental building block of networkwide synchronization. Typically, multihop synchronization is achieved by synchronizing neighboring nodes lying along the path.

The existing approaches for synchronizing a pair of nodes can be broadly classified as sender-receiver or receiver-receiver. The classical approach of sender- receiver synchronization is based on a simple handshake between the pair of nodes being synchronized. In contrast, receiver-receiver synchronization is used to synchronize a set of receivers. In this approach, the two synchronizing nodes compare the timestamps of the receipt of a packet from a common sender. All the existing time-synchronization protocols use some variant of these two basic approaches. For example, NTP and TPSN are based on sender-receiver synchronization, whereas RBS and FTSP are based on receiver-receiver synchronization. In this section, we analyze and contrast these two approaches in detail. We derive the analytical expression for the synchronization error for both of the approaches.

5.4.1   Sender-Receiver Synchronization

The approach of using a simple handshake to synchronize the sender of a message with its receiver was first used by NTP. Figure 5.2 shows this handshake between nodes A and B. Notice that here $T1$, $T4$ represents the time measured by the local clock of A; similarly, $T2$, $T3$ represents the time measured by the local clock of B. At time $T1$, A sends a packet to B. Node B receives this packet at $T2$, which can be expressed in terms of $T1$ as follows:

$$T2 = T1 + \Delta^{A\text{->}B} + \delta \ldots \qquad (5.1)$$

Here, δ represents the packet delay. At time $T3$, B sends back a packet, which is received at $T4$ at A. $T4$ can be expressed in terms of $T3$ as

$$T4 = T3 - \Delta^{A\text{->}B} + \delta \ldots \qquad (5.2)$$

Using (5.1) and (5.2), A can calculate the offset and propagation delay as

$$\hat{\Delta}^{A\text{->}B} = \frac{(T2 - T1) - (T4 - T3)}{2}; \quad \hat{\delta} = \frac{(T2 - T1) + (T4 - T3)}{2} \ldots \qquad (5.3)$$

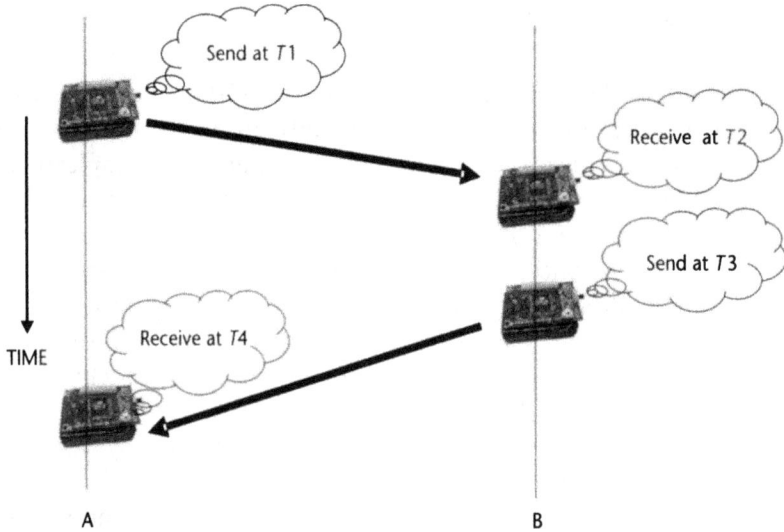

Figure 5.2 Sender-receiver synchronization.

Equations (5.1) to (5.3) implicitly assume that the packet delay is the same for both the packet exchanges. However, as mentioned in the previous section, the packet delay is made up of several variable components. Therefore, (5.1) is more correctly expressed as

$$T2 = T1 + \Delta^{A->B} + S_A + P_{A->B} + R_B \cdots \qquad (5.4)$$

Here S_A, $P_{A->B}$, and R_B refer to the time taken to send a packet (send time + access time + transmission time) from node A, the propagation time between node A and B, and the time taken to receive packet (receive time) at node B, respectively. Analogously, (5.1) to (5.3) also assume that the clock offset remains the same over the span of time in which the handshake is initiated and completed. Let us analyze the validity of this assumption.

The clocks in sensor nodes are crystal based and can deviate from the nominal frequency specified on the crystal (ideal clock). This is shown in Figure 5.3, whereby the local clock of node A evolves at a different rate than the ideal clock. Moreover, as shown in Figure 5.3, the operational frequency can also vary over different nodes. The local clocks of A and B evolve at different rates. This implies that the clock offset between the nodes will vary with time. In light of this observation, from now on, the clock offset will only be defined at a specific time. For example, $\Delta_t^{A->B}$ will represent the clock offset between nodes A and B at time t. Just like the packet delay, variability in the clock offset will also contribute to the synchronization error. Equation (5.4) now gets modified to

$$T2 = T1 + \Delta_{t1}^{A->B} + S_A + P_{A->B} + R_B \cdots \qquad (5.5)$$

We use lowercase letters to represent the equivalent time in the ideal clock, corresponding to the time in the local clock of a node. For example, $t1$ stands for the time in the ideal clock, whereas $T1$ stands for the time in the local clock of node A. Similarly, (5.2) gets modified to

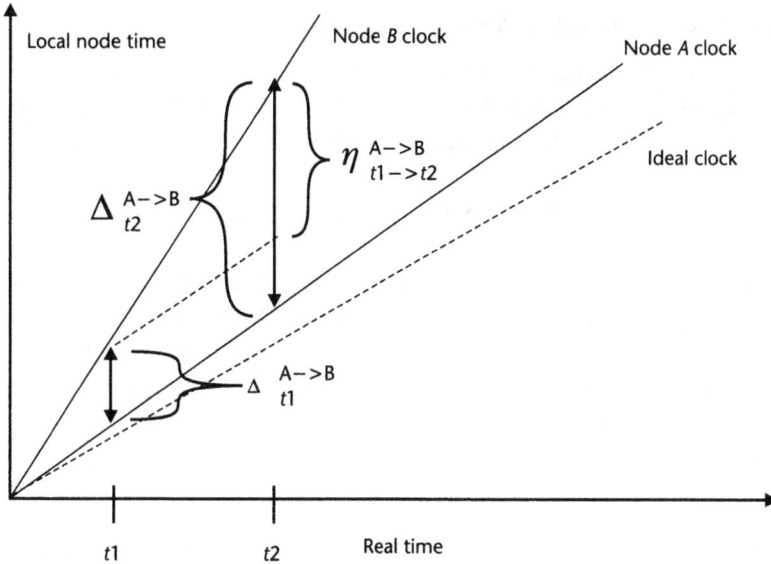

Figure 5.3   Drift among the local node clocks.

$$T4 = T3 - \Delta_{t4}^{A->B} + S_B + P_{B->A} + R_A \cdots \tag{5.6}$$

As mentioned earlier, the objective of the time synchronization algorithm is to estimate the clock offset between the two nodes after the completion of the algorithm (i.e., $\Delta_{t4}^{A->B}$). Notice that $\Delta_{t1}^{A->B} \neq \Delta_{t4}^{A->B}$. We represent the change in the clock offset between two nodes over time by η. As shown in Figure 5.3, the following relationship can be established:

$$\Delta_{t1}^{A->B} = \Delta_{t4}^{A->B} - \eta_{t1->t4}^{A->B} \cdots \tag{5.7}$$

Here, $\eta_{t1->t4}^{A->B}$ represents the change in the clock offset between A and B from time *t1* to *t4*. Subtracting (5.6) from (5.5) and using (5.3) and (5.7), we obtain

$$\left(2 * \hat{\Delta}^{A->B}\right) = \left| S^{UC} + P^{UC} + R^{UC} + \eta_{t1->t4}^{A->B} \right| + \left(2 * \Delta_{t4}^{A->B}\right) \cdots \tag{5.8}$$

Here S^{UC}, R^{UC}, and P^{UC} stand for the indeterminism at sender, at receiver, and in propagation time, respectively. They are given by the following equations:

$$S^{UC} = S_A - S_B; R^{UC} = R_A - R_B; P^{UC} = P_{A->B} - P_{B->A} \cdots \tag{5.9}$$

The synchronization error can then be derived from equation (5.8) as

$$\varepsilon_{S-R} = \hat{\Delta}^{A->B} - \Delta_{t4}^{A->B} = \frac{\left| S^{UC} + P^{UC} + R^{UC} + \eta_{t1->t4}^{A->B} \right|}{2} \cdots \tag{5.10}$$

As mentioned earlier, if there is no indeterminism in packet delay ($S^{UC} = P^{UC} = R^{UC} = 0$) and the clocks have been drifting at the same rate from the ideal clock ($\eta_{t1->t4}^{A->B} = 0$), synchronization error ε_{S-R}, will be zero.

5.4.2   Receiver-Receiver Synchronization

Receiver-receiver synchronization is used to synchronize a set of receivers with one another. In this approach, the two synchronizing nodes compare the timestamps upon the receipt of a packet from a common sender. This approach eliminates the sender from the critical path. This observation was first made by P. Verissimo and L. Rodrigues in [16] and later became central to their CesiumSpray system [17]. Figure 5.4 shows the essence of this approach.

Suppose the two nodes A and B receive the common packet at $T2$ and $T3$, respectively. Let us denote the time at which this packet was sent by the common sender, C, at $T1$. At some later stage, node B will send the timestamp information ($T2$) to A. Suppose node A receives this packet at $T4$, after which it can calculate the clock offset as follows:

$$\hat{\Delta}^{A->B} = T3 - T2... \tag{5.11}$$

As in Section 5.4.1 , the variability in the packet delay and the clock offset introduces synchronization error. A similar breakup of packet latency as in Section 5.4.1 can be used to represent the packet transfer from C to A and B, respectively.

$$T2 = T1 + \Delta_{t1}^{C->A} + S_C + P_{C->A} + R_A ... \tag{5.12}$$

$$T3 = T1 + \Delta_{t1}^{C->B} + S_C + P_{C->B} + R_B \tag{5.13}$$

Subtracting (5.13) from (5.12), we obtain

$$\hat{\Delta}_{A->B} = T3 - T2 = \left(P_{C->B} - P_{C->A}\right) + \left(R_B - R_A\right) + \left(\Delta_{t1}^{C->B} - \Delta_{t1}^{C->A}\right)... \tag{5.14}$$

The offset estimation is done at $T4$; hence, the objective is to estimate $\Delta_{t4}^{A->B}$. The last term in (5.14) can be further decomposed as follows:

$$\Delta_{t1}^{C->B} - \Delta_{t1}^{C->A} = \Delta_{t1}^{A->B} = \Delta_{t4}^{A->B} - \eta_{t1->t4}^{A->B}... \tag{5.15}$$

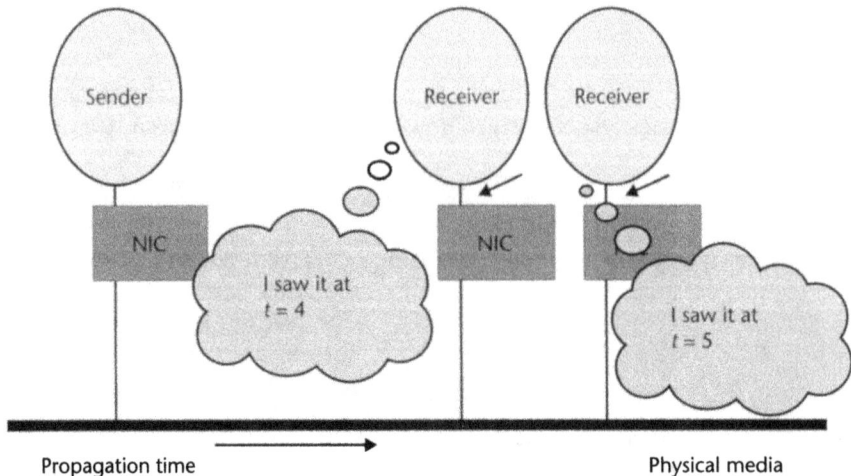

Figure 5.4   Receiver-receiver synchronization.

Using (5.14) and (5.15), the synchronization error can be derived as

$$\varepsilon_{R-R} = \hat{\Delta}_{t4}^{A->B} - \Delta_{t4}^{A->B} = \left| P^{UC} + R^{UC} + \eta_{t1->t4}^{A->B} \right| ... \qquad (5.16)$$

5.4.3 Analysis

As can be seen from (5.10) and (5.16), receiver-receiver synchronization completely eliminates the uncertainty on the sender side, whereas sender-receiver synchronization reduces the impact of uncertainty on the receiver side, in propagation, and due to the variation in clock offset by half. Thereby, no scheme can completely overshadow the other. Receiver-receiver synchronization will give better results when the radio and its driver make up a closed black box. Such is the case with typical wireless LANs (WLANs). The uncertainty in MAC delay (at the sender side) in these networks is so large that it completely overshadows the effects of other factors. However, in the case of sensor networks, there exists a strong coupling between the radio and the application layer. In fact, available sensor networking platforms such as motes [18] do most of the radio processing at the application layer. This provides a huge amount of flexibility in sensor networks. Packets can be time-stamped at the physical layer (i.e., after the MAC layer), leveraging the fact that the operating system has direct access to the underlying MAC layer. As a result, much of the normal nondeterminism of channel access is eliminated. Based on this observation, it can be argued that the classical approach of doing sender-receiver synchronization will give a performance roughly two-times better than the receiver-receiver approach in sensor networks for synchronizing a pair of nodes. Reference [14] provides detailed experimental results over Berkeley motes verifying this claim.

5.5 Protocols

5.5.1 Timing-Sync Protocol for Sensor Networks

TPSN [14] uses a two-way message exchange between a pair of nodes (as shown in Figure 5.2) to synchronize them. TPSN time-stamps the packets in the MAC layer of the radio stack, thereby reducing the effect of sender-side indeterminism from contributing to the synchronization error. TPSN has been implemented on Berkeley motes and is available as a part of the TinyOS [19] distribution operating system for motes, in the subdirectory *tinyos-1.x/contrib/Timesync-NESL-UCLA*. In their original paper, the author S. Ganeriwal, et. al report an accuracy of 17 μs and 8 μs on the Mica1 and Mica2 motes, respectively. The latest implementation of TPSN exploits the hooks provided by the communication stack of TinyOS to synchronize the nodes within an error of 5 μs.

In order to establish a global timescale, TPSN first creates a hierarchical structure in the network and then performs pairwise synchronization along the edges of this structure. At the onset of the network, a reference node is chosen, which acts as the root of this hierarchical topology. The algorithm makes provisions for choosing the next reference node after the failure of the initially chosen reference node. It also provides hooks for adapting to the runtime dynamics of the system, such as maintenance of the hierarchical topology, handling packet drops, and so forth. The performance of this algorithm has been evaluated in NESLsim [20], a PARSEC-based simulation platform for sensor networks; see [21] for details.

The handshake between two neighboring nodes in the hierarchical topology does not depend on the presence of other nodes in the vicinity. Therefore, the synchronization error between two neighboring nodes should not vary with the total number of nodes in the network. The results obtained in simulations were consistent with this speculation. Even the energy consumed by every node to synchronize itself remains the same with the increase in the total number of nodes. This clearly proves that TPSN is scalable with the number of nodes in the network.

However, the synchronization error of a node with respect to the root node depends on its hop distance from it. This is because error adds up over multihop. Although the randomness in the sign and magnitude of the synchronization error and the drift prevents it from becoming arbitrarily high with an increase in hop distance, the synchronization error is still a nondecreasing function of the hop distance. Hence, TPSN is not scalable with the diameter of the network. This clearly highlights the fact that the networkwide performance of TPSN depends on the efficiency of the hierarchical structure. Optimal results will be obtained if the path in the hierarchical topology between a node and the reference node corresponds with the actual shortest-distance path between them.

5.5.2 Reference Broadcast Synchronization

RBS is based on the approach of receiver-receiver synchronization and leverages the broadcast property inherent in wireless communication. In this scheme, nodes periodically send a message to their neighbors using the network's physical-layer broadcast. A reference broadcast does not contain an explicit timestamp; instead, receivers use its arrival time as a point of reference for comparing their clocks. RBS removes sender's nondeterminism from the critical path and, in this way, produces high-precision clock agreement. Reference [22] compares the performance of RBS vis-à-vis NTP. For this, RBS was implemented as a UNIX daemon using User Datagram Protocol (UDP) datagrams and a commodity hardware platform (Strong-ARM-based Compaq IPAQs with Lucent Technologies 11-Mbit 802.11 wireless Ethernet adapters). To make the comparison as fair as possible, RBS was implemented under the same constraints as NTP: a pure user-space application with no special kernel or hardware support or special knowledge of the MAC layer (other than that it supports broadcasts). In the light traffic scenario, RBS performed more than eight times better than NTP, an average of $6.29 \pm 6.45\,\mu$s error, compared to $51.18 \pm 53.30\,\mu$s for NTP. A more efficient implementation of RBS can be achieved by using the timestamps acquired in the network interface's interrupt handler. This optimized version of RBS achieves an accuracy of $1.85 \pm 1.28\,\mu$s.

In [13], an extension of the basic RBS scheme to synchronize nodes on a multihop path was proposed. The algorithm federates clocks across broadcast domains without losing the critical receiver-to-receiver property of RBS. The analysis shows that the multihop experiences a slow decay in precision; the standard deviation grows as $O(sqrt\{n\})$ after n hops. An error of $3.68 \pm 2.57\,\mu$s was measured over a four-hop distance of IPAQ devices. Similar to the results obtained in [14], the multihop algorithm allows a coordination of any number of nodes; the error incurred is only as large as the distance between the nodes being synchronized.

Post facto synchronization, proposed in [23], is a technique that allows significant energy savings in networks where time synchronization is needed occasionally and unpredictably. Instead of keeping clocks synchronized all the time, clocks run undisciplined (at their natural rate) and are reconciled after an event of interest occurs. In [13], J. Elson, L. Girod, and D. Estrin propose an algorithm where RBS is used to achieve post facto synchronization. They show that even 60 seconds after an event, RBS can reconcile clocks to a phase error that is below the mean jitter of the detector.

5.5.3   Other Protocols

Several other proposed protocols try to improve upon the basic foundations of RBS and TPSN to achieve better accuracy in sensor networks.

- *FTSP.* FTSP [15] utilizes broadcast messages to obtain time-synchronization reference points between a sender and its neighbors. It uses a unique combination and efficient implementation of two key ideas from TPSN and RBS to yield an order of magnitude better performance: MAC-layer time-stamping and skew compensation with linear regression.

- *LTS.* In [24], J. V. Greunen and J. Rabaey propose a Lightweight Time Synchronization (LTS) scheme specially geared toward applications with lower accuracy requirements and that trades off accuracy for gains in energy efficiency. It uses the same approach of pairwise synchronization as TPSN. LTS has not been implemented, and the efficacy of the scheme has been only tested over simulations. Greunen and Rabaey provide several hooks to make the algorithm robust and efficient in face of runtime dynamics such as channel fading and dynamic links.

- *Tiny/Mini Sync.* This protocol, proposed in [25], is based on the assumption that the clock drifts of nodes are of the linear form; under this assumption, the offset between two nodes is also linear. In order to perform pairwise synchronization, Tiny/Mini-Sync nodes exchange similar time-stamped packets as in TPSN. These exchanged packets are used to estimate the best-fit offset line between the two nodes. As more packets are exchanged, the computational complexity required for calculating the best-fit line increases. The linear constraint is used to identify redundant packets that are discarded to lower computation complexity. Tiny/Mini-Sync has not been implemented, and its performance comparison with other protocols still remains an open problem.

- *Optimal time synchronization.* A recent work [26] has some promising results: in an RBS-like protocol, optimal time synchronization can be achieved by distributing complete information to all neighbors. R. Karp et al. also propose a mostly localized implementation of their idea that can achieve nearly optimal results. These assertions are still in need of quantitative justification or practical verification on real systems.

5.6   Discussion

While there has been significant work in sensor-network synchronization, many system-level and practical challenges have not been addressed. In this section, we discuss some of these issues, highlighting some recent efforts and laying down basic foundations for future work.

5.6.1 Long-Term Synchronization

Current approaches for timing synchronization, such as RBS [13], TPSN [14] and FTSP [15], guarantee error to be within a few microseconds per minute but do not comment about error behavior over an hour after synchronization. These studies are based on short-term datasets (a few tens of minutes) and, therefore, do not enable us to understand either how error accumulates over long periods into the future or how to use past beacons to better estimate future error. Due to the unavailability of good methods for predicting clock drift over long timescales, the approach taken by current sensor-network deployments, such as James Reserve [27], Great Duck Island [28], and the Shooter Localization system [29], has been to run either one of these approaches in a periodic manner with an arbitrarily decided short period to bound the synchronization error.

Better synchronization can lead to significantly lower radio duty cycles and enable an order of magnitude greater lifetime than current techniques. Sensor MAC (S-MAC) [30] can afford to sleep significantly longer (from the current periods of a few seconds to many tens of minutes) yet remain tightly synchronized within tens of microseconds; the preamble lengths required to wake up neighbors in Berkeley MAC (BMAC) [31] can be significantly reduced (from 100 ms to a few minutes); synchronized sampling with errors within 100 μs can often be done with an overhead of just two to five time synchronization packets per hour. The periodic beaconing time of 5 minutes in James Reserve (RBS based), 10 minutes in Great Duck Island, and 1 minute in the Shooter Localization system (FTSP based) can be scaled appropriately, while still meeting the desired requirements for synchronization accuracy, translating into significant improvements in network lifetime.

5.6.1.1 Rate Adaptive Time Synchronization

Reference [32] performs a detailed empirical study of long-term time synchronization across sensor nodes in many different environments, both indoors and outdoors. A thorough empirical and analytical study reveals complex relationships between the sampling rate, window of past samples, and the estimation scheme. The authors show that there is an optimal time window of historical synchronization data that provides best estimation and error prediction. They also provide both analytical techniques to bound the prediction error and an empirical verification of this bound. Finally, a multiplicative-increase, multiplicative-decrease synchronization protocol is proposed that learns the operational parameters and adapts to different environments seamlessly. For a given user-defined synchronization error bound, Rate Adaptive Time Synchronization (RATS) consumes 1.5 to 12 times less energy than the best possible periodic synchronization scheme and outperforms the schemes based on the calibration parameters in the clock crystal datasheet by three orders of magnitude. Although, RATS is a promising effort in this direction, several future challenges need to be resolved before an efficient time-synchronization mechanism for long-lived sensor networks will be realized.

- *Multihop synchronization.* RATS has focused on long-term synchronization of adjacent nodes. As with single-hop synchronization, current multihop experiments have also focused on very short timescales. Thus, many questions addressed in [32] for the single-hop case can also be extended to the multihop case. For instance, is the linear estimator and use of history appropriate when a node needs to synchronize with a base station that is multiple hops away (perhaps for synchronized sampling). A related question concerns the propagation of error in a multihop setting, which, in the Gaussian case [13], has a simple *sqrt(n)* growth across *n* hops. The error analysis in [32] shows that unlike the short-timescale studies, errors can be quite non-Gaussian at large timescales; hence, error propagation requires further study. Other protocol-related issues emerge in the multihop case as well; for instance, how can the broadcast channel be exploited for overhearing or group synchronization.

- *Better estimators.* In [32], the authors show that the rate of change of drift is highly correlated with changes in ambient conditions such as temperature. However, they do not explicitly consider temperature and other ambient environmental variables in the estimator in order to keep it simple. On larger sensor platforms, such modeling would be feasible and can potentially improve the results.

5.6.1.2 Secure Time Synchronization

Since time synchronization forms a critical building block for signal-processing applications as well as for the efficient design of several networking layers, a malicious adversary can bring detrimental effects to the network by compromising a node and then abusing the underlying time-sync protocol. For example, a parent node in TPSN can easily mislead all of the nodes lying in its subtree on a different timescale than the rest of the network. Analogous attacks can be imagined for the rest of the protocols, such as RBS and FTSP. In fact, every existing time-synchronization protocol for sensor networks will break down under such malicious attacks, as they have been developed on the premise that all nodes will abide by the rules of the protocol. Realizing the inadequacy of existing protocols, there are broadly two approaches for countering such attacks. First, we can build security mechanisms into the time-synchronization protocol itself. Such approaches are also being investigated to secure other basic services of sensor networks, such as localization [33] and calibration [34]. Second, we can use a higher-level reputation-based framework [35] on top of existing time-synchronization protocols to make sure that nodes collaborate with only other trustworthy (neither malicious nor faulty) nodes while running these protocols.

5.7 Conclusion

Time synchronization is a critical piece of infrastructure in any distributed system, and WSNs make particularly extensive use of synchronized time. Almost any form of sensor data fusion or coordinated actuation requires synchronized physical time for reasoning about events in the physical world. However, while the requirements along axes such as tolerable error are often stricter in sensor networks than in traditional distributed systems, energy and channel constraints limit the resources available to meet these goals. New approaches to time synchronization can better support the broad range of application requirements seen in sensor networks, while meeting the unique resource constraints found in such systems.

An important step toward understanding the problem domain is the articulation of the relevant metrics. We conclude that any time-synchronization approach to sensor networks has to be scalable, energy efficient, multimodal, and flexible enough to be tuned to meet desired user-level requirements. The hindrance to precise network time synchronization is nondeterminism; to gain greater insight into this, we performed a decomposition of the packet latency as it traversed a wireless link to find the sources of error.

All the existing protocols for time synchronization can be broadly classified as either sender-receiver based or receiver-receiver based. We have analyzed each of these two approaches in detail, coming up with exact analytical expressions for the synchronization error for both of them. TPSN and RBS are the two most widely used protocols in the domain of sensor networks, each being a representative example of the two diverse approaches to synchronization. While TPSN has been able to achieve accuracy within a few microseconds on Berkeley motes, RBS is equally efficient in providing synchronization on Linux-based platforms such as IPAQs, Stargates, and the like. Both of these protocols are available as stand-alone implementations on their respective platforms with clearly specified application programmer interfaces (APIs) exposed to the application writer, who can tune them to meet the desired user-level requirements.

We note that although time synchronization has been one of the most widely researched areas in the realm of sensor networks, there are still some challenges left for the development of a complete solution. First, current approaches to time synchronization concentrate on establishing pairwise timing relationships between nodes at a given instant in time. However, the requirements of most of the applications are beyond this basic snapshot case; event notification requires nodes to synchronize along a routing path (line), tracking requires a cluster of nodes to be synchronized, and synchronized sampling of a physical process requires networkwide timing synchronization. Thereby, the problem unfolds as follows: if we have error bounds that need to be achieved for a specific long-lived sensor-network application, how do we exploit temporal and spatial broadcast patterns to achieve them? A recent effort, RATS, exploits the temporally correlated behavior of clock drift for developing an adaptive and efficient scheme for resynchronization. Besides the fact that this scheme still needs to be implemented and verified in a practical setting, it does not exploit the spatially correlated drift among proximate nodes. This is an open area of future research in this domain.

Second, since time synchronization forms a critical building block for signal-processing applications as well as for the efficient design of several networking layers, a malicious adversary can wreak detrimental effects on the network by compromising a node and then abusing the underlying time-sync protocol. All of the existing protocols will completely break down under such attacks, as they were never designed with maintaining security in mind. Thus, there is a need to make time-synchronization protocols secure and robust.

References

[1]   Girod, L., and. Estrin D, "Robust Range Estimation Using Acoustic and Multimodal Sensing," *Proc. IEEE/RSJ International Conference on Intelligent Robots and Systems (IROS 2001)*, Maui, HI, March 2001.

[2] Wang, H., et al., "A Wireless Time-Synchronized COTS Sensor Platform Part II: Applications to Beamforming," *Proc. IEEE CAS Workshop on Wireless Communications and Networking*, Pasadena, CA, September 2002.

[3] Asada, G., et al., Wireless Integrated Network Sensors: Low Power Systems on a Chip," *Proc. European Solid State Circuits Conference*, The Hague, Netherlands, 1998.

[4] Cerpa, A., et al., "Habitat Monitoring: Application Driver for Wireless Communications Technology," *Proc. 2001 ACM SIGCOMM Workshop on Data Communications in Latin America and the Caribbean*, San Jose, Costa Rica, April 2001.

[5] Intanagonwiwat, C., Govindan R., and Estrin D., "Directed Diffusion: A Scalable and Robust Communication Paradigm for Sensor Networks," *Proc. Sixth Annual International Conference on Mobile Computing and Networking*, Boston, MA, August 2000, pp. 56–67.

[6] DARPA Advanced Technology Office (ATO), Self-healing Minefield, at http://www.darpa.mil/ato/programs/SHM.

[7] Cristian, F., "Probabilistic Clock Synchronization," *Distributed Computing*, Vol. 3, 1989, pp. 146–158.

[8] Gusell, R., and Zatti S., "The Accuracy of Clock Synchronization Achieved by TEMPO in Berkeley UNIX 4.3 BSD," *IEEE Transactions on Software Engineering*, Vol. 15, 1989, pp. 847–853.

[9] Srikanth, T. K., and Toueg S., "Optimal Clock Synchronization," *J-ACM*, Vol. 34, No. 3, July 1987, pp. 626–645.

[10] Mills, D. L., Internet Time Synchronization: The Network Time Protocol," In Zhonghua Yang and T. Anthony Marsland, (eds.), *Global States and Time in Distributed Systems*, IEEE Computer Society Press, 1994.

[11] Lamport, L., "Time, Clocks, and the Ordering of Events in a Distributed System," *Communications of the ACM*, Vol. 21, No. 7, 1978, pp. 558–565.

[12] Romer, K., "Time Synchronization in Ad Hoc Networks," *Proc. MobiHoc 2001*, Long Beach, CA, October 2001.

[13] Elson, J., Girod L., and Estrin D., "Fine-Grained Network Time Synchronization Using Reference Broadcasts," *Proc. Fifth Symposium on Operating Systems Design and Implementation (OSDI 2002)*, Boston, MA, December 2002.

[14] Ganeriwal, S., Kumar R., and Srivastava M. B., "Timing-Sync Protocol for Sensor Networks," *Proc. 1st ACM Conference on Embedded Networked Sensor Systems (SenSys)*, Los Angeles, CA, 2003.

[15] Maroti, M., et al., "The Flooding Time Synchronization Protocol," *Proc. Second ACM Conference on Embedded Networked Sensor Systems (SenSys)*, Baltimore, MD, 2004.

[16] Verissimo, P., and Rodrigues L., "A Posteriori Agreement for Fault-Tolerant Clock Synchronization on Broadcast Networks," *Proc. 22nd Annual International Symposium on Fault-Tolerant Computing (FTCS '92)*, Boston, MA, July 1992, p. 85.

[17] Verissimo, P., Rodrigues L., and Casimiro A., "CesiumSpray: A Precise and Accurate Global Time Service for Large-Scale Systems," *Tech. Rep. NAV-TR-97-0001*, Universidade de Lisboa, 1997.

[18] Hill, J., and Culler D., "A Wireless Embedded Sensor Architecture for System-Level Optimization," technical report, University of California, Berkeley, 2001.

[19] TinyOS, at http://webs.cs.berkeley.edu/tos.

[20] Ganeriwal, S., et al., "NESLsim: A Parsec Based Simulation Platform for Sensor Networks," *NESL*, Technical Report, 2002.

[21] Ganeriwal, S., et al., "Network-wide Time Synchronization in Sensor Networks," *NESL Technical Report*, March 2003.

[22] Elson, J., and Estrin D., "Time Synchronization for Wireless Sensor Networks," *Proc. 2001 International Parallel and Distributed Processing Symposium (IPDPS), Workshop on Parallel and Distributed Computing Issues in Wireless and Mobile Computing*, San Francisco, CA, April 2001.

[23] Elson, J., and Romer K., "Wireless Sensor Networks: A New Regime for Time Synchroniza-tion," *Proc. 1st Workshop on Hot Topics in Networks (HotNets-I)*, Princeton, NJ, October 2002.

[24] Greunen, J. V., and Rabaey J., "Lightweight Time Synchronization for Sensor Networks," *Proc. 2nd ACM International Workshop on Wireless Sensor Networks and Applications (WSNA)*, San Diego, CA, 2003.

[25] Sichitiu, M. L., and Veerarittiphan C., "Simple, Accurate Time Synchronization for Wire-less Sensor Networks," *IEEE Wireless Communications and Networking Conference, WCNC*, New Orleans, LA, 2003.

[26] Karp, R., et al., "Optimal and Global Time Synchronization in Sensornets," Technical Report 0009, Center for Embedded Networked Sensing, University of California, Los Angeles, April 2003.

[27] CENS Habitat Sensing Group at James Reserve, at http://www.jamesreserve.edu.

[28] Habitat Monitoring on Great Duck Island, at http://www.greatduckisland.net.

[29] Sallai, J., et al., "Acoustic Ranging in Resource Constrained Sensor Networks," Technical Report, 2004.

[30] Ye, W., Hiedemann J., and Estrin D., "An Energy-Efficient MAC Protocol for Sensor Net-works," *IEEE Infocomm*, New York, NY, 2002.

[31] Polastre, J., Hill J., and Culler D., "Versatile Low Power Media Access on Wireless Sensor Networks," *Proc. Second ACM Conference on Embedded Networked Sensor Systems (SenSys)*, Baltimore, MD, 2004.

[32] Ganeriwal, S., et al., "Rate-Adaptive Time Synchronization for Long-lived Sensor net-works," *CENS Technical Report*, November 2004.

[33] Capkun, S., and Hubaux J. P., "Secure Positioning of Wireless Devices with Application to Sensor Networks," *Proc. 24th IEEE INFOCOM*, Miami FL, 2005.

[34] Koushanfar, F., Potkonjak M., and Sangiovanni-Vincentelli A., "On-line Fault Detection of Sensor Measurements," *IEEE Sensors*, October 2003, pp. 974–980.

[35] Ganeriwal, S., and Srivastava M. B., "Reputation-based Framework for High Integrity Sen-sor Networks," *Proc. ACM Security for Ad Hoc and Sensor Networks, SASN*, Washington, D.C., 2004.

CHAPTER 6

Energy Conservation in Sensor Networks at the Link and Network Layers

John Heidemann and Wei Ye

6.1 Introduction

Sensor networks promise to place sensors in the physical world to gather information, communicate, and act. All of these steps consume energy. With limited battery capacity, sensor networks are characterized by the situation where "each bit sent brings that node closer to death" [1]. Some sensor networks today add energy harvesting with solar panels or other more experimental methods as discussed in Chapter 9, but even there, careful use of energy is essential to an operational system.

Given a limited amount of energy or a limited recharge rate, energy conservation becomes a goal. A successful sensor network will minimize energy consumption at all levels of the system, from the application down to the hardware itself. This chapter considers network-level opportunities for energy conservation, with emphasis on the MAC level, topology-control protocols, and routing-level issues.

6.2 Radio Transmission Power Control

We begin our survey of energy conservation by considering radio transmission power control. Transmission power is often integrated with the MAC protocol or routing protocol, or sometimes it is set externally to the system.

Transmission power control is important for several reasons: First, adjusting power can be important to guarantee connectivity. Second, since transmission power indicates a "radio's footprint," controlling power is essential to managing density and encouraging spatial reuse of spectrum. Finally, minimizing transmission power can reduce energy consumption, both directly by requiring that less power be sent and indirectly by reducing contention with other transmitting nodes.

Guaranteeing connectivity and managing density are related problems. By balancing connectivity and density, wireless networks maximize spatial reuse of the spectrum. Power control is a key component in this process. There is a very, very large body of literature around analysis and protocol design for wireless power control. We describe only two examples of early work here. R. Ramanathan and R. Rosales-Hain have demonstrated how to select transmission power to balance connectivity and energy consumed [2]. Many approach the subject from the MAC layer. A representative MAC protocol that considers power control is Power Control Media Access (PCMA) [3]. It focuses on optimizing spatial channel reuse and extends an Request To Send (RTS)/Clear To Send (CTS) mechanism to support variable power. They demonstrate about 50% better throughput when nodes and traffic are clustered and power control is enabled.

Efficient spatial use also affects the fundamental performance limits of the sensor network. For example, P. Gupta and P. R. Kumar's work establishes a theoretical bound on the capacity of a network, indicating that wireless network capacity tracks $\Omega(\sqrt{n})$ as the number of nodes increases, assuming optimal transmission power and uniform distributions of sources and sinks [4]. Selection of optimal transmission power is necessary for their results.

Fewer researchers focus on power control to reduce energy consumption. The focus is most often on connectivity and spatial reuse because those are more pressing issues in systems design, particularly at longer ranges. The benefits of short-range transmission have been observed by G. J. Pottie and W. J. Kaiser, both due to the d^2 cost of longer-distance transmission and because of the opportunity to trade local processing for transmission [5]. Radio transmission power can be a significant part of energy consumption at short ranges, but without care, other component costs can dominate. For example, the CC1000 radio is widely used in sensor networks on platforms such as Mica2 motes, and its output power ranges over a factor of five (from 5 to 27 mA) [6]. However, the fixed cost of listening makes transmission power differences insignificant at low duty cycles. If 2% of time is spent transmitting, for example, the maximum energy savings is only 8%. Avoiding collisions by spatial reuse doubles the savings by comparison since, after a collision, both parties must retransmit.

Figure 6.1 illustrates these concepts by considering two transmission powers, r and R, where $R \approx 3r$. For communication from node A to node D, one can either transmit in one hop at full power (R) or in three hops a-b-c-d, each at reduced power r. Using a simple d^2 energy model, the relative costs of these transmissions are $1 \times 3^2 = 9$ for one hop with R, and $3 \times 1^2 = 3$ for three hops with r, demonstrating

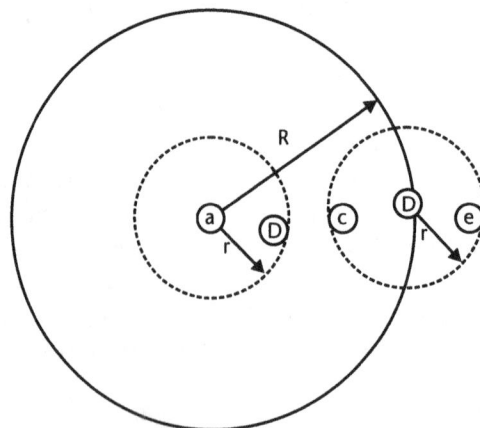

Figure 6.1 Control of transmission power to promote spatial reuse and reduce energy requirements.

the possible energy conservation from shorter, multihop communication. This example also shows the possibility for spatial reuse and reduced contention enabled by lower-power transmission. With strength-r transmissions, concurrent communications are possible between nodes a and b and nodes D and E, while if node A communicates directly with node D at strength R, node D must be silent to avoid interference. Of course, these examples are greatly simplified compared to the real world, where radio propagation is not spherical or symmetric, and listening and other costs must be considered (as described in Section 6.3). However, it illustrates the principles of power control.

Figure 6.1 also shows the reduced contention and increased spatial reuse with short-range communications as a result of less interfering nodes.

When power control is considered for energy savings, it is often viewed as part of the routing layer. An example protocol from this domain is Low Energy Adaptive Clustering Hierarchy (LEACH) [7]. Rather than sending data directly to a central site, nodes form clusters. Data is sent via a short hop to the cluster head, then via a long hop to the sink. By rotating cluster heads over time, energy consumption is reduced and distributed evenly, allowing a fivefold increase in network lifetime.

Systematic studies of the interactions between power control and routing protocols indicate the importance of considering interactions to ensure a reliable overall system [8].

6.3   Medium Access Control

We next consider energy conservation opportunities at the MAC level. For our purposes, we will assume that transmission power has been fixed. This leaves four areas of energy consumption that can be avoided: (1) collisions consume energy by corrupting otherwise good packets, (2) idle listening is a major source of energy consumption when the radio is kept powered on for potential incoming transmissions, (3) overhearing transmitting packets consumes energy in a busy network when a node spends effort receiving packets destined to other nodes, and (4) control packets consume energy that is not directly sending useful data. A number of approaches have been proposed to reduce each of these costs: TDMA and contention-based protocols with scheduled contention periods, asynchronous paging channels, and low-power listening. We briefly describe each of these below.

Several MAC-level approaches have been proposed to reduce these costs. The first class is schedule-based protocols. TDMA protocols can avoid collisions, idle listening, and overhearing by scheduling transmit and listen periods. TDMA protocols require strict time synchronization, often provided by infrastructure such as a base station.

The infrastructure mode of IEEE-802.11 incorporates a contention-free interval, which adopts a TDMA-like structure coordinated by the access point [9], avoiding all three kinds of overhead. Bluetooth behaves similarly in a cluster, called a *piconet*, where a master polls each slave for possible transmissions. Intercluster communication and interference are handled by Code Division Multiple Access (CDMA). K. Sohrabi and G. J. Pottie have proposed a peer-to-peer transmission-scheduling protocol for sensor networks [10].

Their approach avoids base stations, but it depends on assigning different channels, such as CDMA or Frequency Division Multiple Access (FDMA), to any interfering links to allow concurrent transmissions; as a result, it has lower channel utilization. Contention-based protocols are a second class of MAC protocols. They relax the tight synchronization requirements of TDMA protocols and use carrier-sense multiple access (CSMA) techniques to provide more flexibility in multihop communications and better robustness to topology changes. However, because these protocols contend to access the channel, collisions occur, and basic protocols in this class have costs for idle listening and overhearing. IEEE-802.11 ad hoc mode is a very widely used contention-based protocol. It uses carrier-sensing and randomized back-offs to reduce the likelihood of collisions [9]. To reduce idle listening, it defines a power-save mode (PSM), allowing nodes periodically to enter sleep state. The PSM assumes a single-hop network, and so, time synchronization is easy. In multihop operation in sensing applications such as [11], it may have problems in clock synchronization, neighbor discovery, and network partitioning [12].

Overhearing is another source of energy waste. PAMAS first observed the costs of overhearing and suggested using two channels, one for control traffic and the other for data traffic [13]. By keeping the data channel open when packets are exchanged between other nodes overhearing can be avoided.

Scheduled-contention protocols are a subset of contention-based protocols. Besides the PSM in 802.11 ad hoc mode, S-MAC is a second protocol in this class [14, 15]. In S-MAC each node adopts a listen/sleep cycle. Contention occurs only during a brief listen period, reducing the cost of idle listening. Figure 6.2 shows how two nodes exchange packets with the listen/sleep cycles. When there is no data, nodes enter the sleep mode after a brief listening period. Otherwise, they use their sleep time to transmit data packets. During the data transmission, nodes other than the source and destinations sleep to avoid consuming energy due to overhearing (a generalization of PAMAS to in-channel signaling).

S-MAC maintains loose time synchronization between nodes to synchronize schedules, and it allows nodes to adopt multiple schedules, if necessary, to support distributed and multihop operation. Recently, adaptive listen [15] and Time-out-MAC (T-MAC), MAC stands for Medium Access Control [16] have been proposed to improve multihop transmission with sleep-cycled MAC protocols. Asynchronous schemes are a fourth class of MAC protocols. Y. C. Tseng, C. S. Hsu, and T. Y Hsieh [12] have proposed asynchronous wake-up schemes to extend the 802.11 Power Saving (PS) mode into multihop operations. Their basic idea is to design wake-up patterns that guarantee that neighboring nodes have overlapping

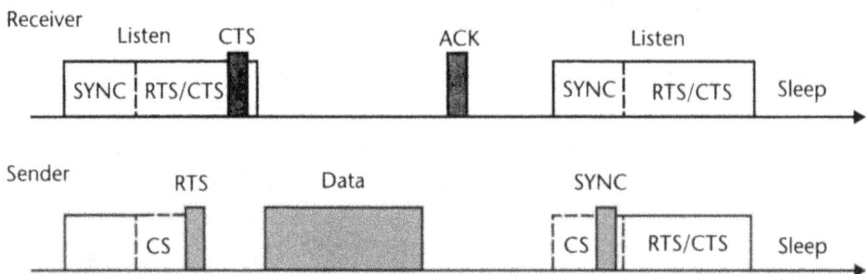

Figure 6.2 Packet exchanges in S-MAC with listen/sleep cycles. CS stands for carrier sense.

listen intervals no matter how large their clock differences are. R. Zheng, J. C. Hou, and L. Sha [17] have proposed an optimal design of the asynchronous sleep patterns to minimize wake-up time by formulating the problem as the block design in combinatorics. Asynchronous wake-up schemes completely remove the requirement for time synchronizations. Their major drawback is their inefficiency in broadcasting since all nodes wake up independently.

Paging channels are another approach to reduce energy consumption: the primary radio is left off when there is no traffic, and a secondary, low-power radio (the paging channel) is used to wake up nodes when data needs to be sent. Sparse Topology and Engery Management (STEM) [18] is an on-demand wake-up protocol using a second radio as a paging channel. In addition to using a low-power paging radio, STEM further reduces energy consumption by letting the paging radio periodically poll the medium for traffic. A sender needs to send a wake-up signal that is at least the length of the period. An advantage of using a second radio is the ability to completely avoid interference with possible transmissions on the main radio.

This approach for low-power listening has been generalized to operate as the primary energy-conservation mechanism with a single radio [19, 20]. A sleeping node periodically wakes up and briefly polls the medium. It stays in active mode only when activity is detected. A sender wakes up a receiver by sending packets with a preamble that is as long as the polling period. Figure 6.3 shows the packet exchanges in low-power listening. The benefit of low-power listening is that very brief polling is possible—as little as 3 ms on Mica2 motes [10]—with most of the delay involving time for the radio's crystal to stabilize. The disadvantage is that transmitting nodes must precede packets with extremely long preambles. This increases control overhead and reduces channel utilization, especially when traffic is heavy. On-demand wake-up offers the most aggressive reduction in listening time. For very low-duty cycle networks (less than a few percent) and light traffic, it appears quite attractive.

In summary, schedule-based MAC protocols, such as TDMA, avoid collisions and easily reduce idle listening and overhearing. However, they can be a poor match to multihop networks because of uneven energy usage due to clustering and the need for strict time synchronization. Contention protocols do not have these disadvantages, but basic protocols consume energy in collisions, idle listening, and overhearing. Versions of contention protocols reduce each of these costs, with four techniques to reduce idle listening: scheduled contention periods, asynchronous paging channels, and low-power listening, each with its own advantages and disadvantages.

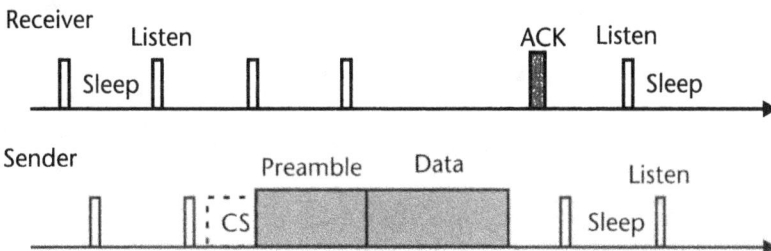

Figure 6.3 Packet exchanges in low-power listening. CS stands for carrier sense.

6.4   Topology-Control Protocols: Between MAC and Routing

Although MAC protocols may put the radio to sleep, they provide the illusion of continuous connectivity to all nodes, buffering and delaying transmission of packets if necessary. Topology-control protocols are a class of protocols set between the MAC layer and routing that violate this abstraction by turning nodes off for longer periods of time.

Topology-control protocols turn off as many nodes as possible to conserve energy, aiming to leave only enough on to keep a connected topology. This constraint assures that data can transit the network and that any node that attempts to send data can connect to the network. However, since some nodes are off, these nodes cannot be destinations for data.

Topology-control protocols complement MAC-level sleep/wake-up protocols for two reasons. First, they typically operate at much coarser timescales, cycling radios on the order of minutes rather than seconds. Coarser granularity reduces mode-switch costs and allows clocks to be less closely synchronized. Second, by relaxing the assumption that all nodes are reachable, sleeping nodes have no need to poll for traffic. The main disadvantage of topology-control protocols is that edge nodes will be sleeping and unreachable for long periods of time. If individual nodes are considered important or explicitly addressed, this constraint may be a problem, requiring backbone or source nodes to cache and resend data. On the other hand, in a data-centric sensor network where queries are made for classes of data rather than specific end nodes, this restriction will likely have little impact. For example, the query "Find seismic sensors covering region x" can be satisfied by whatever sensors are currently awake.

We briefly consider two classes of topology-control protocols: geographic-based protocols, exemplified by Geographic Adaptive Fidelity (GAF) [21], and topology-based protocols, such as Span [22]. Both aim to construct a connected backbone network; they differ in how to select nodes to form that network.

Geographic-based protocols such as GAF use physical location to infer network coverage [21]. Given a nominal radio range, nodes impose a logical grid over the network such that a node in any grid cell is guaranteed to be able to reach any node in any neighboring grid. One node in each grid is then elected to remain on to guarantee a connected network, while other nodes sleep to conserve energy. Figure 6.4(a) shows an

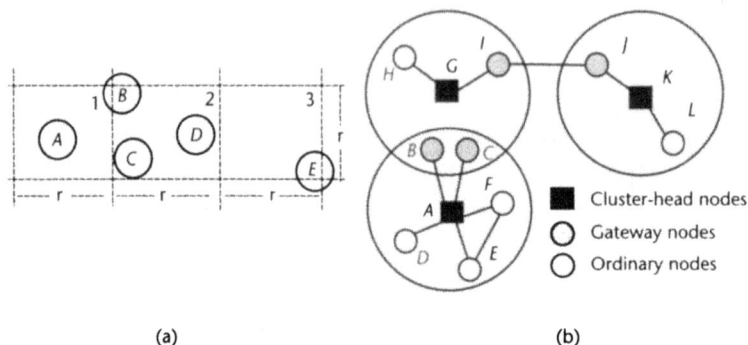

(a)   (b)

Figure 6.4   Node equivalence in dense sensor networks (examples from [29]): (a) geographical node equivalence, and (b) topological node equivalence.

example virtual grid where r is the nominal radio range, and one of nodes B, C, and D needs to remain awake. Geographic topology-control protocols can be fairly simple, but they require a source of node location and depend on a lower-bound estimate of radio range.

By contrast, topology-based protocols directly measure network connectivity, electing coordinators to guarantee coverage. In Span [22], a node becomes a coordinator if any of its neighbors cannot reach each other directly or via one or two coordinator nodes. This election algorithm requires that all nodes share their neighbor information with each other, and it measures local connectivity directly by this information. Figure 6.4(b) shows a sample cluster topology where black nodes (A, G, and K) are cluster heads, gray nodes form gateways between clusters (I, J, and either B or C), and the remainder are edge nodes. Backbone election can also preferentially select nodes with wall power, as proposed in ReOrg [23], or to minimize traffic and congestion [24].

Performance of topology protocols is affected by network density and node mobility. Energy savings and network lifetime are proportional to density. If density is defined as the mean number of neighbors each node has, a basic network requires a density of 6 to 10 to be connected if all nodes are awake. Topology protocols have been evaluated at densities of 20 to 100 neighbors. Densities of 20 to 40 exhibit network lifetimes three to four times that of a comparable network without topology control, both with GAF and Span. While one would expect lifetime to increase linearly with density, simulation results for the protocols suggest that efficiency decreases at higher densities due to overhead from electing and switching backbone nodes.

Although many sensor networks are static, topology-control protocols have been evaluated with mobile nodes. Since topology-control protocols select a minimum backbone of nodes, all are important to maintain connectivity. This balance can easily be disrupted by mobility. Both GAF and Span probe more frequently to cope with mobility, reducing their efficiency. Since GAF presumes that nodes know their location, it can use predicted movement to predict when backbone nodes must be reselected; CEC adds this capability to a Spanlike protocol [25].

Topology-control protocols can interact or be independent of MAC and routing protocols. GAF and CEC are independent of each, but problems can occur if topology control puts a node on a route to sleep [25]. These problems can be avoided by a fast-repairing routing layer or by explicit signaling between topology control and other layers. STEM is an example of topology control integrated with a paging-channel-based MAC protocol [26], using analysis to suggest MAC-level energy conservation can add a tenfold savings over GAF alone. Other researchers have explored the integration of topology control with transmission power control [27].

6.5   Routing

Routing is the highest protocol layer we review in this chapter. Since the routing protocol defines the interactions between many nodes as data travels in a multihop network, it is not surprising that there are several different goals that routing may optimize. S. Singh, M. Woo, and C. S. Raghavendra describe five possible goals; we summarize four here: "minimizing total energy consumed, maximizing time until network partition, and minimizing variance in energy at each node, or minimizing the cost" per packet [28].

Directly minimizing energy consumed can be at odds with the middle two goals since the minimum-energy path may concentrate on certain nodes, unevenly draining their batteries. The final goal seeks to balance these trade-offs, defining cost as a function of remaining battery lifetime. However, [28] reports that optimizing the cost is NP-complete. (J. H. Chang and L. Tassiulas formulated energy-aware routing as a linear programming problem, allowing polynomial solutions [29].)

Traditional routing algorithms such as the Bellman-Ford or Dijkstra's algorithm optimize some metric, such as the shortest path. In wireless networks, this metric is typically hop count or some other measure of latency, as in Destination-Sequenced Distance-Vector (DSDV) Routing [30] and Dynamic Source Routing (DSR) [31]. It is straightforward to generalize this metric to optimize for energy consumed at each hop, or for per-hop costs.

By considering a per-hop cost that is a function of remaining node lifetime, it becomes possible to handle routing in heterogeneous networks where some nodes have larger batteries or even wall power. Although most topology-control protocols described above do not explicitly consider routing, by selecting which nodes remain active, they implicitly influence routing. Another factor that affects the energy consumption in routing is the link quality. Link-level retransmissions could largely increase the energy cost. A shorter path with many retransmissions may be even worse than a longer path with more hops but fewer retransmissions. One way to handle this problem is to exclude bad links from route selection so that the hop-count metric can still be used. Another way is to add the link quality into the routing metric. S. Banerjee and A. Misra have considered both the variable transmission power and link quality at each hop [32]. They define a link cost that combines the link distance and the link error rate. A. Woo, T. Tong, and D. Culler have examined the interaction of routing with link quality on Mica2 motes [33]. They propose using the number of total transmissions (including retransmissions) as the routing metric.

In summary, research on energy-efficient routing is mainly focused on two aspects: minimizing energy cost per packet and balancing energy consumption in the network. The underlying MAC and topology-control protocols can influence the design of the routing layer.

6.6   Energy Conservation in Today's and Tomorrow's Applications

Having considered opportunities to conserve energy at each of these layers of the system, we conclude by placing them in the context of sensor networks that are being deployed today and that we expect may be deployed in years to come.

Habitat monitoring is a representative of the current state of the art for sensor-network applications today [11]. Several dozen Mica2 motes are placed to monitor a 500m × 500m area, augmented by a few computers with additional electrical and computing power and connectivity to the Internet. Deployment is done with some care to insure sufficient radio and sensing coverage.

It is informative to compare energy conservation in such an application. Radio transmission power is selected off-line, with deployment density and configuration in mind, to insure connectivity. On-line radio power control is not necessary. Since target lifetimes are several months or an entire season, MAC-level power control is critical, using either S-MAC or low-power listening. The network is not dense enough to warrant on-line topology control, and with only a single extraction point for data, routing options are limited.

While this application indicates current practice, it is in many ways limited by current cost and deployment constraints. Today's sensors cost a few hundred dollars per node for hardware, and remote deployment, ongoing debugging, and development make total costs higher still. As sensor prices fall and the infrastructure matures, denser deployments will become easier, making on-line use of transmission power control and topology control more feasible. Deployment of applications in less remote areas will motivate multiple connections to traditional networks, opening room for energy-conserving routing.

6.7 Conclusion

Energy conservation is an active area of research in sensors networks. In this chapter, we have briefly surveyed current work in transmission power control, MAC-protocol design, topology control, and routing. As these areas move from theoretical studies to laboratory experiments to fielded systems, we are beginning to see the deployment of long-lived sensor networks and the fruition of this research.

References

[1] Pottie, G. J., personal communication, 1999.

[2] Ramanathan, R., and Rosales-Hain R., "Topology Control of Multi-hop Wireless Networks Using Transmit Power Adjustment," *Proc. IEEE Infocom*, Tel Aviv, Israel, March 2000.

[3] Monks, J., Bharghavan V., and Hwu W., "A Power Controlled Multiple Access Protocol for Wireless Packet Networks," *Proc. IEEE Infocom*, Anchorage, AK, April 2001.

[4] Gupta, P., and. Kumar P. R, 'The Capacity of Wireless Networks," *IEEE Transactions on Information Theory*, Vol. IT-46, No. 2, March 2000, pp. 388–404.

[5] Pottie, G. J., and. Kaiser W. J. "Wireless Integrated Network Sensors," *Communications of the ACM*, Vol. 43, No. 5, May 2000, pp. 51–58.

[6] Chipcon, Cc1000 Single Chip Very Low Power RF Transceiver. Data sheet at http://www.chipcon.com/files/CC1000_Data_Sheet_2_1.pdf, April 2002.

[7] Heinzelman, W. R., Chandrakasan A., and Balakrishnan H., "Energy-Efficient Communication Protocols for Wireless Microsensor Networks," *Proc. Hawaii International Conference on Systems Sciences*, Big Island, HI, January 2000, pp. 3,005–3,014.

[8] Son, D., Krishnamachari B., and Heidemann J., "Experimental Study of the Effects of Transmission Power Control and Blacklisting in Wireless Sensor Networks," *Proc. 1st IEEE Conference on Sensor and Ad Hoc Communication and Networks*, Santa Clara, CA, October 2004.

[9] LAN MAN Standards Committee of the IEEE Computer Society, "Wireless LAN Medium Access Control (MAC) and Physical Layer (PHY) Specification," IEEE Std 802.11, 1999.

[10] Sohrabi, K., and Pottie G. J., "Performance of a Novel Self-organization Protocol for Wireless Ad Hoc Sensor Networks," *Proc. IEEE 50th Vehicular Technology Conference*, Amsterdam, Netherlands, 1999, pp. 1222–1226.

[11] Szewczyk, R., et al., "Application Driven Systems Research: Habitat Monitoring with Sensor Networks," *Communications of the ACM*, Vol. 47, No. 6, June 2004, pp. 34–40.

[12] Tseng, Y. C., Hsu C. S., and Hsieh T. Y, "Power Saving Protocols for IEEE 802.11-based Multi-hop Ad Hoc Networks," *Proc. IEEE Infocom*, New York, June 2002, pp. 200–209.

[13] Singh, S., and Raghavendra C. S., "PAMAS: Power Aware Multi-access Protocol with Signaling for Ad Hoc Networks," *ACM Computer Communication Review*, Vol. 28, No. 3, July 1998, pp. 5–26.

[14] Ye, W., Heidemann J., and Estrin D., "An Energy-Efficient MAC Protocol for Wireless Sensor Networks," *Proc. IEEE Infocom*, New York, June 2002, pp. 1,567–1,576.

[15] Ye, W., Heidemann J., and Estrin D., "Medium Access Control with Coordinated, Adaptive Sleeping for Wireless Sensor Networks," *IEEE/ACM Transactions on Networking*, Vol. 12, No. 3, June 2004, pp. 493–506.

[16] van Dam, T., and Langendoen K., "An Adaptive Energy-Efficient MAC Protocol for Wireless Sensor Networks," *Proc. 1st ACM International Conference on Embedded Networked Sensor Systems (SenSys)*, Los Angeles, CA, November 2003, pp. 171–180.

[17] Zheng, R., Hou J. C., and Sha L., "Asynchronous Wakeup for Ad Hoc Networks," *Proc. ACM International Symposium on Mobile Ad Hoc Networking and Computing*, Annapolis, MD, June 2003.

[18] Schurgers, C., et al., "Optimizing Sensor Networks in the Energy-Latency-Density Space," *IEEE Transactions on Mobile Computing*, Vol. 1, No. 1, 2002, pp. 70–80.

[19] Hill, J. L., and Culler D. E., "Mica: A Wireless Platform for Deeply Embedded Networks," *IEEE Micro*, Vol. 22, No. 6, November–December 2002, pp. 12–24.

[20] El-Hoiydi, A., Decotignie J. D., and Hernandez J., "Low Power MAC Protocols for Infrastructure Wireless Sensor Networks," *Proc. Fifth European Wireless Conference*, Barcelona, Spain, February 2004.

[21] Xu, Y., Heidemann J., and Estrin D., "Geography-Informed Energy Conservation for Ad Hoc Routing," *Proc. ACM International Conference on Mobile Computing and Networking*, Rome, Italy, July 2001, pp. 70–84.

[22] Chen, B., et al., "Span: An Energy-Efficient Coordination Algorithm for Topology Maintenance in Ad Hoc Wireless Networks," *Proc. ACM International Conference on Mobile Computing and Networking*, Rome, Italy, July 2001.

[23] Conner, W. S., et al., "Experimental Evaluation of Synchronization and Topology Control for In-Building Sensor Network Applications," *Proc. Second ACM Workshop on Sensor Networks and Applications*, San Diego, CA, 2003, pp. 38–49.

[24] Cerpa, A., and Estrin D., "ASCENT: Adaptive Self-configuring Sensor Networks Topologies," *Proc. IEEE Infocom*, New York, June 2002.

[25] Xu, Y., et al., "Topology Control Protocols to Conserve Energy in Wireless Ad Hoc Networks," Technical Report 6, Center for Embedded Networked Sensing, University of California, Los Angeles, January 2003.

[26] Schurgers, C., et al., "Optimizing Sensor Networks in the Energy-Latency-Density Design Space," *IEEE Transactions on Mobile Computing*, Vol. 1, No. 1, January 2002, pp. 70–80.

[27] Pan, J., et al., "Topology Control for Wireless Sensor Networks," *Proc. of the ACM International Conference on Mobile Computing and Networking*, San Diego, CA, September 2003, pp. 286–299.

[28] Singh, S., Woo M., and Raghavendra C. S., "Power-Aware Routing in Mobile Ad Hoc Networks," *Proc. ACM International Conference on Mobile Computing and Networking*, Dallas, TX, October 1998, pp. 181–190.

[29] Chang, J. H., and L. Tassiulas, "Energy Conserving Routing in Wireless Ad Hoc Networks," *Proc. IEEE Infocom*, Tel-Aviv, Israel, March 2000.

[30]   Perkins, C. E., and Bhagwat P., "Highly Dynamic Destination-Sequenced Distance-Vector Routing (DSDV) for Mobile Computers," *Proc. ACM SIGCOMM Conference*, London, U.K., August 1994.

[31]   Johnson, D. B., and Maltz D. A., "Dynamic Source Routing in Ad Hoc Wireless Networks." In Tomasz Imielinski and Hank Korth, (eds.), *Mobile Computing*, Kluwer Academic Publishers, Hingham, MA, 1996.

[32]   Banerjee, S., and Misra A., "Minimum Energy Paths for Reliable Communication in Multi-hop Wireless Networks," *Proc. ACM International Symposium on Mobile Ad Hoc Networking and Computing*, Lausanne, Switzerland, June 2002, pp. 146–156.

[33]   Woo, A., T. Tong, and Culler D., "Taming the Underlying Challenges of Reliable Multihop Routing in Sensor Networks," *Proc. ACM International Conference on Embedded Networked Sensor Systems (SenSys)*, Los Angeles, CA, November 2003, pp. 14–27.

CHAPTER 7

Multihop Routing

Nirupama Bulusu

In large WSNs, multihop communication optimizes energy consumption and enables reuse of the spectrum. WSNs require scalable, energy-efficient routing protocols to (1) disseminate networkwide queries and code updates, (2) enable robust data collection from the sensor network, and (3) enable distributed coordination among sensor nodes. Wireless communication with low-power radio transceivers found in common wireless sensor devices is loss prone and highly dynamic. To adapt to such dynamics, routing must be coupled with underlying issues such as link-quality estimation and neighborhood management. In this chapter, we explore some of these issues.

7.1 Introduction

In large WSNs, the energy required for communication increases as a power law over distance. In other words, sending large amounts of data over long distances must be avoided. The use of multihop communication has been promoted in WSNs to conserve energy [1] and to increase network capacity by enabling spectrum reuse [2].

In order to support self-organizing, multihop communication, one of the essential components is routing [3]. *Routing* is the act of moving information across a network from a source to a destination. Along the path, at least one intermediate node is encountered. The topic of routing has been covered in computer science literature for more than two decades, but routing achieved prominence as late as the mid-1980s with the growth of the Internet.

Routing involves two basic activities: *path determination*, or determining optimal routing paths, and *packet switching*, or transporting information groups, typically called *packets*, through a network. Although packet switching is relatively straightforward, path determination can be very complex. Routing protocols use metrics to evaluate what path will be the best for a packet to travel. A *metric* is a standard of measurement, such as path bandwidth, that is used by routing algorithms to determine the optimal path to a destination. To aid the process of path determination, routing algorithms initialize and maintain routing tables, which contain route information. Route information varies depending on the routing algorithm used. The design goals for a routing algorithm include selection of optimal routes, simplicity, low operational overhead, robustness and stability in the face of unusual and unforeseen circumstances, rapid convergence or agreement on routes, and flexibility in quickly and accurately adapting to a wide variety of network circumstances.

Beginning in the mid 1990s, several new routing protocols were developed for mobile, ad hoc networks of wireless devices [4]. These protocols dynamically discover routes from sources to the destination and enable the network to self-organize through local operations. Route discovery is either initiated by sources or on demand. In its simplest incarnation, on-demand routing involves no periodic exchanges of route information but, instead, establishes a route when needed by flooding route requests into the network.

Sensor-network routing protocols derive many of their principles of self-organization from the rich body of routing literature in ad hoc networks. We do not review this literature here ([4] provides an excellent survey) but, instead, focus on the several factors that distinguish sensor-network routing requirements from those of ad hoc networks [5]. Most significantly, sensor devices are resource constrained with limited processing, bandwidth, energy, and storage. Moreover, they are densely connected and employ low-power radios. Finally, they operate in aggregate over multiple hops with application-specific communication patterns.

In sensor networks, routing protocols are typically required to support several different communication models, illustrated in Figure 7.1. These include the following:

- *Dissemination (one-to-many communication)*. Routing is used to disseminate queries and interests for particular data in the network, as well as to reprogram and monitor sensor nodes via code propagation and maintenance.

- *Data gathering (many-to-one communication)*. Routing is used to gather data from distributed sensor sources, which detect events of interest and transmit them back to a data sink. This typically involves a many-to-one communication model. Moreover, the sensors can generate either a continuous stream of data or a sporadic stream only when interesting events occur. In the latter case, data may flow concurrently from several sensors that observe the event. One important feature in sensor networks is that redundant data can be aggregated along the path, thereby reducing the total amount of information transmitted [5].

- *Any-to-any communication*. Routing is used to support distributed coordination among sensor devices, for example, to enable load balancing in distributed storage mechanisms or to enable triggering of special devices. This involves an any-to-any communication model.

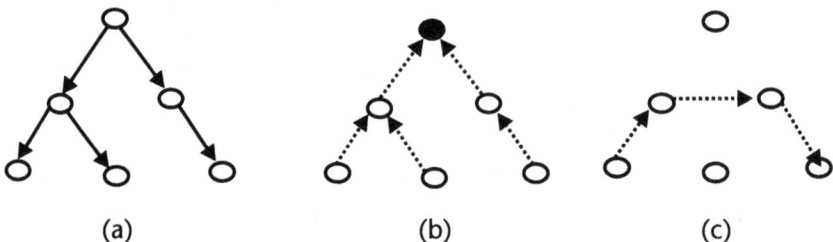

Figure 7.1 Routing used in sensor networks: (a) dissemination (one-to-many communication), (b) data gathering (many-to-one communication), and (c) triggering (any-to-any communication).

What makes sensor-network routing challenging? Several studies have shown that wireless communication links using the low-power radio transceivers found in sensor networks have significant loss rates [6–9]. Moreover, the loss rate varies dynamically with environmental factors or contention arising from the highly correlated behavior of the application. A nearby node may be within communication range most of the time but not always, depending on interference, congestion, and other sources of loss.

For example, as A. Woo, T. Tong, and D. Culler argue in [6], a basic notion for path determination like shortest path is typically defined relative to a connectivity graph describing which nodes can communicate directly over a single hop. In sensor networks, connectivity is not a binary relation but a measure of the likelihood of successful communication. The connectivity graph has to be discovered by nodes observing communication events and sharing this information.

Thus, sensor networks motivate new approaches to routing. Sensor networks cannot afford to abstract all the details of lower-layer protocols and, thus, motivate cross-layer design and optimizations in routing. Routing protocols now take into account the underlying factors and encompass connectivity analysis, neighborhood management with limited storage, and routing on dense sensor networks with simple, low-power radios. They must be evaluated in concert with lower-level estimation mechanisms under realistic loads.

The rest of this chapter is organized as follows. In Section 7.2, we study the various multihop routing protocols developed to support dissemination, many-to-one data gathering, and any-to-any communication. In Section 7.3, we study the cross-layer routing components for sensor networks and why we need them. We conclude in Section 7.4.

7.2   Multihop Routing for Data Gathering

In this section, we use the example of data gathering for our discussion. In data gathering, a large collection of nodes route sampled data over multiple hops to an individual sink.

7.2.1   Application Scenario: Data Gathering

Sensor nodes, or *sources*, gather data such as temperature, salinity, acoustic, or vibration measurements. The data gathered may be periodic, event driven, or location specific. The end goal is for the data to arrive at some base station, or *sink*, to be consumed. The base station is assumed to have persistent storage or global memory.

A simple, easy-to-deploy topology choice to support this operation is a star topology, wherein all nodes communicate directly with the base station, which constitutes the hub of a topology. However, this topology has restricted communications coverage and makes inefficient use of wireless and energy spectrum. In contrast, a tree topology rooted at the base station allows coverage of a greater area. Multihop communication is more energy efficient, not only because it avoids the energy overhead of power-law communication but because it enables in-network processing and aggregation. Tree-based routing is the underlying basis of many practical data-gathering schemes [10–12].

7.2.2 Building the Data-Gathering Tree

Building the data-gathering tree consists of two logically distinct operations: (1) creating and maintaining a topology, and (2) tasking nodes with when and what to sense and report.

In practice, both of these operations are performed jointly. One of the most widely using mechanisms to accomplish this is *flooding*. A sink floods out interest. A node reforwards interest and chooses a previous node as parent in the data-gathering tree. Therefore, flooding serves both to set up the tree via reverse paths, as well as to disseminate tasks and commands. This is illustrated in Figure 7.2.

Flooding is a reactive technique and does not involve topology maintenance and complex route-discovery algorithms. However, it has several performance drawbacks, including an implosion of duplicate messages and resource blindness to the available energy of the sensor nodes [13].

Gossiping [14] is a variant of flooding wherein each node forwards incoming packets to a randomly selected neighbor instead of broadcasting them to all neighbors. Gossiping avoids the implosion problem by having just one copy of a packet at each node. However, it takes a very long time to propagate the message to all sensor nodes. Consequently, numerous routing protocols for sensor networks have been proposed that try to find an optimal balance between flooding and gossiping.

Directed Diffusion [10] and Sensor Protocols for Information Dissemination via Negotiation (SPIN) [13] are two of the first proposed data-centric routing protocols for WSNs and work in a somewhat similar fashion. In Directed Diffusion, the data generated by the source is named using attribute-value pairs. The sink node requests the data by periodically broadcasting an *interest* in the named data. Each node in the network will establish a gradient toward its neighboring nodes from which it receives the interest. The gradient specifies both the data rate and the direction toward which the data should be sent. Once the source detects an interest, it will

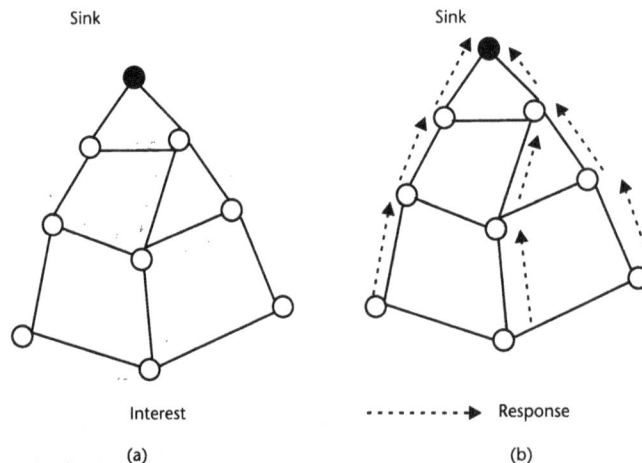

Figure 7.2 Building the data-gathering tree: (a) interests are flooded from the sink into the network, and (b) each node selects a parent node, forming a reverse path to the sink.

send exploratory packets toward the sink, possibly along multiple paths. As soon as the sink begins to receive exploratory packets from the source, it will reinforce one particular neighbor from whom it chooses to receive the rest of the data. The data will then flow back toward the sink along the reinforced path. The path reinforcement packets are also used for local path repairs in case of the failure of some nodes during the data-delivery phase.

Two Tier Data Dissemination (TTDD) uses a more hierarchical approach to improve scalability [15]. It uses a grid structure to divide the topology into cells. Only sensors located at a cell boundary need to forward the data. The sink actively builds the grid structure through the network and sets up forwarding points the *dissemination nodes* (DNs), the sensors closest to the grid boundary called. One tier is the cell at the sink's current location, and the other is the DN at the cell boundaries. The sink only floods the query within its own cell. When the nearest DN hears the query, it forwards it to its adjacent DNs. This process continues until the query reaches the source or one of the DNs with the corresponding data. During the query-propagation period, the network establishes the reverse path toward the sink so that it can enable the data path to be the same as that of the query propagation.

In Gradient Broadcast (GRAB) [16], a cost-based routing approach is used. A node on deployment sets its cost to reach the sink at infinity. As soon as the sink node starts up, it broadcasts the advertisement message containing its initial cost. Each intermediate node that hears the advertisement will calculate the receiving cost of the message. At the end of the cost-field setup period, each working node will have calculated the minimum cost for it to reach the sink. Each message carries a *credit* in its header. Data packets can flow along multiple paths if the credit is set higher than the minimum cost. Each intermediate node will make its own decision regarding the forwarding of a packet based on the amount of credit in the data message, its own minimum cost value, and the remaining ratio.

Directed Diffusion and TTDD incur a relatively higher overhead compared to GRAB in route discovery, but they minimize redundant transmissions.

7.2.3   Clustering: From Tree to Forest

A single-tree topology has several drawbacks. The root of the tree is a unique point of failure. As the network grows, paths become longer and longer, leading to increased end-to-end packet loss and congestion and decreased network lifetime. Moreover, it has very poor load balancing as nodes near the root carry more traffic.

We can anticipate that the sensor networks of the future will deploy multiple, dispersed base stations, known as *microservers*. Data needs to arrive only at one of the sinks (preferably the closest). This approach decreases the mean number of hops to closest sink and, consequently, decreases the energy burn rate while increasing the delivery ratio and spreading traffic relay load more evenly over sensors.

A forest-based routing scheme works by restricting the scope of interest flooding, using techniques such as *Voronoi scoping* [12]. In Voronoi scoping, a node n receiving interest from a sink rebroadcasts it if, and only if, (1) it has never seen this packet (verified using packet sequence number), and (2) the distance (number of hops) that a packet has traversed from the sink is smaller than the distance to any of the other sinks. The key idea is to form Voronoi clusters Vk rooted at each sink k, containing all nodes whose closest sink is k. Ideally, each sink only floods the nodes in its Voronoi cluster.

With Voronoi scoping, the scoping decision is entirely distributed. Clusters adaptively grow and shrink as links come and go. No coordination is required between sinks. Some redundancy can be retained among trees to ensure robustness for weak links.

Likewise, for the complementary process of data collection, LEACH is a clustering-based protocol that minimizes energy dissemination in sensor networks by partitioning the network into dynamic clusters [17]. All nodes are assumed to be capable of direct communication with a base station; however, only the cluster head is required to do so. In LEACH, cluster heads are randomly selected and rotated so as to balance the energy dissipation across the network nodes. LEACH has a *setup phase* and a *steady phase*.

During the setup phase, a node selects a random number between zero and one. If this number lies above a threshold (carefully chosen to ensure the required density of cluster heads in the network), it is a cluster head and advertises itself to the other sensor nodes in the network. Sensor nodes select which cluster head they must belong to based on proximity information (e.g., the signal strength of a cluster head's advertisement) and communicate this information to the appropriate cluster head. The cluster head assigns it time slots for transmission.

During the steady phase, sensors transmit their data to the cluster head, which forwards it to a base station, perhaps aggregating the data. After a period, nodes go back to the setup phase and rotate cluster heads.

One drawback of LEACH is that it assumes the base station is close to the sensor nodes.

It works best if the base station is positioned in the middle of the sensor network, but this is not always possible.

7.2.4 Geographic Routing

On-demand routing techniques, such as those described in Section 7.2.2, work well for small and moderately sized sensor networks and for large sensor networks with relatively stable routes and limited destination locality. In large sensor networks, where triggering or activation of sensors creates bursty any-to-any communication patterns, the overhead and latency of route discovery can be significant [18–20].

Geographic routing uses the locations of nodes as their addresses and forwards packets to their destination, when possible, in a greedy manner. Geographic routing is particularly suitable to sensor networks because data is often named with location attributes, and sensor nodes are expected to be location aware. Greedy Perimeter Stateless Routing (GPSR) [18] is probably the most widely known scheme for geographic routing. One of the key challenges in geographic routing is how to deal with dead ends, where greedy routing fails because a node has no neighbor close to the destination. GPSR uses the perimeter to circumnavigate voids. GPSR can be used to support any-to-any communication for distributed storage mechanisms, as will be discussed in Chapter 13.

The performance of geographic routing schemes is sensitive to errors in location information [21]. More recently, several techniques have emerged to facilitate geographic routing in the absence of precise location information [19, 20]. These techniques work by creating virtual coordinates for nodes based on their network connectivity. These virtual, rather than true, coordinates are used for routing.

Greedy geographic routing does not account for the energy available at sensor nodes along the path. Consequently, the shortest path may also be the most energy-depleted path. This can cause premature partitioning of the sensor network [22]. Recent techniques [22, 23] have developed forwarding heuristics that balance greedy and energy-efficient forwarding to optimize the overall system lifetime.

7.3   Routing Components

In a study of wireless links across multiple environments, J. Zhao and R. Govindan [7] found the following:

- In harsh indoor environments, 40% of the links have quality less than 70%.
- When mapping packet loss as a function of distance, the connectivity region can be distributed into three regions: (1) an *effective region*, wherein all nodes essentially have good connectivity, (2) a *transitional region*, wherein the average link quality falls off smoothly, but individual pairs exhibit high variation, and (3) a *no connectivity* region. This is illustrated in Figure 7.3.
- The effective range increases with the transmission power level.

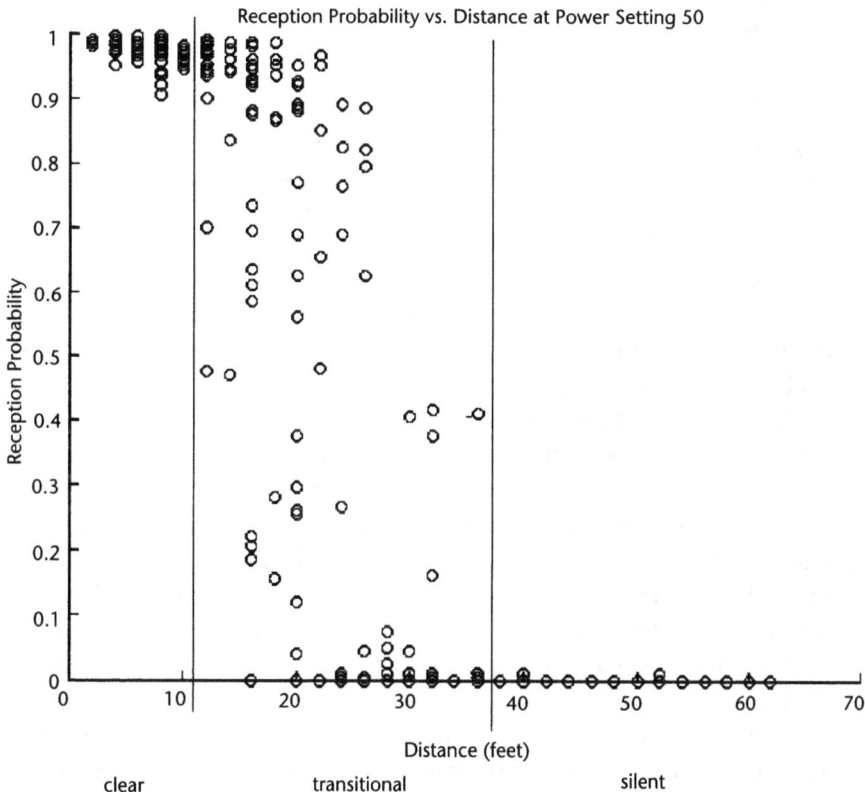

Figure 7.3   Reception probability of all links in a linear topology. Each link pair appears twice to indicate link quality on both directions. (*Source:* Alec Woo, used with permission.)

- The transitional range is quite significant. It ranges from 20% to 50% of the communication range.
- Higher transmit power does not eliminate transitional range.
- Packet loss increases as network load increases.
- Asymmetric links are quite common. More than 10% of the links have a quality difference greater than 50%. The environment alone does not cause asymmetry, as swapping nodes swaps the asymmetry.

An ad hoc routing protocol creates a many-to-one spanning tree topology, is self-organizing through local operations, is simple and efficient, and explores quality or reliable routing paths to the base station. There are several pitfalls with applying traditional routing-protocol design to sensor networks [6, 7]. Foremost is the assumption that links are inherently good. This assumes that the link-layer abstraction provides good links and pays little attention below the network layer. The second assumption is that the reverse link quality is as good as that of the forward links used in setting up the reverse-path tree in protocols such as Directed Diffusion. Finally, a small, fixed, consecutive number of packet delivery failures are treated as link failures. Thus, "semigood" links are easily treated as failures, creating network instability.

In order to take the underlying issues into account, Woo [6] argues that routing must encompass (1) link-quality estimation, (2) neighborhood management, and (3) new route-selection metrics. We review Woo's methodology in detail in the rest of this section.

7.3.1 Link-Quality Estimation

The goal of link-quality estimation is to estimate the rate of successful reception from neighboring nodes. Received-signal strength indication (RSSI) alone may not work well to characterize link quality. Instead, neighbors must exchange estimations to derive bidirectional link quality.

There are two techniques for link estimation—*passive snooping* and *active probing*. Because the wireless communication medium is a broadcast medium, we can employ passive techniques wherein nodes snoop on data packets sent by their neighbors to estimate link quality.

There are several considerations when estimating link quality. For a given error bound and agility, good estimator yields the most stable distribution with the smallest memory footprint and the simplest algorithm.

How can we infer packet loss? One simple scheme would be to use the packet sequence number for inferring packet loss. With this scheme, we cannot infer loss until we have heard the subsequent packets, which is not possible with dead nodes or mobility. Another scheme would be to infer losses based on time. This assumes that the minimum data rate is known, which is likely to be true in periodic data collection.

Because agility and stability are at odds with each other, getting an estimate may require a large number of samples. The Window mean with exponentially weighted moving average (WMEWMA) estimator computes an average success rate over time and smoothes with an exponentially weighted moving average. The tuning parameters are T and α, the history size of the exponentially weighted moving average (EWMA).

The WMEWMA estimator yields agile and stable estimations, uses constant memory, and is very simple.

7.3.2   Neighborhood Management under Limited Memory

The neighbor table maintains link-estimation statistics and routing information for each neighbor. The size of the table increases linearly with the network density (i.e., number of neighbors of a node). In WSNs, the density can be very high, but memory is limited. Moreover, at high density, many links are poor or asymmetric.

The goal of neighborhood management is to use constant memory to maintain the set of good neighbors of a node regardless of cell density. When the table becomes full, it must decide whether to add a new neighbor and, if so, which old neighbor to evict.

What should the metric for neighbor goodness be? Link quality is a good metric, but it is unknown. Signal strength provides a hint. The approach is to rely on the frequency of packet reception, assuming periodic data packets or beacons. This is similar to frequency estimation of data streams or the classical cache policy. The pseudocode for the management algorithm is described below:

```
When we hear a node, if
    In table: increment a counter for this node
    Not in table
            Insert if table is not full
            down-sample if table is full
                    If successful, insert only if some nodes can be evicted
Eviction: (FREQUENCY)
    Decrement counter for each table entry
    Nodes with counter = 0 can be evicted
            Otherwise, all nodes stay in the table
```

The FREQUENCY algorithm can effectively utilize 50% to 70% of the table space to maintain a set of good neighbors, while being adaptive to neighborhood changes.

To avoid maintaining sibling nodes based on routing-cost difference, neighborhood goodness is augmented with routing simulation.

7.3.3   Route Selection

Routing algorithms have used many different metrics to determine the best route. Sophisticated routing algorithms can base route selection on multiple metrics, combining them in a single (hybrid) metric. Metrics include the following:

- Path length;
- Reliability;
- Delay;
- Bandwidth;
- Load;
- Communication cost;

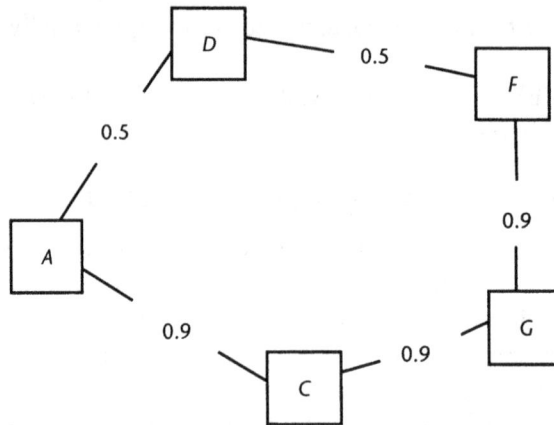

Figure 7.4 Sensor-network connectivity graph. Edges are weighted with link quality. The shortest hop-count path *A-D-F* between the nodes *A* and *F* also requires more retransmissions for successful reception.

- Energy cost.

Path length is the most common routing metric. Some routing protocols allow network administrators to assign arbitrary costs to each network link. In this case, path length is the sum of the costs associated with each link traversed.

In traditional networks, the metric for path length is hop count. In WSNs, if link quality is not considered in route selection, the real cost of packet delivery can exceed the hop count. For example, consider the sensor-network connectivity graph shown in Figure 7.4. Edges are weighted with link quality (i.e., the ratio of packets transmitted with successful reception). Using hop count as the metric for path length, the shortest path from *A* to *F* is the path *A-D-F*. However, the link quality of *A-D* and *D-F* is only 0.5; so, on average, a packet has to be transmitted $1/(0.5 \times 0.5) = 4$ times to be received successfully at *D*. On the other hand, the link qualities of *A-C*, *C-G*, and *G-F* are 0.9 each; so, on average, a packet only has to be transmitted $1/(0.9 \times 0.9 \times 0.9) = 1.37$ times to be received successfully at *D*. This example illustrates why the shortest path between two nodes according to hop count is not necessarily the better path.

Woo proposes two possible routing metrics that take link quality into account.

- *Shortest path with threshold [SP(t)]*. Shortest-path routing can still be useful in unreliable networks, given that poor links are filtered out from route selection. A simple technique is to apply shortest-path routing only to links whose estimated link quality exceeds a specified threshold *t*. This increases the depth of the network since reliable links are likely to be shorter. However, network density and physical deployment may result in a connectivity graph that is weakly connected.

- *Minimum expected transmissions (MT)*. With links of varying quality, a longer path with fewer retransmissions is preferable to a shorter path with many retransmissions. An alternate approach is to use the expected number of transmissions as the cost metric for routing. The best path minimizes the total number of transmissions in delivering a packet over multiple potential hops to the destination.

7.3.4 Evaluation

Woo's evaluation of these routing metrics on the MICA motes showed that MT might be the preferable metric. SP(70%) fails to construct a viable routing tree, while SP(40%) does not adapt well to high network congestion. MT avoids these problems by picking the best available paths without the arbitrary threshold. The lesson here is that by tracking the link quality, routing protocols can successfully avoid routing over asymmetrical links.

No cycles were detected, suggesting that simple cycle avoidance mechanisms are sufficient for relatively static networks. Even at low power settings, the number of potential neighbors is quite large, reinforcing the need for neighborhood management.

7.5 Conclusion

In large sensor networks, multipath communication conserves energy and enables spectrum reuse. Routing is a prerequisite to support multihop communication and must support different types of communication models. In this chapter, we reviewed some of the major routing protocols for sensor networks and the key concepts behind them, such as data naming, clustering, and geographic forwarding. Data-centric communications, storage, and programming will be studied in greater depth in Chapters 12 to 14. Security aspects of routing protocols will be discussed in Chapter 15.

Various real-world studies show that because of the dynamic, loss-prone nature of wireless communication in sensor networks, routing protocols for sensor networks should take into account underlying link-level and topology features and should be evaluated in concert with lower-level estimation mechanisms under realistic loads. Although newer generations of radios and applications will have correspondingly different connectivity and traffic characteristics, the three-region structure is expected to persist. Therefore, link estimation, neighborhood table management, and reliability-based routing metrics are likely to remain core issues for future generations of routing protocols.

References

[1] Pottie, G. J., and Kaiser W., "Wireless Integrated Network Sensors," *Communications of the ACM*, Vol. 43, No. 5, 2000, pp. 51–58.

[2] Ramanathan, R., and Rosales-Hain R., "Topology Control of Multihop Wireless Networks Using Transmit Power Adjustment," *Proc. IEEE Infocom*, Tel-Aviv, Israel, March 2000.

[3] Perlman, R., *Interconnections: Bridges, Routers, Switches, and Internetworking Protocols*, Boston, MA: Addison-Wesley, 2000.

[4] Royer, E. M., and Toh C. K., "A Review of Current Routing Protocols for Ad Hoc Mobile Wireless Networks," *IEEE Personal Communications*, Vol. 6, No. 2, 1999.

[5]   Estrin, D., et al., "Next Century Challenges: Scalable Coordination in Sensor Networks," *Proc. ACM MobiCOM '99*, Seattle, WA, August 1999.

[6]   Woo, A., T. Tong, and Culler D., "Taming the Underlying Challenges of Reliable Multihop Routing in Sensor Networks," *Proc. ACM SENSYS*, Los Angeles, CA, November 2003.

[7]   Zhao, J., and Govindan R., "Understanding Packet Delivery Performance in Dense Wireless Sensor Networks," *Proc. ACM SENSYS 2003*, Los Angeles, CA, November 2003.

[8]   Couto, D. D., et al., "Performance of Multi-hop Wireless Networks," *Proc. 1st Workshop on Hop Topics in Networks (HotNetS-I)*, Princeton, NJ, October 2002.

[9]   Cerpa, A., Busek N., and Estrin D., "SCALE: A Tool for Simple Connectivity Assessment in Lossy Environments," CENS Technical Report 21, September 2003.

[10] Intanagonwiwat, C., Govindan R., and Estrin D., "Directed Diffusion: A Scalable and Robust Communication Paradigm for Networked Sensors," *Proc. ACM MOBICOM*, Boston MA, August 2000, pp. 56–67.

[11] Madden, S. R., et al., "TAG: A Tiny AGgregation Service for Ad Hoc Sensor Networks," *Proc. OSDI 2002*, Boston MA, December 2002.

[12] Dubois-Ferriere, H., and Estrin D., "Efficient and Practical Query Scoping in Sensor Networks," CENS Technical Report 39, April 2004.

[13] Heinzelman, W. R., Kulik J., and Balakrishnan H., "Adaptive Protocols for Information Dissemination in Wireless Sensor Networks," *Proc. ACM Mobicom '99*, Seattle, WA, 1999, pp. 174–185.

[14] Hedetniemi, S., Hedetniemi S., and Liestman A., "A Survey of Gossiping and Broadcasting in Communication Networks," *Networks*, Vol. 18, 1988.

[15] Cheng, J., et al., "Two-Tier Data Dissemination Model for Large-Scale Wireless Sensors Networks," *Proc. ACM/IEEE International Conference on Mobile Computing and Networking (MobiCom 2002)*, Atlanta, GA, September 2002.

[16] Lu, S., et al., "Gradient Broadcast: A Robust Data Delivery Protocol for Large Scale Sensor Networks," *Proc. IEEE IPSN*, Palo Alto, CA, April 2003.

[17] Heinzelman, W., Chandrakasan A., and Balakrishnan H., "Energy-Efficient Communication Protocols for Wireless Sensor Networks," *Proc. 33rd Annual Hawaii International Conference on System Sciences (HICSS)*, Big Island, HI, January 2000, pp. 3005–3014.

[18] Karp, B., and Kung H. T., "Greedy Perimeter Stateless Routing," *Proc. Sixth Annual ACM/IEEE International Conference on Mobile Computing and Networking (Mobicom 2000)*, Boston MA, 2000.

[19] Rao, A., et al., "Geographic Routing without Location Information," *Proc. ACM MobiCom*, Boston MA, September 2003, pp. 96–108.

[20] Newsome, J., and Song D., "GEM: Graph Embedding for Routing and Data-centric Storage in Sensor Networks without Geographic Information," *Proc. ACM Sensys*, Los Angeles, CA, 2003.

[21] Seada, K., Helmy A., and Govindan R., "On the Effect of Localization Errors on Geographic Face Routing in Sensor Networks," *Proc. IEEE/ACM 3rd International Symposium on Information Processing in Sensor Networks (IPSN)*, Berkeley, CA, April 2004.

[22] Yu, Y., Govindan R., and Estrin D., "Geographical and Energy Aware Routing: A Recursive Data Dissemination Protocol for Wireless Sensor Networks," USC technical report, 2001.

[23] Seada, K., et al., "Energy-Efficient Forwarding Strategies for Geographic Routing in Lossy Wireless Sensor Networks," *Proc. ACM SenSys*, Baltimore, MD, November 2004.

Reliable Transmission and Congestion Control

Özgür B. Akan and Mehmet C. Vuran

8.1 Introduction

Variable wireless link errors, node failures, and challenges due to the multihop nature of WSNs necessitate reliable, efficient, and application-specific communication protocols to ensure reliable event-information collection through the collaborative networking of individual sensor nodes. The collaborative effort of these networked wireless sensor nodes may provide sensing capabilities that surpass the capabilities of current sensing and monitoring systems. In order to realize the potential gains of such collaboration, it is important that extracted event features at sensor nodes be reliably communicated to the sink. Efficient communication protocols for sensor networks are required to maintain the reliability levels required by specialized applications [1].

The varying wireless channel conditions constitute the main challenge in ensuring reliable event-information collection in WSNs. Each communication attempt between sensor nodes is prone to errors that degrade overall reliability. Moreover, in contrast to traditional cellular networks and WLANs, the multihop communication nature of WSNs amplifies the adverse effects of wireless channel unreliability on the communication of event information to the sink. Because efficient communication between the sensors and the sink ultimately depends on transmissions between sensor nodes and the sink, both link-level and network-level reliability measures are required. Consequently, reliable communication mechanisms are imperative at the physical, data-link, network, and transport layers, which provide bit-, packet-, route-, and transport-level reliability, respectively.

In addition to wireless channel errors and link failures, communication reliability is also affected by congestion in the network. In a multihop sensor network, every sensor node may act as both source and router at the same time. Consequently, the memory, processing, and communication constraints of sensor nodes may well lead to congestion in the network. Packet loss due to congestion can impair event detection at the sink, even when enough information is sent out by the sensor nodes. Moreover, the unique characteristics of WSNs, such as dense node deployment, the event-based communication paradigm, and low-energy consumption requirements, necessitate congestion-control protocols specifically tailored to the needs of the WSN.

While there are similarities between WSNs and traditional wireless networks, the communication protocols and algorithms proposed for traditional wireless networks may not be adequate in providing reliable communication in WSNs. Unlike wireless networks, sensor networks are deployed to achieve a specific application objective via the collaborative effort of numerous sensor nodes. Hence, the application-specific objectives of a sensor network also influence the design requirements of the communication protocols. Hence, new reliability and congestion-control mechanisms tailored to the specific needs of the WSN are required.

In this chapter, the challenges and possible solution strategies for the design and development of reliable communication protocols are presented. The protocol requirements to ensure reliable transmission are discussed in Section 8.2. More specifically, physical, link, and network layer reliability issues are presented in Sections 8.2.1 to 8.2.3. Transport-layer reliability problems and the existing solution approaches are discussed in Section 8.2.4. Congestion control in WSNs is explored in Section 8.3, and concluding remarks are given in Section 8.4.

8.2 Reliable Transmission in WSNs

In this section, the challenges and the solution approaches for reliable transmission in WSNs are investigated, considering physical-, data-link-, network-, and transport-layer reliability mechanisms.

8.2.1 Physical-Layer Reliability

The physical layer is responsible for the conversion of bit streams into signals that are best suited for communication across the wireless channel. More specifically, the physical layer is responsible for frequency selection, carrier-frequency generation, signal detection, modulation, and data encryption. The reliability of the communication depends also on the hardware properties of the nodes, such as antenna sensitivity and transceiver circuitry.

The wireless medium used in WSNs is one of the most important factors since the unique properties of different media constraint the capabilities of the physical layer. The unreliability and varying nature of wireless communication channels necessitate efficient error-control strategies to be implemented according to the properties of the specified wireless medium. The wireless links can be formed by radio or optical media. For radio links, one option is to use industrial, scientific, and medical (ISM) bands, which offer license-free communication in most countries. Some of the ISM frequency bands are already being used for communication in cordless phone systems and WLANs. Much of the current hardware for sensor nodes is based upon RF circuit design. The μAMPS wireless sensor node [2] uses a Bluetooth-compatible 2.4-GHz transceiver with an integrated frequency synthesizer. In addition, the low-power sensor device [3] uses a single-channel RF transceiver operating at 916 MHz. The Wireless Integrated Network Sensors architecture [4] also uses radio links for communication. Although there exist many advantages to using the ISM bands, such as free radio, huge spectrum allocation, and global availability, these bands are prone to interference from different sources that operate in the same frequency. Hence, careful selection of operation bands in conjunction with the location of the WSN and sophisticated interference-cancellation hardware is required to provide reliable communication.

The ultrawideband (UWB), or impulse radio, has also been used as communication technology in WSN applications, especially in indoor wireless networks [5]. The UWB employs baseband transmission and, thus, requires no intermediate or radio carrier frequencies. Generally, pulse position modulation (PPM) is used. The main advantage of UWB is its resilience with regard to multipath fading [6, 7]. Hence, increased reliability is possible by exploiting UWB techniques in sensor networks along with low transmission power and simple transceiver circuitry.

An optical medium can also be used for communication among sensor nodes. An example is the Smart Dust mote [8], which is an autonomous, sensing, computing, and communication system that uses an optical medium for transmission. While an optical medium can enable ultra-low-power communication with the help of passive devices and mirrors on the sensor nodes, line-of-sight requirements and robustness problems against node-position changes constitute significant challenges to WSN deployment. As a specific case of the use of an optical medium, infrared (IR) communication is also used for internode communication in sensor networks. IR communication is license-free and, in contrast to the RF links, robust to interference from electrical devices. Although the IR medium provides low error rates and prevents interference, the main drawback is, again, the requirement of line of sight between the sender and receiver. This makes IR a less optimal choice of transmission medium in sensor-network-deployment scenarios.

Application-specific requirements also constrain the capabilities of sensor nodes. For instance, marine applications may require the use of an underwater transmission medium. Hence, acoustic waves that can penetrate the water are a favorable choice. However, high error rates and low data rates make assuring reliable communication challenging with underwater channels. Inhospitable terrain or battlefield applications might encounter error-prone channels and greater interference. Moreover, due to low cost requirements, the antenna of the sensor nodes might not have the height, sensitivity, and radiation power of those in traditional wireless devices. Hence, the choice of transmission medium must be supported by robust coding and modulation schemes that efficiently model these vastly different channel characteristics.

Channel-coding schemes have long been investigated in the context of wireless communication theory. There exist many powerful channel codes, such as Reed Solomon (RS), convolutional, and Bose-Chaudhuri-Hocquenghem (BCH) codes. However, recently, the efficiency of distributed source-channel coding has been investigated in the context of WSNs [9]. Since the information gathered by sensor nodes follows the physical properties of the sensed phenomenon, the characteristics of the source can be closely matched with the channel characteristics. It has been shown that in distributed networks, where the information about an event is more important than the individual readings of each sensor, joint source-channel coding yields optimal results [10]. Moreover, a recent work has shown that uncoded transmission achieves a scaling-law optimal performance [9]. Hence, exploiting the intrinsic properties of the sensed phenomenon provides additional advantages to channel coding. Based on the joint source-channel coding strategies and uncoded transmission principles, many networking protocols extending the physical layer have also been proposed in the literature. In [11], possible approaches to transport and MAC-layer protocols, exploiting the spatial and temporal correlation and uncoded transmission in WSNs, have been discussed. The source-channel coding has been exploited in [12] in order to investigate the reliability-versus-efficiency trade-off in data gathering. Furthermore, an application-level error-correction algorithm is presented that exploits the spatiotemporal properties of the physical phenomenon in [13] as an alternative to RS codes.

Modulation is also a critical building block for reliable wireless communications since it performs the conversion of digital information to analog waveforms. Generally, binary or M-ary modulation schemes are used in WSNs. The general structure of binary modulation is given in Figure 8.1. In this figure, the frequency synthesizer is integrated with the modulation circuitry [2]. Moreover, the M-ary modulation is illustrated in Figure 8.2. Using M-ary modulation, multiple bits can be sent through the channel. This is accomplished by parallelizing the input data and using these parallel data as inputs to a digital-to-analog converter (DAC). As a result, the parallel input levels provide the in-phase and quadrature components of the modulated signal.

Binary and M-ary modulation schemes are compared in [2]. While an M-ary scheme can reduce the transmit on-time by sending multiple bits per symbol, it results in complex circuitry and increased radio-power consumption. E. Shih et al. [2] formulate these trade-off parameters and conclude that under startup-power-dominant conditions, the binary modulation scheme is more energy efficient. Hence, M-ary modulation gains are significant only for low-startup-power systems. The choice of a good modulation scheme depends on the types of waveforms, characteristics of the channels, and the nature of the interference, as well as the energy

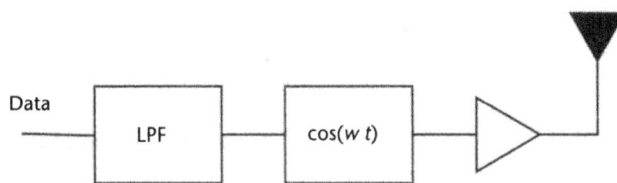

Figure 8.1 Binary modulation block.

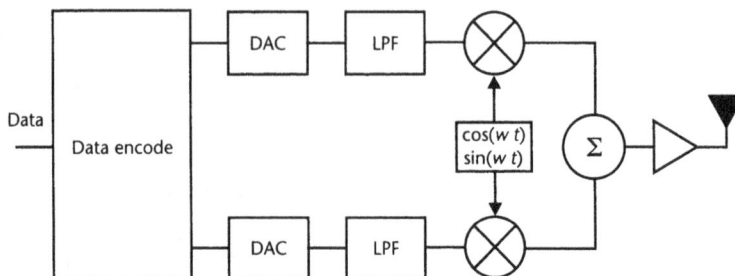

Figure 8.2 M-ary modulation block.

efficiency of the modulation scheme, as discussed above. Hence, all of these factors should be considered when choosing a modulation scheme for reliable communication in WSNs.

It is well known that long-distance wireless communication can be expensive, both in terms of energy and cost. While designing the physical layer for sensor networks, energy minimization assumes significant importance, over and above the decay, scattering, shadowing, reflection, diffraction, multipath, and fading effects. For instance, multihop communication in a sensor network can effectively overcome shadowing and path-loss effects if the node density is high enough. Similarly, while propagation losses and channel capacity limit data reliability, this very fact can be used for spatial frequency reuse. Moreover, network-layer protocols are usually developed to provide shortest-hop-count routes to the packet. Although these routes may seem optimal in the network layer, due to high error rates caused by the increased transmission range, a penalty will be paid [14]. Energy-efficient, reliable physical-layer solutions are currently being pursued by researchers.

8.2.2   Link-Layer Reliability

The main objectives of the data-link layer are multiplexing/demultiplexing of data, data-frame detection, medium access, and error control. While fulfilling these objectives, the data-link layer should provide reliable and energy-efficient, point-to-point and point-to-multipoint communication throughout the network. An overview of the data-link components is shown in Figure 8.3, which is implemented in [15].

In WSNs, where correlation between sensors can be exploited in terms of aggregation, collaborative source coding, or correlation-based protocols, error control is of extreme importance. Since the above-mentioned techniques aim to reduce the redundancy in the traffic by filtering correlated data, it is essential for each packet to be transmitted reliably. Moreover, the multihop features of the WSN require a unique definition of reliability other than the conventional reliability metrics which

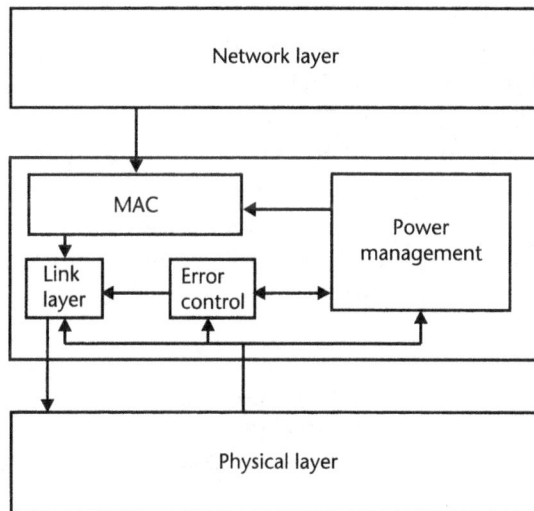

Figure 8.3   An overview of the data-link layer providing reliability in WSNs.

focus on point-to-point reliability. More specifically, in a WSN, when a packet is injected into the network, each node along the path to the sink consumes a certain amount of its scarce resources to relay the packet. Each packet has a different importance due to the path it has already traversed. Hence, packets in different network locations require different reliability measures. Furthermore, in WSNs, the applications are interested in the collaborative information from sensors about a specific event rather than individual readings of each sensor. Consequently, the reliability notion considered in WSNs differs from the approach in traditional wireless networks in terms of both multihop reliability and event-based reliability.

In general, the error-control mechanisms in communication networks can be categorized into three main approaches: power control, automatic repeat request (ARQ), and forward error correction (FEC).

- *Power control.* Controlling the transmission power can be used to achieve desired error rates. Higher transmission power reduces the packet error rate by improving the signal-to-noise ratio. As a result, however, energy consumption is increased, as is interference with the other nodes. Power control necessitates that sophisticated protocols be implemented, which requires additional memory footprint for implementation. Another drawback of power control in error control is that the radio should support different power levels, which may not be applicable to many WSNs, where low node cost is of critical importance.

- *ARQ.* ARQ-based error control mainly depends on retransmission for the recovery of lost data packets/frames. ARQ protocols enable transmissions of failed packets by sending explicit notification upon reception and detection of missing acknowledgments. The main ARQ strategies can be summarized as go-back-n, selective repeat, and stop-and-wait [16–18]. It is clear that such ARQ-based error-control mechanisms incur significant additional retransmission cost and overhead. Although ARQ-based error-control schemes are utilized at the data-link layer for conventional wireless networks, the efficiency of ARQ in sensor-network applications is limited due to the scarcity of energy and processing resources for the sensor nodes.

- *FEC.* FEC adds redundancy to the transmitted packet such that it can be received error-free, even if a limited number of bits are received in error. More specifically, in an (n, k) FEC code, as shown in Figure 8.4, $(n - k)$ redundant FEC symbols are added to the k-symbol FEC payload to improve the error resilience of the wireless communication at the cost of increased bandwidth consumption. As a result, the overall probability of error is decreased. There exist various FEC codes, such as BCH, linear-block, and RS codes, which are optimized for specific packet sizes, channel conditions, and reliability notions.

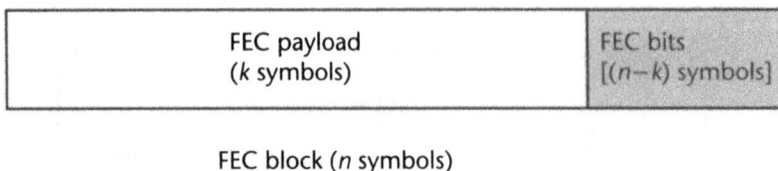

FEC payload (k symbols)	FEC bits [$(n-k)$ symbols]

FEC block (n symbols)

Figure 8.4 An illustration of FEC in WSNs.

For the design of efficient FEC schemes, it is important to have good knowledge of the channel characteristics and implementation techniques.

Energy consumption is the most important performance metric in the design of communication protocols. Error-control protocols should also consider energy efficiency as the first design goal. Because transmission of a bit is more expensive than processing, protocols that exploit the onboard processing capabilities of sensor nodes are desirable. Consequently, use of FEC is the most efficient solution given the constraints of the sensor nodes.

The use of FEC codes can decrease transmit power due to the increased redundancy in the constructed packets [2]. However, the required processing power due to the encoding and decoding of the packet increases the overall energy consumption. The required energy during encoding and decoding of BCH codes is shown in Figure 8.5, while the incurred encoding and decoding latency is shown in Figure 8.6 [19]. Moreover, the increase in packet length also incurs additional energy cost due to the longer packet transmission times and, hence, the increased packet collision rate.

Although FEC can achieve significant reduction in the bit-error rate (BER) for any given value of transmit power, the additional processing power consumed during encoding and decoding must be considered when designing an FEC scheme. FEC is a valuable asset to the sensor networks if the additional processing power required is less than the transmission power savings. Thus, the trade-off between this additional processing power and the associated coding gain needs to be optimized in order to have powerful, energy-efficient, low-complexity FEC schemes for error control in sensor networks. Furthermore, powerful FEC codes incur additional decoding latency, which should also be considered in choosing error-control schemes.

Along with the discussions presented above, an adaptive error-control scheme is presented in [20]. The transmit power and the coding rate are increased as the required range between sensor nodes is increased. Given the BER and latency

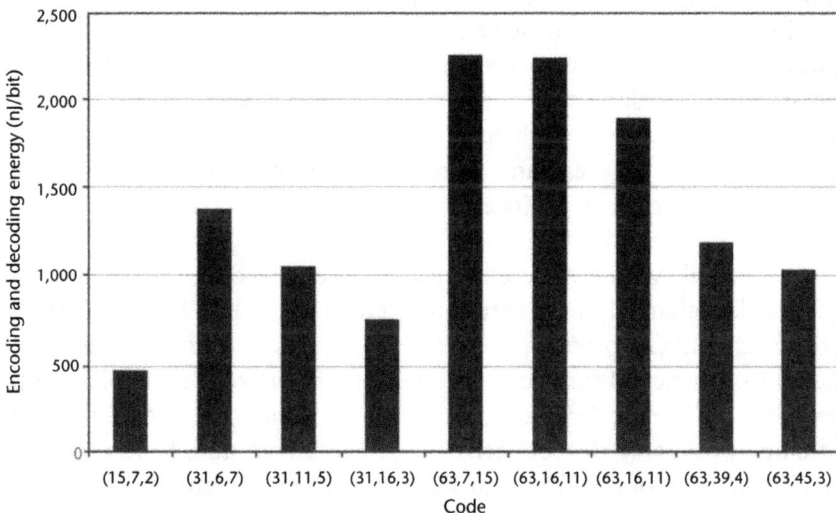

Figure 8.5 Total energy consumption for encoding and decoding using BCH codes.

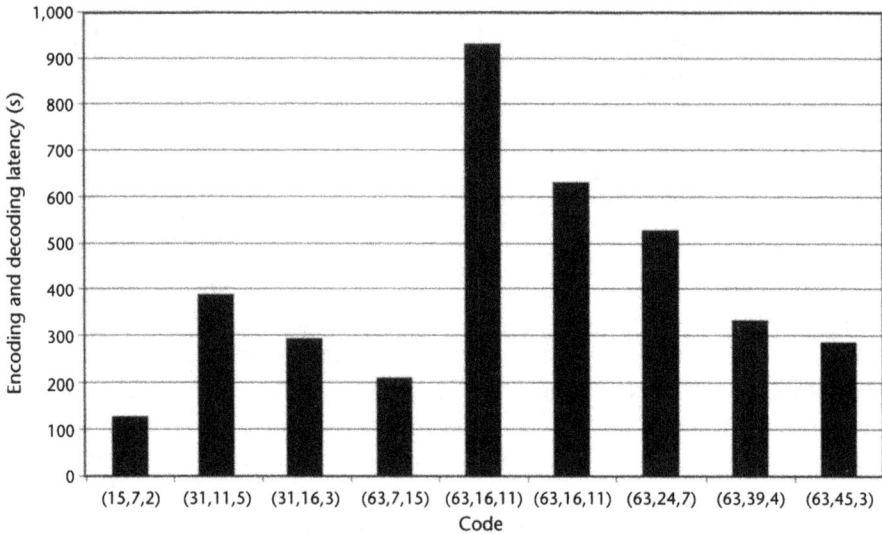

Figure 8.6   Total encoding and decoding latency using BCH codes.

requirements, the lowest-power FEC code that satisfies them is continuously chosen. It has been found that, using this protocol, energy is scalable over nearly two orders of magnitude, realizing range scalability to well over 100m and BER scalability across several decades. Therefore, adaptive error control can also be a promising approach as part of low-power WSN protocols [19].

Although FEC codes have been shown to provide flexible error-control capabilities over a wide variety of ranges between nodes, such an advantage is limited in scenarios where limited error probabilities are acceptable. More specifically, no coding provides better energy efficiency than convolutional codes for asimilar probability of error $P_b > 10^{-4}$ [2] because the encoding/decoding energy is small at high P_b, and output power is limited. As a result, the transceiver energy dominates overall energy consumption. Since the packet length is increased due to coding, overall energy consumption increases. Consequently, if lower P_b is not required by an application for individual packets from sensors, FEC coding can be inefficient.

In addition to the various error-control schemes explained above, hybrid solutions also exist. Data-aware error-control schemes aim to exploit the unique properties of the sensed data, such as spatial and temporal correlation [21]. The correlation among individual packets helps implementation of hybrid error-control protocols that require less overhead and energy consumption [13]. Moreover, a channel-quality-based error-control protocol is also proposed that selects an FEC rate depending on the channel quality [5].

While link-layer error-control protocols aim to overcome the errors incurred by the wireless channel, more care can be taken at the link layer through MAC protocols. Since MAC protocols govern the procedures to access the shared wireless channel, the efficiency of the access scheme helps improve the reliability of the WSN. By reducing collisions, less energy can be consumed for a specific communication attempt, preserving the connectivity of the network and, hence, the overall reliability. The density of the WSN can also be exploited to provide more reliable communication. As the network density increases, the number of nodes contending with each other also increases, resulting in higher collision probability. On the other hand, the connectivity of the network can be provided without compromising the overall energy efficiency due to a high number of neighbor nodes. In addition, due to the high density of sensor nodes, the information gathered by each node is highly correlated. Exploiting the correlation between sensor nodes at the MAC layer as well can be a promising approach to improve overall network reliability further.

8.2.3   Network-Layer Reliability

The network layer is responsible for determining efficient routes for the transmission of event information from sensor nodes to the sink in WSNs. Since the overall route the packets traverse from the sensor field to the sink is determined by network-layer algorithms, reliability should also be guaranteed at this layer in terms of route determination and dynamic route management. Due to the stringent energy capacity of sensor nodes, the topology of WSNs may change continuously due to energy depletion and node failures. Moreover, low-duty-cycle MAC protocols switch sensor nodes to sleep state in order to save energy during low-traffic periods. Hence, the sensor-network topology varies continuously. This necessitates that information-flow routes in the sensor network be formed dynamically in order to provide reliable data delivery in WSNs.

Wireless channel errors, congestion at the sensor nodes along the path, and continuous changes in the network topology decrease the reliability of a route after it is established by the routing protocol. Moreover, the hop count between sensor nodes and the sink has a direct effect on the reliability of a route. A naive approach to improving the reliability of a route would be to send multiple copies of a packet through a single route, which increases the probability that the packet will be received at the sink [22, 23]. However, such an approach does not address reliability issues due to topology changes or congestion along the path. Hence, recently, routing solutions based on a multipath approach have been proposed. Multipath routing protocols assign multiple routes for the transmission of event information in order to increase the redundancy and, hence, amplify the probability of successful delivery of the packets to the sink. Generally, the multipaths formed fall into two classes [24]:

- *Disjointed multipaths*. In this case, alternate routes that are disjointed from the other paths are constructed. In this formation, the failure of one path does not effect the performance of the others. However, the overall latency of the alternate paths increases.
- *Braided multipaths*. In this case, alternate paths may have sections that overlap with other paths as shown in Figure 8.7. As a result, all the paths are not independent of each other. The latency incurred by the additional paths may be decreased by constructing them using certain portions of the shortest path.

In [23], multiple copies of the same packet are sent through multiple paths. It has been shown that in the case of severe channel conditions (i.e., error probability of 60%), multipath routing achieves a reliability level equivalent to controlled flooding. In [24], a multipath routing scheme using k disjointed paths is proposed. S. Dulman et al. propose a combination of FEC and multipath routing in order to

Disjointed multipath routing

(a)

Braided multipath routing

(b)

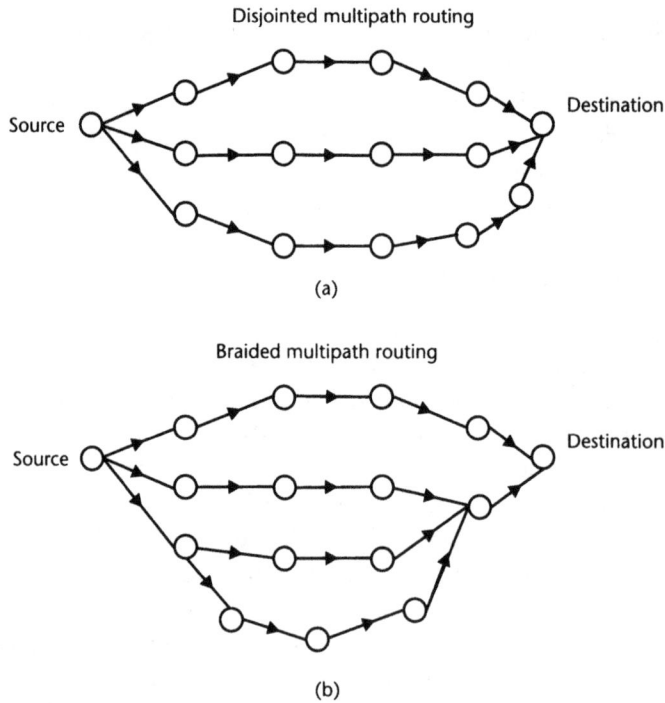

Figure 8.7 (a) Disjointed multipath routing, and (b) braided multipath routing in WSNs.

provide reliable data transmission. Each packet is partitioned into k parts and transmitted through k paths. The subpackets are created such that a fraction of k packets is enough to reconstruct the original packet.

Similarly, the Multipath On-Demand Routing (MOR) protocol [25] makes use of all possible paths to the sink in WSNs. The MOR protocol resides at sensor nodes as a reliability layer for sensor-network routing. Each node has, wherever possible, a choice of next hops for a given destination. The main approach followed in this scheme is to use a different node on each retransmission, while keeping track of the successful transmissions. Therefore, the objective is to increase the probability that a packet will be successfully delivered even if a given neighbor is temporarily unavailable.

In addition to pure single-path and pure multipath routing schemes, an adaptive multipath routing scheme is proposed in [26] to further improve the probability of reliable delivery of packets in WSNs. In this scheme, data is forwarded along a preestablished single path to save energy, and a high delivery ratio is achieved by path repair whenever a break is detected. Moreover, a simple, quick, local path-repairing approach is incorporated in this protocol. This allows a designated pivot node to skip over a path break by using existing routing information in its neighborhood and, hence, utilizing possible multipath routes adaptively.

Many routing protocols proposed in the literature [27] mainly consider energy efficiency, route distance, and latency as the primary objectives. The multihop nature of sensor-network communications, frequent node and link failures, and congestion due to sensor-node resource constraints impose reliability as one of the most important requirements for network-layer protocol design. Consequently, network-layer protocols, which can address reliability along with the energy efficiency, latency, and application-specific requirements, must be developed for reliable and efficient communication in WSNs.

8.2.4   Transport-Layer Reliability

In addition to robust modulation, physical- and link-layer error-control mechanisms, and reliable network-layer algorithms, a reliable transport mechanism is imperative. This reliable transport mechanism must be able to accommodate unique characteristics of the sensor-network paradigm. Conventional end-to-end retransmission-based error control adopted by the widely used TCP implementations in the Internet domain may not be feasible for the sensor-network domain and, hence, may lead to the waste of scarce resources.

Sensor networks are deployed with a specific sensing-application objective, such as event detection, event identification, location sensing, and local control of actuators. This specific objective also influences the design requirements of transport-layer reliability mechanisms.

Due to the application-oriented and collaborative nature of the sensor networks, the main data flow takes place in the forward path, where the source sensor nodes transmit their data to the sink. The reverse path, on the other hand, carries data originating from the sink such as programming/retasking binaries, queries, and commands to sensor nodes for the operation and maintenance of the WSN. Therefore, different functionalities are required to handle the transport reliability needs of the forward and reverse paths. Hence, transport-layer reliability issues pertaining to these distinct cases are investigated separately in the following sections.

8.2.4.1   Event-to-Sink Transport Reliability

The need for a transport-layer reliability mechanism for data delivery in sensor networks was also questioned in [28] under the premise that data flows from source to sink are generally loss tolerant. While the need for an end-to-end reliability measure may not exist due to the sheer number of correlated data flows, an event in the sensor field needs to be tracked with a certain accuracy at the sink. Hence, unlike traditional communication networks, the sensor-network paradigm necessitates an event-to-sink reliability notion at the transport layer [29]. This involves a reliable communication of the event features to the sink rather than conventional packet-based reliable delivery of the individual sensing reports or packets generated by each sensor node in the field. Such an event-to-sink reliable-transport notion based on the collective identification of data flows from the event to the sink is illustrated in Figure 8.8.

In [30], the Reliable Multisegment Transport (RMST) protocol is proposed to address the requirements of reliable data transport in WSNs. RMST is mainly based on the functionalities provided by directed diffusion [31]. Furthermore, RMST utilizes in-network caching and provides guaranteed delivery of the data packets generated by the event flows. However, as discussed above, event detection and tracking does not require guaranteed end-to-end data delivery since the individual data flows are correlated, hence loss tolerant to a certain extent. Moreover, such guaranteed reliability via in-network caching may bring significant overhead for sensor networks with power and processing limitations.

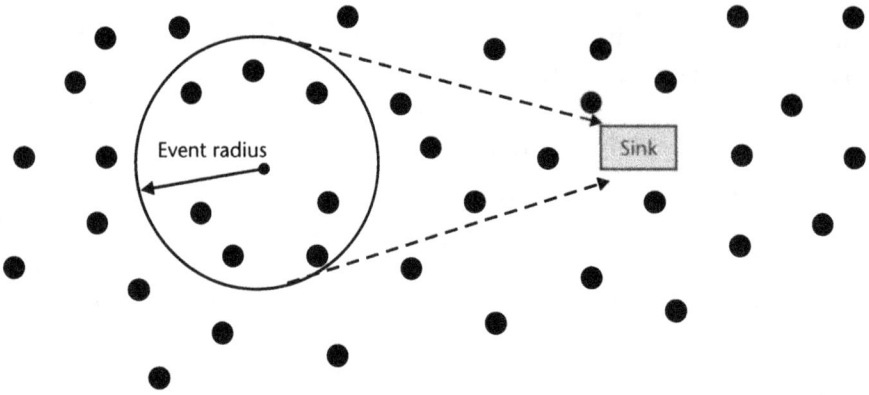

Figure 8.8 Typical sensor-network topology with event and sink. The sink is only interested in the collective information of sensor nodes within the event radius and not in their individual data.

Still, although the transport-layer solutions in conventional wireless networks are relevant, they are simply inapplicable for Event-to-Sink Reliable Transport (ESRT) in WSNs. These solutions mainly focus on reliable data transport following end-to-end TCP semantics and are proposed to address the challenges posed by wireless link errors and mobility [32]. The primary reason for their inapplicability is their notion of end-to-end reliability, which is based on acknowledgments and end-to-end retransmissions. Due to inherent correlation in the data flows generated by the sensor nodes, however, these mechanisms for strict end-to-end reliability are significantly energy-draining and superfluous. Furthermore, all these protocols bring considerable memory requirements to buffer the transmitted packets until they are acknowledged by the receiver. In contrast, sensor nodes have limited buffering space (< 4 KB in MICA motes [33]) and processing capabilities.

In order to provide reliable event detection at the sink, possible congestion in the forward path should also be addressed at the transport layer. Once an event is sensed by a number of sensor nodes within the coverage area of the phenomenon (i.e., event radius), a significant amount of traffic is triggered, which may easily lead to congestion in the forward path.

In [29], a case study is performed to observe the event transport reliability r for varying sensor reporting frequency f. As shown in Figure 8.9, the event reliability r shows a linear increase (note the log scale) with sensor reporting rate f, until a certain $f = f_{max}$, beyond which the event reliability drops. This is because the network is unable to handle the increased injection of data packets, and packets are dropped due to congestion. Hence, event-to-sink reliability also necessitates efficient congestion-control mechanisms, which are discussed in detail in Section 8.3.

In contrast to the transport-layer protocols for conventional end-to-end reliability, ESRT [29] is based on the event-to-sink reliability notion and provides reliable event detection without any intermediate caching requirements. ESRT is a novel transport solution developed to achieve reliable event detection in sensor networks with minimum energy expenditure. It mainly exploits the fact that the sheer number of data flows generated by the sensor nodes toward the sink are correlated due to spatial and temporal correlation among the individual sensor readings [11]. Consequently, ESRT achieves application-specific, desired, transport-reliability levels via collective effort of resource-constrained wireless sensor nodes.

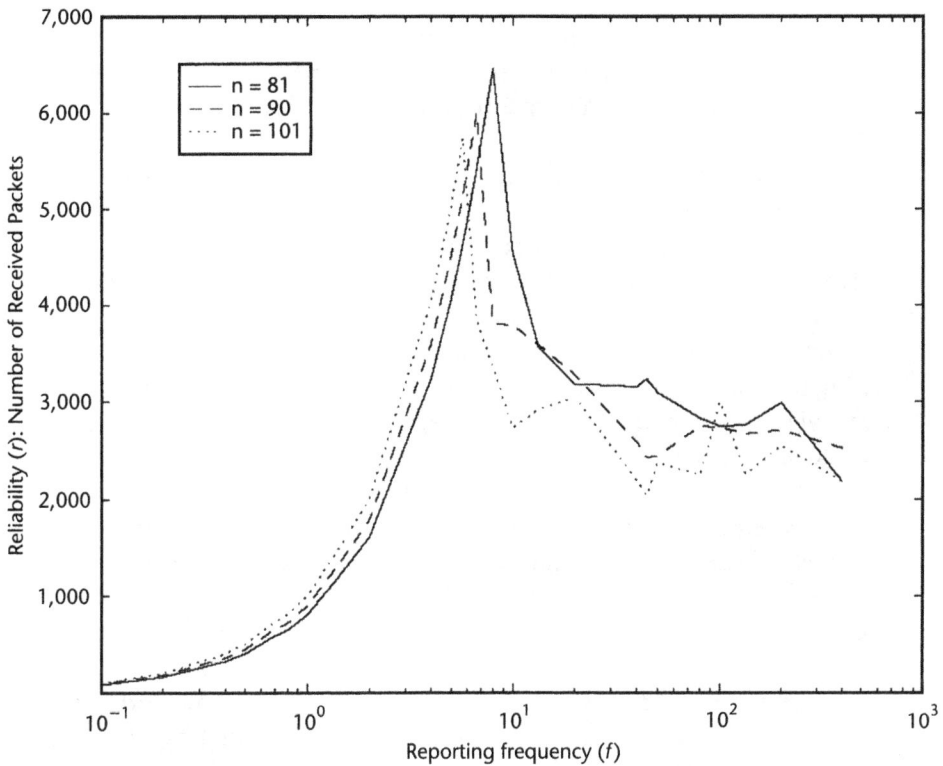

Figure 8.9   The effect of varying reporting rate f of source nodes on event reliability r observed at the sink. The number of source nodes is denoted by n.

8.2.4.2. Sink-to-Sensors Transport Reliability

While the data flows in the forward path carry correlated sensed or detected event features, the flows in the reverse path mainly contain data transmitted by the sink for operational or application-specific purposes. This may include operating system binaries, programming or retasking of configuration files, application-specific queries, and commands for the routine operation of the sensor network. Dissemination of this type of data mostly requires 100% reliable delivery. Therefore, the event-to-sink reliability approach introduced in Section 8.2.4.1 will not suffice to address the tighter reliability requirement of the flows in the reverse paths.

Such a strict reliability requirement for the sink-to-sensors transport of operational binaries and application-specific queries and commands necessitates a certain level of retransmissions as well as acknowledgment mechanisms. However, these mechanisms should be incorporated cautiously in order not to compromise scarce sensor-network resources. In this respect, local retransmissions and negative acknowledgment (NACK) approaches would be preferable over end-to-end retransmissions and acknowledgments to maintain minimum energy expenditure.

In [28], the Pump Slowly, Fetch Quickly (PSFQ) mechanism is proposed for reliable retasking or reprogramming in sensor networks. PSFQ is based on slowly injecting packets into the network but performing aggressive hop-by-hop recovery in case of packet loss. The pump operation in PSFQ simply performs controlled flooding and requires each intermediate node to create and maintain a data cache to be used for local loss recovery and in-sequence data delivery. Although this is an important transport-layer-reliability solution for sink-to-sensor communication, PSFQ does not address packet loss due to congestion in the sensor network, which may severely hamper the ultimate reliable-communication objective as discussed in Section 8.3.

A new framework called GARUDA for providing sink-to-sensors reliability in WSNs is introduced in [34]. This framework incorporates an efficient pulsing-based solution, which informs the sensor nodes of an impending, reliable, short-message delivery by transmitting a specific series of pulses at a certain amplitude over a certain period. A virtual infrastructure called the *core*, which approximates a near-optimal assignment of local, designated servers, is instantaneously constructed during the course of a single packet flood. In case of packet loss detected by a core node via out-of-sequence packet reception, a core node initiates a two-stage NACK-based packet-recovery process that performs out-of-sequence forwarding to assure the reliable delivery of the original message. GARUDA also supports other reliability semantics that might be required for sink-to-sensors communication, such as (1) reliable delivery to all nodes within a subregion of the sensor network, (2) reliable delivery to a minimal number of sensors required to cover an entire sensing area, and (3) reliable delivery to a probabilistic subset of the sensor nodes in the network.

In summary, the physical-, link-, network-, and transport-layer reliability mechanisms that can address the unique challenges posed by the sensor-network paradigm are essential to achieve the potential gains of collaborative sensing applications. The solutions discussed above need to be exhaustively evaluated under real sensor-network deployment scenarios to reveal their shortcomings and, hence, to determine necessary modifications in order to obtain a complete reliability solution for WSNs.

8.3 Congestion Control in WSNs

Packet loss due to congestion may severely impair event detection at the sink, even when enough information is sent out by the sensor nodes. Furthermore, congestion control not only increases network efficiency but also helps conserve scarce wireless sensor resources. In addition to the reliability mechanisms developed for sensor networks as discussed in Section 8.2, efficient congestion-control mechanisms are desirable. The main objectives and desired essential features of congestion-control solutions for WSNs are as follows:

- *Self-configurability.* The congestion-control mechanisms must be adaptive to dynamic topologies caused by node mobility/failure/temporary power-down, as well as spatial variation of events and random node deployment in various smart environment scenarios.

- *Energy awareness.* The congestion-control algorithms should be energy aware (i.e., the objectives must be achieved with minimum possible energy expenditure).

- *Biased implementation.* The algorithms must be designed such that they mainly run on the sink with minimum functionalities at sensor nodes in order to conserve limited sensor resources and shift the burden to the high-powered sink.

- *Constrained routing and addressing.* Unlike protocols such as TCP, the congestion-control protocols for WSNs should not assume the existence of end-to-end global addressing. WSNs are more likely to have attribute-based naming and data-centric routing, which call for different transport-layer approaches.

In addition to the results of the case study in [29] presented in Section 8.2.4.1, the need for transport-layer congestion control to assure reliable event detection at the sink is also revealed by the results in [35]. It has been shown in [35] that exceeding network capacity can be detrimental to the observed goodput at the sink. Moreover, although event-to-sink reliability may be attained even in the presence of packet loss due to network congestion, thanks to correlated data flows, a suitable congestion-control mechanism can also help conserve energy, while maintaining desired accuracy levels at the sink.

A novel transmission-control scheme for use at the MAC layer in WSNs is proposed in [3] with the main objective of per-node fair bandwidth share. Energy efficiency is maintained by controlling the rate at which the MAC layer injects packets into the channel. Although such an approach can control the transmission rate of a sensor node, it neither considers congestion control for event flows nor addresses reliable event detection in the entire sensor network.

The Congestion Detection and Avoidance (CODA) protocol for sensor networks is presented in [36]. CODA mainly aims to detect and avoid on the forward path in WSNs via receiver-based congestion detection. It uses open-loop hop-by-hop back-pressure signaling to inform the source about the congestion and closed-loop multisource regulation for persistent and larger-scale congestion conditions. The simulation results presented in [36] show that CODA can increase network performance via congestion avoidance. However, the CODA protocol does not address reliable event transport in sensor networks. On the contrary, it has been observed in the experiment results in [36] show that the congestion control performed at the sensor nodes, without considering reliability, impairs event-to-sink transport reliability.

ESRT [29] also includes a congestion-control component that serves the dual purpose of achieving reliability and conserving energy. Its congestion-control approach is tightly coupled with the event-to-sink reliability objective; hence, it relieves congestion in the sensor network without compromising the achieved event transport reliability. ESRT uses a congestion-detection mechanism based on local buffer-level monitoring in sensor nodes. Any sensor node whose routing buffer overflows due to excessive incoming packets is said to be congested and informs the sink of the same.

In the event-to-sink model of ESRT [29], the traffic generated during each reporting period (i.e., $1/f$) mainly depends on the reporting frequency f and the number of source nodes n. The reporting frequency f does not change within one reporting period since it is controlled periodically by the sink at the end of each decision interval with period of $\tau > 1/f$. Assuming n does not significantly change within one reporting period, the traffic generated during the next reporting period will have negligible variation. Therefore, the amount of incoming traffic to any sensor node in consecutive reporting intervals is assumed to stay constant. This, in turn, signifies that the increment in the buffer-fullness level at the end of each reporting interval is expected to be constant.

Let b_k and b_{k-1} be the buffer-fullness levels at the end of the k_{th} and $(k-1)_{-th}$ reporting intervals, respectively, and let B be the buffer size, as in Figure 8.10. For a given sensor node, let Δb be the buffer-length increment observed at the end of last reporting period; that is,

$$\Delta b = b_k - b_{k-1}$$

Thus, if the sum of the current buffer level at the end of the k_{th} reporting interval and the last experienced buffer-length increment exceeds the buffer size (i.e., $b_k + \Delta_b > B$,), then the sensor node infers that it is going to experience congestion in the next reporting interval. Hence, it sets the congestion notification (CN) bit in the header of the packets it transmits, as shown in Figure 8.11. This notifies the sink of the upcoming congestion condition to be experienced in next reporting interval.

Accordingly, ESRT operation is determined by the current network state based on the reliability achieved and congestion condition in the network. If the event-to-sink reliability is lower than required, ESRT adjusts the reporting frequency f of the source nodes aggressively in order to reach the target reliability level as soon as possible. If the reliability is higher than required, then ESRT reduces the reporting frequency conservatively in order to conserve energy, while still maintaining reliability. This self-configuring nature of ESRT makes it robust to the random, dynamic topology of WSNs.

ESRT also does not require individual sensor identification. An event ID suffices for the congestion-control operation. Importantly, the congestion-control algorithms of ESRT mainly run on the sink, with minimal functionality required at resource-constrained sensor nodes.

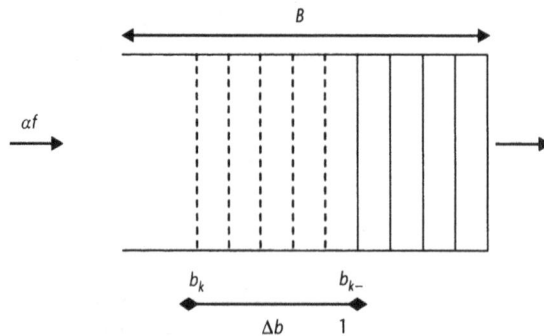

Figure 8.10   An illustration of buffer-level monitoring in sensor nodes.

Event ID	CN (1 bit)	Destination	Timestamp	Payload	FEC

Figure 8.11   A typical data packet with a CN field, which is marked to alert the sink of congestion.

For the reverse communication path (i.e., the sink-to-sensors direction), the sink has greater involvement in the dissemination of operational, control, and query data packets. The sink, with plentiful energy and communication resources, can broadcast the data with its powerful antenna. This reduces the amount of forwarding traffic in the multihop network and, hence, helps sensor nodes conserve energy. Therefore, data flows in the reverse path may experience less congestion in contrast to those in the forward path, which depend entirely on multihop communication. This calls for less aggressive congestion-control mechanisms, which have yet to be developed for the reverse path, as compared to the forward path in the sensor networks.

8.4 Conclusion

The realization of the potential of sensor networks requires efficient and application-specific communication protocols to assure reliable communication of the features of events sensed through the collective efforts of distributed sensor nodes. This chapter has explored the challenges and possible solution strategies for the design and development of reliable transmission and congestion-control mechanisms for sensor networks. There exist many physical-, link-, network-, and transport-layer reliability and congestion-control mechanisms specifically tailored to sensor networks. An exhaustive evaluation of real sensor-network applications must be performed to reach a satisfactory reliable-transmission and congestion-control solution.

References

[1] Akyildiz, I. F., et al., "Wireless Sensor Networks: A Survey," *Computer Networks Journal (Elsevier)*, Vol. 38, No. 4, March 2002, pp. 393–422.

[2] Shih, E., et al., "Physical Layer Driven Protocol and Algorithm Design for Energy-Efficient Wireless Sensor Networks," *Proc. ACM Mobicom 2001*, Rome, Italy, July 2001, pp. 272–286.

[3] Woo, A., and Culler D., "A Transmission Control Scheme for Media Access in Sensor Networks," *Proc. ACM MOBICOM 2001*, Rome, Italy, July 2001, pp. 221–235.

[4] Pottie, G. J., and Kaiser W. J., "Wireless Integrated Network Sensors," *Communications of the ACM*, Vol. 43, No. 5, May 2000, pp. 551–558.

[5] Mireles, F. R., and Scholtz R. A., "Performance of Equi-correlated Ultra-wideband Pulse-Position-Modulated Signals in the Indoor Wireless Impulse Radio Channel," *Proc. IEEE Communications, Computers and Signal Processing 1997*, Vol. 2, Victoria, B.C., Canada, August 1997, pp. 640–644.

[6] Cramer, R. J., Win M. Z., and Scholtz R. A., "Impulse Radio Multipath Characteristics and Diversity Reception," *Proc. IEEE ICC 1998*, Vol. 3, Atlanta, GA, 1998, pp. 1650–1654.

[7] Lee, H., et al., "Multipath Characteristics of Impulse Radio Channels," *Proc. IEEE VTC 2000*, Vol. 3, Tokyo, Japan, May 2000, pp. 2487–2491.

[8] Kahn, J. M., Katz R. H., and Pister K. S. J., "Next Century Challenges: Mobile Networking for Smart Dust," *Proc. ACM Mobicom 1999*, Seattle, WA, August 1999, pp. 271–278.

[9] Gastpar, M., "Distributed Source-Channel Coding for Wireless Sensor Networks," *Proc. IEEE ICASSP 2004*, Vol. 3, May 2004, Montreal, Quebec, Canada, pp. 829–832.

[10] Gastpar, M., and B. Rimoldi, "Source-Channel Communication with Feedback," *Proc. IEEE Information Theory Workshop 2003*, Paris, France, March 2003, pp. 279–282.

[11] Vuran, M. C., et al., "Spatio-temporal Correlation: Theory and Applications for Wireless Sensor Networks," *Computer Networks Journal (Elsevier)*, Vol. 45, No. 3, June 2004, pp. 245–261.

[12] Marco, D., and Neuhoff D. L., "Reliability vs. Efficiency in Distributed Source Coding for Field-Gathering Sensor Networks," *Proc. ACM IPSN 2004*, Berkeley, CA, April 2004, pp. 161–168.

[13] Mukhopadhyay, S., Panigrahi D., and Dey S., "Data Aware, Low Cost Error Correction for Wireless Sensor Networks," *Proc. IEEE WCNC 2004*, Vol. 4, Atlanta, GA, March 2004, pp. 2492–2497.

[14] De Couto, D. S. J., et al., "A High-Throughput Path Metric for Multi-hop Wireless Routing," *Proc. Mobicom 2003*, Vol. 1, San Diego, CA, September 2003, pp. 134–146.

[15] Zhong, L. C., Rabaey J. M., and Wolisz A. "An Integrated Data-Link Energy Model for Wireless Sensor Networks," *Proc. IEEE ICC 2004*, Paris, France, June 2004.

[16] Chlamtac, I., and Petrioli C., "Energy-Conserving Selective Repeat ARQ Protocols for Wireless Data Networks," *Proc. IEEE PIMRC 1998*, Vol. 3, Boston, MA, September 1998, pp. 836–840.

[17] Chlamtac, I., Petrioli C., and Redi J., "Energy-Conserving Go-Back-N ARQ Protocols for Wireless Data Networks," *Proc. IEEE ICUPC 1998*, Florence, Italy, October 1998.

[18] Liu, H., et al., "Error Control Schemes for Networks," *Mobile Networks and Applications*, Vol. 2, No. 2, October 1997, pp. 167–182.

[19] Min, R., et.al., "Low-Power Wireless Sensor Networks," *Proc. 14th International Conference on VLSI Design*, West Lafayette, ID, January 2001, pp. 205–210.

[20] Min, R., and Chandrakasan A., "A Framework for Energy-Scalable Communication in High-Density Wireless Networks," *Proc. ISLPED 2002*, Monterey, CA, August 2002, pp. 36–41.

[21] Zhao, J., and Govindan R., "Understanding Packet Delivery Performance in Dense Wireless Sensor Networks," *Proc. ACM SenSys '03*, Vol. 1, Los Angeles, CA, November 2003, pp. 1–13.

[22] Shih, E., et al., "Energy-Efficient Link Layer for Wireless Microsensor Networks," *Proc. IEEE Workshop on VLSI*, Orlando, FL, April 2001.

[23] Deb, B., Bhatnagar S., and Nath B., "ReInForM: Reliable Information Forwarding Using Multiple Paths in Sensor Networks," *Proc. IEEE LCN 2003*, Bonn, Germany, October 2003, pp. 406–415.

[24] Dulman, S., et al., "Trade-off between Traffic Ovehead and Reliability in Multipath Routing for Wireless Sensor Networks," *Proc. IEEE WCNC 2003*, Vol. 3, New Orleans, LA, March 2003, pp. 1918–1922.

[25] Biagioni, E., and Chen S. H., "A Reliability Layer for Ad Hoc Wireless Sensor Network Routing," *Proc. 37th Annual Hawaii International Conference on System Sciences 2004*, Big Island, HI, January 2004, pp. 300–307.

[26] Tian, D., and Georganas N. D., "Energy Efficient Routing with Guaranteed Delivery in Wireless Sensor Networks," *Proc. IEEE WCNC 2003*, Vol. 23, New Orleans, LA, March 2003, pp. 1923–1929.

[27] Akkaya, K., and Younis M., "A Survey of Routing Protocols in Wireless Sensor Networks," *Ad Hoc Networks Journal (Elsevier)*, Vol 13, 2005, pp. 325–349.

[28] Wan, C. Y., Campbell A. T., and Krishnamurthy L., "PSFQ: A Reliable Transport Protocol for Wireless Sensor Networks," *Proc. WSNA 2002*, Atlanta, GA, September 2002.

[29] Sankarasubramaniam, Y., Akan O. B., and Akyildiz I. F., "ESRT: Event-to-Sink Reliable Transport for Wireless Sensor Networks," *Proc. ACM MOBIHOC 2003*, Annapolis, MO, June 2003, pp. 177–188.

[30] Stann, F., and Heidemann J., "RMST: Reliable Data Transport in Sensor Networks," *Proc. IEEE SNPA 2003*, Anchorage, AK, May 2003, pp. 102–112.

[31]   Intanagonwiwat, C., Govindan R., and Estrin D., "Directed Diffusion: A Scalable and Robust Communication Paradigm for Sensor Networks," *Proc. ACM MOBICOM 2000*, Boston, MA, August 2002, pp. 56–67.

[32]   Balakrishnan, H., et al., "A Comparison of Mechanisms for Improving TCP Performance over Wireless Links," *IEEE/ACM Trans. Networking*, Vol. 5, No. 6, December 1997, pp. 756–769.

[33]   MICA motes and sensors, at http://www.xbow.com.

[34]   Park, S. J., et al., "A Scalable Approach for Reliable Downstream Data Delivery in Wireless Sensor Networks," *Proc. ACM MOBIHOC 2004*, Tokyo, Japan, May 2004, pp. 78–89.

[35]   Tilak, S., Abu-Ghazaleh N. B., and Heinzelman W., "Infrastructure Trade-offs for Sensor Networks," *Proc. WSNA 2002*, Atlanta, GA, September 2002.

[36]   Wan, C.-Y., Eisenman S. B., and Campbell A. T., "CODA: Congestion Detection and Avoidance in Sensor Networks," *Proc. ACM SENSYS 2003*, Los Angeles, CA, November 2003.

CHAPTER 9

Energy-Harvesting-Aware Power Management

Aman Kansal and Mani B. Srivastava

9.1 Introduction

The true autonomy of WSNs depends on their reliable operation for extended times without human intervention. Energy supply is a critical factor in this design. Wireless, ad hoc deployment, which is essential in some scenarios and cost-effective in others, precludes the use of a wired energy infrastructure. The sensor nodes are thus forced to operate on limited battery reserves, and low-power design is an important consideration [1, 2].

Unlike human-carried devices such as hand-helds or cell phones, which can be returned to charging docks periodically, sensor-node batteries are limited in supply. They cannot be replaced in large numbers of nodes as the embedded nature of deployment makes it hard to access each individual node. A limited amount of energy supply, however, is not sufficient to ensure uninterrupted operation for the several-year-long lifetimes typically expected from an embedded deployment. The small node size puts constraints on the maximum battery size. Batteries already dominate the node volume in prototype sensor nodes. The energy density for common battery technologies varies in the range of 1,200 J/cm^3 (alkaline) to 3,780 J/cm^3 (zinc-air). At such an energy density, assuming a sensor node operating at 1 mW (average consumption after power management) and assuming the full battery capacity can be utilized, a year-long operation requires a battery size on the order of 10 cm^3, which is rather large. Thus, batteries alone cannot be expected to supply a sensor-network deployment reliably for several years.

A viable alternative, then, is to endow the sensor nodes with the appropriate energy-harvesting technologies such as solar, vibrational [3], wind or water flow, thermal-gradient scavenging [4–6], electromagnetic direct conversion [7], and others [8–11]. These sources can supplement or even entirely replace the battery energy supply. A fundamental difference between environmental-energy and battery supply is that the environmental energy can be scavenged for as long as desired and, if efficiently utilized, can enable a system to last eternally (until its hardware is outdated).

However, the introduction of harvesting components into sensor nodes requires design changes spanning the hardware of the node, the node-level power management,

119

and networkwide energy scheduling. Consider for instance that the power consumption at a node is to be matched to the environmental energy available to it. Then, the node must be equipped with additional hardware to measure the environmental energy input rather than just the residual battery measurement. Further, power-management decisions in a network differ when harvested energy is available. For instance, consider a solar-energy-harvesting network. At a particular instance, the network may have two alternative data routes available to satisfy the immediate data-transfer requirement. However, the harvested-energy available at nodes along the two routes may be different, say, due to the presence of a shadow on part of the network. In such a case, the network requires a harvesting-aware power-management strategy, along with the environmental-energy-input measurement, which allows it to choose the route passing through nodes outside the shadow.

This chapter addresses the resulting issues in harvesting-aware system design, including both the hardware and power-management software.

9.2   Harvesting Technologies

We define a *harvesting node* to be any sensor node with at least one form of environmental-energy harvesting as part of its power supply. Typically, such a device will also have an energy-storage mechanism, such as a battery or an ultracapacitor, to allow energy harvesting and consumption to occur without total synchrony. However, the storage device is not essential in all scenarios, and harvesting nodes (see [10]) use the energy generated from the press of a button immediately to transmit a packet and are inactive otherwise.

A network of such devices will be referred to as a *harvesting network*. The design of the network involves further considerations than just the individual nodes since the nodes in such a network may be heterogeneous and the environmental energy available at each node may be different.

We consider the harvesting node first. Figure 9.1 shows the new modules that are part of a harvesting node, in addition to the usual sensor-node components. The various blocks are discussed in subsequent sections, including an example implementation. While most of the blocks shown are implemented as hardware circuits,

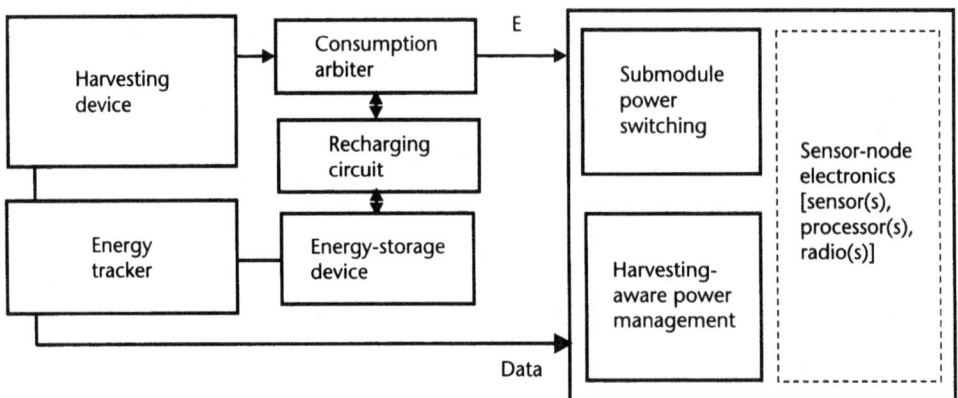

Figure 9.1   Block diagram of a harvesting node.

the block labeled "Harvesting-Aware Power Management" is best implemented as a set of algorithms on the sensor-node processor itself. A design component not shown in the figure but that influences the hardware and software design is knowledge about the application behavior with respect to energy consumption; this knowledge should be exploited to the extent available.

The key distinguishing characteristic of the harvesting node appears at the top left of the block diagram—the harvesting device. It is based on one of the harvesting technologies relevant to the deployment environment and outputs energy. The key harvesting technologies of interest to embedded systems are discussed below.

9.2.1 Solar Energy

Solar or other light energy can be converted to electric power using solar cells. The magnitude of energy generated varies from approximately $15 \, \mu\text{W/cm}^2$ in noon-time sunlight to $10 \, \mu\text{W/cm}^2$ in indoor, incandescent lighting. The energy output depends on the material used. Crystalline materials such as silicon and gallium arsenide have moderate absorption efficiency and high conversion efficiency (15% to 30%), while thin film materials such as cadmium telluride have high absorption efficiency and lower conversion efficiency ($\leq 10\%$). The choice of material also depends on its spectral response and the light source of interest.

For the purposes of circuit design, the solar cell may be modeled as a voltage source with an internal resistance. The output voltage is fairly constant in the useful operating range, and the supply current varies with light intensity. A single solar cell output is 0.6V, but panels with series of such cells can generate any required voltage for the circuit.

9.2.2 Vibration Energy

Vibrations are available in many environments of interest, including commercial buildings, parking structures, aircrafts, trains, industrial facilities, and even residential buildings. Preliminary analysis and experiments presented in [3] show that 300 $\mu\text{W/cm}^3$ are available in such environments. The sources of vibrations, which may be heavy machinery, home appliances, Heating Ventilation and Air Conditioning (HVAC) vents, movement of people or vehicles, or other movements, vary a great deal in their acceleration characteristics and frequency spectra [3]. Methods to convert this energy to electricity can be classified as electromagnetic, electrostatic, and piezoelectric [12–16].

Electromagnetic conversion uses vibration to move a conductor in a magnetic field. Existing prototypes [15, 12] generate a voltage output too low to be usable. Electrostatic conversion uses vibration energy to move the conductors of a charged capacitor. The disadvantage of this approach is that a separate voltage source is required to charge the capacitor. An advantage, however, is that the output voltage is in the usable range of two to several volts. Piezoelectric conversion uses materials that when mechanically deformed, generate an electric potential. The piezoelectric method combines the advantages of electromagnetic and electrostatic conversion but is difficult to implement at the microscale. With the current technology, piezoelectric conversion has have the greatest available energy density among the three methods.

9.2.3 Other Sources

Wind or water flow can be converted to energy. While macroscale generators based on these flows are widely used, compact technologies to extract such energy are lacking. A sensor-network application of wind energy is also for locomotion, such also employed to enable locomotion, as used in NASA's Tumbleweeds [17]. These are inflatable spheres that can roll along the deployment surface using wind energy and are aimed at Martian and polar exploration.

Thermoelectric generation using the Seebeck effect (flow of current in a loop made of two wires of certain metals when a temperature difference is applied to the wire junctions) and other methods have been demonstrated to yield 10 to 40 μW/cm2 using a 5°C to 10°C temperature gradient [4, 6, 18].

Pressure variations, such as the pressure of a fluid or gas in an enclosed space changing with time of day, can also be used to generate energy. In the Atmos clock, for example, invented in 1928 by Jean-Leon Reutter, a mixture of gas and liquid enclosed in a sealed capsule expands as the temperature rises and contracts as it falls, moving the capsule back and forth, providing sufficient motion to run the clock. Methods to convert such limited pressure variation or motion into electricity are not readily available, however.

The exact choice of the harvesting technology depends on several factors, including the achieved energy density from a particular technology, sensor-node form factor, and, most importantly, availability of the source in the deployment scenario. For instance, a sensor-network deployed for environmental monitoring or precision agriculture in outdoor settings may have ready access to sunlight and, hence, could use solar energy, while a sensor network operating indoors for industrial applications such as machine-health monitoring may have plenty of vibrations available for harvesting energy.

9.3 Descriptions of the Components of a Harvesting Node

Figure 9.1 alludes to various new components that are required when a sensor node is converted into a harvesting node. This section discusses the reasons why these blocks are needed and their design requirements.

The *recharging circuit* receives part of the energy output of the harvesting device and stores it. This block is needed since the current and voltage characteristics of the harvesting device output may not be directly suitable for charging the storage device. For example, a rechargeable battery must be charged with a voltage equal to or higher than its output, and if the harvesting device outputs current at a lower voltage than this, a voltage conversion circuit may have to be used.

The recharging circuit also depends on the energy-storage mechanism and, in case of batteries, on the battery-chemistry used. For a battery charged from a solar cell, the simplest implementation is to directly connect a solar panel to the battery through a diode to prevent the battery's being drained through the solar cell.

However, this circuit would yield low longevity for the battery as the battery could be repeatedly damaged from overcharging (continued charging of battery even when it is charged to its full capacity) and undercharging (continued discharging after the battery voltage has dropped below a battery-chemistry-specific threshold). In general, to ensure battery longevity, a recharging circuit is required to produce an acceptable charging profile and to disconnect the battery at its undercharge limit.

The *consumption arbiter* uses a combination of the energy from the harvesting source and the storage device to supply the power requirements of the harvesting node. Three reasons necessitate this arbiter. First, the harvested energy may not always be sufficient to power the load; hence, a mechanism is needed to share the load between the harvested and stored energy. Storage efficiency is always less than one. The arbiter must ensure that the environmental energy is supplied to the load directly and that only the deficit is drawn from the storage. When the energy available from the harvesting device exceeds the consumption, the excess energy is routed to the recharging circuit, which may store it if the storage is not already full.

Second, for a given environmental stimulus, the energy output of the harvesting device depends on its own internal resistance and the characteristics of the load presented to it. The consumption arbiter may also modify the load characteristics appearing at the output of the harvesting device to maximize the scavenged energy.

Third, the storage mechanism used may not supply the energy at the voltage required by the sensor node; the arbiter regulates the supply voltage.

The arbiter may also be designed to prioritize storage of energy in certain cases, such as when the battery has been drained to a very low level. In such a case, if the harvested-energy level is just sufficient to power the sensor node but no excess is available to charge the battery, then the arbiter may shut off the node and charge the battery first in order to ensure system availability in emergency situations.

The *energy tracker* monitors the energy available from the harvesting device, as well as the current state of the energy store. Such data may be used by the harvesting-aware power-management algorithms for learning the energy environment. The instantaneous battery status, environmental-energy availability, and long-term availability behavior are used by several networkwide scheduling algorithms to distribute workload as discussed later in sections on power management.

The *submodule power-switching block* is typical to any device that provides for shutting down parts of its circuit for power management. This is very crucial in sensor networks where sensors and minimal-direct-memory-access peripherals may be kept active for much longer durations than the energy-intensive communication, processing, and larger-memory modules. In a harvesting node, this block is useful for turning on or off various components as the amount of environmental availability varies. This switching functionality is controlled by the harvesting-aware power-management algorithms.

A prototype harvesting node, implemented at the Networked and Embedded Systems Laboratory, UCLA, is shown in Figure 9.2. This node uses a pair of solar cells connected in parallel as the harvesting device. The storage is a pair of AA-sized NiMH rechargeable batteries. The recharging circuit provides the charging profile required for NiMH batteries. The arbiter supplies the solar-cell output energy to the load, while the excess or deficit flows through the battery. The heliomote arbiter also regulates the voltage output of the NiMH battery pack, which varies between 2.2V and 2.8V to the stable 3V required by the sensor node. The energy-tracker module monitors the total energy charged into the battery, the instantaneous battery current, and the voltage. The measured data is provided to the sensor-node processor via a one-wire interface. The detailed circuit schematics are available in [19].

Figure 9.2 Heliomote.

9.4 Harvesting-Aware Power Management

Power-management strategies differ significantly when the power source changes from a fixed battery supply to a harvesting device. There are two reasons that lead to these differences. First, the environmental-energy source is highly variable. Unlike the stored supply, which is simply characterized by the amount of residual energy left in the battery and is reliably available, the environmental energy requires a more sophisticated characterization. Second, the environmental energy is not a limited resource and has the potential to be used eternally. In Section 9.4.1, we present a model to characterize the production and consumption of environmental energy, which can be used to design power-management algorithms for harvesting nodes.

9.4.1 Harvesting Theory

Since the harvesting source is highly variable, an important design consideration is to provide reliable system performance. Another objective is to decide if the environmental energy alone is sufficient for the system to operate at the desired level of performance. If so, the system can operate indefinitely. Otherwise, the designer may want to determine the required harvesting additions. The goal of developing a harvesting theory is to enable analysis of a wide variety of harvesting technologies with respect to system performance. We begin with a source characterization.

Definition 1. $(\rho, \sigma_1, \sigma_2)$ – *source:*   Suppose $E(t)$ is a continuous and bounded function of a continuously varying parameter t. $E(t)$ is said to be a $(\rho, \sigma_1, \sigma_2)$ – source if, and only if, for any finite real number T, it satisfies

$$\int_T E(t)dt \geq \rho T - \sigma_1 \tag{9.1}$$

$$\int_T E(t)dt \leq \rho T + \sigma_2 \tag{9.2}$$

This definition uses only three parameters, keeping it analytically tractable but still allowing it to model a wide variety of variations in energy sources. $E(t)$ models the power output at time t. The model captures the asymptotic rate of availability, which is the maximum power at which the system can operate. Since we are modeling physical energy sources, the restrictions placed on the function $E(t)$ are justified. It may be noted that the unit of ρ is power (e.g., watts) and the units of σ_1 and σ_2 are energy (e.g., joules).

The consumption itself may not be constant, and the following definition can characterize most consumption profiles.

Definition 2. (ρ', σ)   – consumer: A device is said to be a (ρ', σ) consumer if its power consumption, $E_c(t)$, satisfies the constraint

$$\int_T E_c(t)dt \le \rho'T + \sigma \tag{9.3}$$

for any value of T.

With this definition, the following theorem specifies the minimum performance that can be guaranteed. The achievable performance may, in fact, be higher, such as if an application requires node operation only when the environmental energy is available.

Theorem 1.   Sustainable Performance at Eternity (Variable Consumption Profile): If a (ρ', σ) – consumer device is powered by a $(\rho, \sigma_1, \sigma_2)$ – source, has an energy-storage capacity of $\sigma + \sigma_1 + \sigma_2$, and $\rho' < \rho$, then the device can operate forever.

The proof is available in [20]. The following example demonstrates an immediate application of the above theorem.

Example 1.   Figure 9.3 shows the power flowing into the battery observed by a test heliomote, when placed in sunlight, in the month of January in the Northern Hemisphere. Negative values represent battery drain.

Measuring the battery current instead of the current out of the solar cell ensures that only the actual solar power available, and not any power lost due to circuit inefficiencies, is considered. Since the negative portions of the waveform dominate, the batteries suffer a net discharge. For the waveform plotted in Figure 9.3, the source characterization parameter values are given in Table 9.1.

With these values, the solar cell is a $(\rho, \sigma_1, \sigma_2)$ – source in the test environment. We now need a (ρ', σ) classification of the consumer node. The sensor node used in the heliomote has a sleep-mode power drawn $P_{sleep} \le 3\text{mW}$, and the maximum

Table 9.1   Solar Cell Parameters in Experimental Environment

Parameter	Value	Units
ρ	23.6	mW
s_1	1.4639×10^6	J
s_2	1.8566×10^6	J

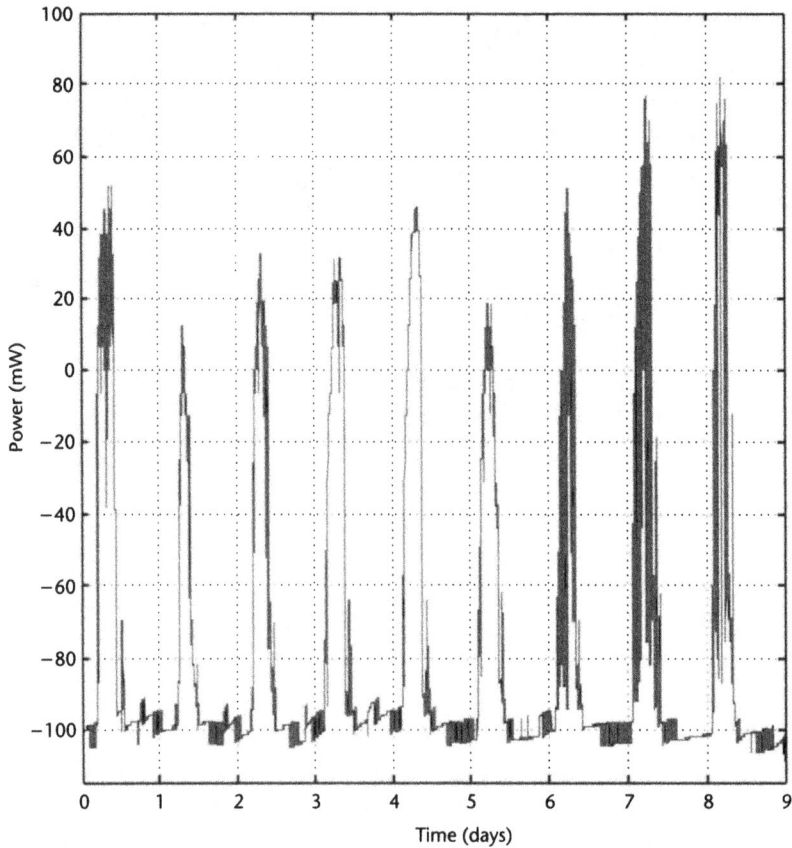

Figure 9.3　Solar-energy-based charging power recorded for nine days.

current is drawn P_{max} = 100 mW. Thus, $(\rho' = 100mW, \sigma = 0)$ is a valid classification. However, to achieve $\rho' = \rho$, we can set the node to sleep for 78.7% of the time. If the minimum wake-up duration is 2 seconds, this leads to $\sigma = 153J$. Theorem 2 implies that a rechargeable battery with capacity $= \sigma + \sigma_1 + \sigma_2 = 3.32 \times 10^6 J$, or 922.43mAh, is required to sustain the node indefinitely with solar energy. This helps the designer to choose a AAA-sized NiMH battery. Using a larger battery does not help improve long-term performance.

The key utility of the above theory is that it enables characterizing the energy availability using a small number of parameters, which can be used for power management as discussed next.

9.4.2　Scheduling Algorithms

In battery-powered systems, the power-management strategies are designed merely to minimize energy consumption. Several power-scaling strategies to minimize battery consumption have been studied [21–24]. Many of these can be used to minimize battery drain in harvesting nodes also. However, these strategies are not sufficient for efficiently utilizing the harvesting source. In harvesting systems, the problem space expands. The aim is not always to minimize consumption but adapt the consumption to the available environmental energy within the performance constraints. For instance, the tasks may be rescheduled when possible to operate directly off the environmental energy. This does not reduce the energy consumed in executing the task, but it reduces the wastage due to battery inefficiency. Research has recently begun in the area of harvesting-aware power management [25, 26].

Various power-scaling mechanisms are typically available to control the power consumption in embedded systems, such as dynamic voltage scaling (DVS), transmit power control, and node duty cycling. For each mechanism, reducing power affects performance. For DVS and duty cycling for instance, using a lower-energy mode increases the latency of executing a task. The algorithms below are illustrated using duty cycling as an example power-scaling mechanism since it is available in most processors.

Theorem 1 gives the asymptotic rate of energy consumption available to a harvesting node. As mentioned, a fundamental difference in harvesting systems compared to battery-powered systems is that if system performance is adapted to the environmental availability, the system can last eternally. To adapt the performance in this manner, the system only needs to estimate the parameters ρ, σ, σ_1, and σ_2 for its deployment scenario. This can be achieved if the environmental behavior over a learning phase is assumed to be representative of the long-term behavior, which is valid for many sources. For instance, solar energy follows a diurnal and annual cycle. Winds have known repetitive patterns. Sustainable performance depends directly on ρ hence, we wish to estimate ρ to within an error margin Δ. The power parameter ρ can be estimated by averaging the energy obtained from the energy source over time. Let the average at time $t, \rho(t)$ be calculated as

$$\rho(t) = \frac{\int_0^t E(t)dt}{t} \tag{9.4}$$

Since the device will sample $E(t)$ at discrete times, the above integral will be evaluated as a discrete summation. Apart from the running estimate of average, also store the most recent local minima and maxima observed in $\rho(t)$. When the difference between the maxima and the minima reaches Δ, we assume that estimation of p is complete. The E(t) waveform can also be used to estimate σ_1 and σ_2 [20].

Once the source has been characterized, the device can adjust its performance to operate at the available rate ρ. Suppose the active-mode power is P_{max} and sleep-mode power is P_{sleep}. Then the duty cycle x satisfies

$$\rho = xP_{max} + (1-x)P_{sleep} \tag{9.5}$$

neglecting the energy consumption of switching between modes. The value of x determined from this equation can be used to decide the sleep duration if the minimum time spent in active mode, T_{min}, is known. Thus, performance can be adjusted at a single node to match the environmental availability.

9.5   Distributed Harvesting

In a distributed system with several harvesting nodes, the harvesting-aware power-management problem is to distribute the workload among the nodes in such a way that the overall performance of the system is maximized. Finding the optimal solution requires complete knowledge of what energy resources are available at every node and the complete set of tasks to be performed by the system. Many of these tasks, such as routing a data packet from one node to another, involve energy consumption at several intermediate nodes, making the scheduling decisions coupled among nodes. Also, in a distributed system, scalability concerns dictate that all the information at every node not be communicated to a central node for scheduling decisions. Moreover, in many distributed systems such as sensor networks, communication itself is the major energy consumer, and distributed decisions are the assumed norm. Distributed power-management solutions are thus used, even though they may not be globally optimal.

Distributed power management is a nascent area of research. Methods have been proposed for distributed battery-powered systems [27–31]. These methods attempt to reduce and balance energy usage across multiple nodes to ensure systemwide sustainability rather than minimize the total energy consumption, which could cause certain nodes to deplete faster and make the overall system useless. In harvesting networks, on the other hand, instead of balancing the energy consumption among nodes, the objective is to share the workload in proportion to the harvesting opportunity available to each node. The distribution of workload may be highly application dependent, and such methods typically require an understanding of the application characteristics. One such method that addresses a typical usage scenario is discussed below.

One of the applications for sensor networks is to monitor a deployment scene for specific events and to report the occurrence of such events with low latency. The sensing transducers consume minimal energy and are kept active, while the power-intensive modules, the processor and the radio, are woken up when an event is triggered by a transducer. The event must then be communicated with low latency to a central base station.

The power-management problem is to provide an energy-efficient communication topology for the network, which adapts the energy consumption at different nodes to their individual harvesting opportunities. The scheme below achieves this in a completely distributed fashion. Also, it can report the expected performance to the base with minimal communication overhead.

9.5.1 Communication Protocol

The communication protocol followed is depicted in Figure 9.4. When a node has data to transmit to its next-hop neighbor along the data path, the node transmits a BEACON packet and listens for response for a period T_{ack}. It repeats this process until an ACK is received. T_{ack} is the time required by an active node to send an ACK after receiving a BEACON packet. Suppose the time required to transmit a BEACON packet is T_{beacon}. Every node in the network wakes up for a duration $T_{min} = 2T_{beacon} + T_{ack}$ to listen for any BEACON messages from nodes attempting to send data to it. After every awake period, a node sleeps for a duration

$$T_{sleep} = \frac{1-x}{x} T_{min} \qquad (9.6)$$

Transmitter

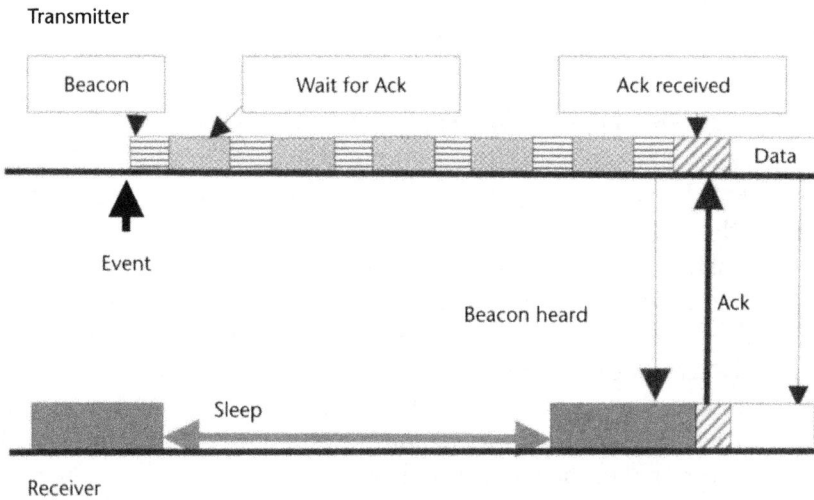

Figure 9.4   Communication with sleep cycle.

where x is the sustainable duty cycle at this node, determined from (9.5). Whenever a node receives a BEACON packet, it transmits an ACK and stays awake until it has received the relevant data and forwarded it to the next hop. With this arrangement, the worst-case delay in receiving an ACK when repeatedly sending a BEACON is $T_{sleep} + T_{min}$. The energy spent in transmitting the event data is negligible compared to the energy spent in periodically entering active mode to listen for potential data when events are infrequent.

The route discovery from all nodes to the base is based on the formation of a data-gathering tree, adopted from [32]. The base node transmits an INIT packet. All nodes that receive this packet treat the sender of the INIT packet as their parent and send an ACK with random back-off delay. They now know the route to the base, which is one hop away. These nodes assign themselves depth of one. They retransmit the INIT message with their own IDs. All nodes that have not already assigned themselves a depth acknowledge this packet. These nodes now know their next hop on the route to the base and assign themselves depth of two. The process continues. When a node transmits an INIT message but does not receive any responses for a time-out duration, it assumes that there are no nodes deeper than itself along the path that passes through it. Such nodes are denoted *leaf nodes*.

9.5.2   Distributed Power Management

The network starts at an arbitrary performance level. Once the estimation of parameters ρ, σ_1 and σ_2 is completed, using the same methods as discussed in the previous section, the nodes will begin the transition to the sustainable performance level. Each node sets its response latency L equal to $T_{min} + T_{sleep}$. If a node is not receiving any energy from the environment, it may have to select a preset value of L for itself. When a leaf node has estimated its L, it sends a LATENCY message to its parent containing its estimate of L. A node that is not a leaf waits for the LATENCY packet from all its children. Let L_i denote the latency value heard from ith child, and let it have N_c children. When it has collected all LATENCY packets from its children and has estimated its own L, it sets

$$L' = \max_{i \in [1,\dots,Nc]} \{L_i\} + 1 \qquad\qquad (9.7)$$

It sends this cumulative value of L' to its parent. L' represents the total worst-case data-transfer latency along the path through this node from the worst leaf. The process repeats. The base then chooses L_{max} equal to the largest L' among those it receives from its children. This L_{max} is the maximum latency of data transfer in the network along the worst path. Each node switches to the duty cycle based on its estimated ρ after having sent the LATENCY message. Thus, the nodes gradually transition to their individual performance levels with no centralized coordination.

The in-network processing of the transmitted L values at each parent reduces the data sent to the parent nodes, and the amount of data sent to the base node is thus proportional to the number of depth-one nodes and not to the total number of nodes in the network. This ensures scalability, while at the same time returning a networkwide performance metric to the base. If the achieved performance is not acceptable, additional resources may have to be added.

The above method provides algorithms to adapt the power consumption on a networkwide basis to the local energy availability at the various network components. Further research is underway to exploit more detailed information about the energy-availability profile, including its temporal variations and spatial patterns to optimize performance in given energy environments.

9.6 Conclusion

Energy harvesting provides a viable alternative to create long-lasting WSNs, using harvested energy to supplement or replace stored battery reserves. Harvesting is not only useful but essential in some usage scenarios, such as space exploration and extraterrestrial sensor-network deployments, where the operating lifetime required is very long and battery replacements are next to impossible. The unmanned vehicles exploring the surface of Mars use solar-energy harvesting for all their energy needs. Different types of harvesting opportunities exist in most deployment scenarios, and the harvested form of energy should be actively exploited for enhanced system performance. The use of harvested energy leads to several research issues that are different from conventional power-management based on only stored battery energy. Not only is the design of the hardware different to account for the harvesting source, but workload scheduling now depends on the nature of the environmental source. The scheduling is also different in the case of distributed systems with multiple harvesting nodes. The harvesting theory described in this chapter can be used to determine sustainable performance levels and the additional resources required for achieving requisite performance. These techniques also facilitate the design of distributed scheduling methods. An ideal system would utilize its starting battery resources and harvested energy to achieve the maximum possible application throughput with the available resources. Such power management for harvesting networks is an open problem with a rich set of research challenges.

References

[1]   Raghunathan, V., et al., "Energy Aware Wireless Microsensor Networks," *IEEE Signal Processing Magazine*, Vol. 19, No. 2, 2002, pp. 40–50.

[2]   Min, R., et al., "An Architecture for a Power-Aware Distributed Microsensor Node," *IEEE Workshop on Signal Processing Systems (SiPS '00)*, October 2000.

[3]   Roundy, S., Wright P. K., and Rabaey J. M., *Energy Scavenging for Wireless Sensor Networks with Special Focus on Vibrations*, Kluwer Academic, Hingham, MA, 2004.

[4]   Stordeur, M., and Stark I., "Low Power Thermoelectric Generator—Self-sufficient Energy Supply for Microsystems," *16th International Conference on Thermoelectrics*, 1997, pp. 575–577.

[5]   Weber, W., "Ambient Intelligence: Industrial Research on a Visionary Concept," *Proc. 2003 International Symposium on Low Power Electronics and Design*, ACM Press, Seoul, South Korea, 2003, pp. 247–251.

[6]   Pescovitz, D., "The Power of Small Tech," *Smalltimes*, Vol. 2, No. 1, 2002.

[7]   DARPA energy harvesting projects, at http://www.darpa.mil/dso/trans/energy/projects.html.

[8]   Kymisis, J., et al., "Parasitic Power Harvesting in Shoes," *Proc. Second IEEE International Conference on Wearable Computing, (ISWC)*, IEEE Computer Society Press, Pittsburgh, PA, October 1998, pp. 132–139.

[9]   Shenck, N. S., and Paradiso J. A., "Energy Scavenging with Shoe-Mounted Piezoelectrics," *IEEE Micro*, Vol. 21, May–June 2001, pp. 30–42.

[10]   Paradiso, J. A., and Feldmeier M., "A Compact, Wireless, Self-powered Pushbutton Controller," in *ACM Ubicomp Third International Conference on Ubiquitous Computing (Ubicomp)*(Atlanta, GA), Berlin Heidelberg, Springer-Verlag, September 2001, pp. 299–304.

[11]   Starner, T., "Human-Powered Wearable Computing," *IBM Systems Journal*, Vol. 35, No. 3–4, 1996.

[12]   Williams, C. B., et al., "Devlopment of an Electromagnetic Micro-Genrator," *IEEE Proceedings—Circuits, Devices and Systems*, Vol. 148, December 2001, pp. 337–342.

[13]   Ottoman, G. K., Hofmann H. F., and Lesieutre G. A., "Optimized Piezoelectric Energy Harvesting Circuit Using Step-Down Converter in Discontinuous Conduction Mode," *IEEE Transactions on Power Electronics*, Vol. 18, No. 2, 2003, pp. 696–703.

[14]   Glynn-Jones, P., et al., "The Modeling of a Piezoelectric Vibration Powered Generator for Microsystems," *Transducers 01/Eurosensors XV, 11th International Conference on Solid-State Sensors and Actuators*, Munich, Germany, June 10–14, 2001.

[15]   Amitharajah, R., and Chandrakasan A. P., "Self-Powered Signal Processing Using Vibration-based Power Generation," *IEEE Journal of Solid State Circuits*, Vol. 33, No. 5, 1998, pp. 687–695.

[16]   Shearwoo, C., and Yates R. B., "Development of an Electromagnetic Microgenerator," *IEE Electronics Letters*, Vol. 33, October 1997, pp. 1883–1884.

[17]   Jet Propulsion Laboratory, California Institute of Technology, "Exploring Mars: Blowing in the Wind," August 10, 2001, at http://www.jpl.nasa.gov/technology/features/tumbleweed.html.

[18]   Whalen, S., et al., "Design, Fabrication and Testing of the p3 Micro Heat Engine," *Sensors and Actuators*, Vol. 104, No. 3, pp. 200–208.

[19]   CVS Repository, "HeliomoteR," http://cvs.nesl.ucla.edu/cvs/viewcvs.cgi/HeliomoteR.

[20]   Kansal, A., Potter D., and Srivastava M., "Performance Aware Tasking for Environmentally Powered Sensor Networks," *ACM Joint International Conference on Measurement and Modeling of Computer Systems (SIGMETRICS)*, New York, NY, June 2004.

[21]   Simunic, T., et al., "Energy Efficient Design of Portable Wireless Systems," *ACM/IEEE International Symposium on Low Power Electronic Design*, Rapallo, Italy, 2000.

[22] Min, R., T. Furrer, and Chandrakasan A., "Dynamic Voltage Scaling Techniques for Distributed Microsensor Networks," *IEEE Computer Society Workshop on VLSI*, Chicago, Il, April 2000, pp. 43–46.

[23] Schuette, M. A., and Barr J. R., "Embedded Systems Design for Low Energy Consumption," *Proc. 1994 IEEE/ACM International Conference on Computer-Aided Design*, IEEE Computer Society Press, Cambridge, MA, 1994, pp. 534–540.

[24] Simunic, T., et al., "Source Code Optimization and Profiling of Energy Consumption in Embedded Systems," IEEE International Symposium on System Synthesis (ISSS), Madrid, Spain, 2000.

[25] Kansal, A., and Srivastava M. B., "An Environmental Energy Harvesting Framework for Sensor Networks," *International Symposium on Low Power Electronics and Design*, ACM Press, Seoul, South Korea, 2003, pp. 481–486.

[26] Voigt, T., Ritter H., and Schiller J., "Utilizing Solar Power in Wireless Sensor Networks," *The 28th Annual IEEE Conference on Local Computer Networks (LCN)*, Bonn/Konigswinter, Germany, October 2003.

[27] Rong, P., and Pedram M., "Extending the Lifetime of a Netork of Battery Powered Mobile Devices by Remote Processing: A Markovian Decision-based Approach," ACM Design Automation Conference (DAC), Anaheim, CA, June 2003.

[28] Chang, J. H., and Tasiulas L., "Maximum Lifetime Routing in Wireless Sensor Networks," *Advanced Telecommunications and Information Distribution Research Program (ATIRP)*, College Park, MD, March 2000.

[29] Li, Q., Aslam J., and Rus D., "Online Power Aware Routing in Wireless Ad Hoc Networks," *ACM SIGMOBILE*, Rome, Italy, July 2001.

[30] Xu, Y., Heidemann J., and Estrin D., "Geography-Informed Energy Conservation for Ad Hoc Routing," *Proc. ACM Mobicom*, ACM Press, Rome, Italy, 2001, pp. 70–84.

[31] Schurgers, C., et al., "Optimizing Sensor Networks in the Energy-Density-Latency Design Space," *IEEE Transactions on Mobile Computing (TMC)*, January–March 2002, pp. 70–80.

[32] Intanagonwiwat, C., Govindan R., and Estrin D., "Directed Diffusion: A Scalable and Robust Communication Paradigm for Sensor Networks," *ACM Mobicom*, Boston, MA, August 2000.

CHAPTER 10

Sensor-Coordinated Actuation

Gaurav S. Sukhatme

This chapter describes representative recent results in sensor-coordinated actuation by focusing on three problems spanning the automated deployment and maintenance of sensor networks, their use in mediating and coordinating mobile robots, and the detection of environmental phenomena. Our focus is on the global behavior of systems whose constituents are mobile and on static nodes capable of limited computation and local communication.

10.1 Introduction

Sensor-coordinated actuation refers broadly to the control of multiple, spatially distributed actuators based on measurements from multiple, spatially distributed networked sensors. Systems of this class are also often called robotic sensor networks [1] or sensor-actuator networks [1]. In contrast to traditional distributed control and multirobot systems, such systems explicitly take into account networking constraints. Nevertheless, there is broad overlap between these areas, and some distinctions are largely a matter of terminology. Sensor-actuator networks harness the superior temporal-sampling capabilities of static-node assemblages and the ability of moving nodes to reposition themselves to provide high-fidelity spatial sampling. The combination (if properly designed) can yield powerful algorithms for extracting information from the environment using sensors. Additionally, the two classes of systems (static nodes and autonomous mobile nodes) exist in tandem. Sensor-network-mediated control of robots leads to simpler robots and works well with heterogeneous robots. The robots in turn can be used for sensor deployment, repair, and network maintenance. In this chapter, we focus on three problems: the automated deployment and maintenance of sensor networks, their use in mediating and coordinating mobile robots, and the detection of environmental phenomena using sensor-coordinated actuation. In each case, we present and analyze algorithms that address the problem. It is worth noting that these problems are a small but representative subset of the sensor-coordinated-actuation area. This is also the case for the algorithms discussed in this chapter since they are drawn from my previously published work. For further reading, see Section 10.5, where examples of related work are discussed.

10.2   Mobile-Robot-Assisted Sensor-Network Deployment and Repair

In this section we present an algorithm for mobile-robot-assisted sensor- network deployment and repair. The algorithm concurrently achieves three objectives: (1) it allows a mobile robot to explore and cover an unknown environment efficiently, (2) it causes a sensor network to be deployed into the environment, and (3) it repairs the network as needed (network maintenance). The algorithm is called least recently visited (LRV), following the policy used by the sensor nodes to guide robot exploration locally. As shown in the algorithm LRV is really the concurrent execution of two algorithms—one on a robot (robot loop) and another on every network node (sensor-node loop). For every node i, denote by $D(i)$ the set of all possible traversable directions from i. Then for all d in $D(i)$, $W(i, d)$ is the weight (cached on a network node) maintaining the number of times d was traversed from i. The function $ANY\_OF(T)$ returns a single element of a set T according to some arbitrary rule (i.e., in order, random). When a robot is deployed into the environment initially, according to the algorithm in Figure 10.1, it deploys a node because there is no sensor node within communication range. Over time, LRV causes a network of nodes to be deployed since every time a new node is deployed, it must be able to communicate with at least one other sensor node in the network. Once deployed, each sensor node starts to emit the locally least recently visited direction (hence the name LRV), which is the direction with the smallest weight W (if there are multiple directions of the same weight, one is arbitrary picked). In practice, the number of directions per node is often bounded and application dependent. The weight W of a direction is incremented in two cases: right before a direction is traversed and on the destination node right after a direction is traversed.

```
Robot Loop:
if no sensor node within communication range threshold then
    deploy sensor node
else
    move in direction suggested by nearest sensor node
    update sensor nodes if necessary
end if

Sensor Node Loop:
Emit least recently visited direction = ANY_OF(argminW(d))
Update sensor nodes weight if necessary: W(i):=W(i)+1
```

LRV is a complete algorithm [2] (i.e., a robot is guaranteed to visit every node in the network within a finite period of time). Using tests in simulation, LRV has also been shown to be moderately efficient. For regular graphs, it is conjectured that LRV is $O(V^{l+\varepsilon})$, but this remains unproved. For a tree with $|E|$ edges, the exploration time of LRV is no more than $2|E|$. For proofs of these results as well as experimental details, the reader is refered to [2].

LRV makes no distinction between network deployment and network repair. If nodes become energy depleted and stop transmitting navigation directives to robots, LRV will simply deploy new nodes in their vicinity with no purposeful network repair plan. Repair is thus a side effect of continual, vigilant deployment.

10.3   Using a Sensor Network to Mediate Robot Task Allocation

In the previous section, an algorithm was presented to deploy and repair a sensor network using a mobile robot. In this section, we discuss ways in which robots may use such a network, once deployed, to navigate an unfamiliar environment and to respond to events in the environment.

10.3.1   Navigation

Navigation, defined here simply as the ability to plan and execute motion from a designated start location to a goal, is a basic prerequisite for robot autonomy. In "traditional" robot systems, this is accomplished either through the use of a map or by following landmarks (in effect, a map). In this parlance, the deployed sensor network is a collection of active landmarks or beacons capable of sensing, computing, and communicating. It is relatively easy for such a network to compute a navigation solution for a robot based on dynamic programming. The sensor node closest to the desired goal is designated the goal node. Its neighbors only need to compute the one-hop navigation solution from their locations to the goal, using some notion of cost associated with the hop. The neighbors of these nodes in turn can compute the local navigation solution toward the best intermediate goals, using the same notion of cost and purely local considerations. This propagation through the network ultimately results in each node in the network's computing a preferred local direction of travel for a robot in its vicinity. This set of directions is called a *navigation field*. As before, robots merely follow local node directives regarding which direction to follow. The behavior of the ensemble is rather like "robot routing," where the network, instead of routing packets, is effectively routing robots to a destination. Some details of the field computation are provided below; for a full description, the reader is referred to [1, 2].

Consider the deployed sensor network as a graph, where the sensor nodes are vertices. Assume a finite set of vertices S in the deployed-network graph and a finite set of actions A a robot can take at each node. Denote by $P(s'|s, a)$ the probability that the robot arrives at vertex s' after starting at vertex s and commanding an action a. Here, a belongs to the set $A(s)$, and $A(s)$ is a subset of A. The vertices s and s' belong to the deployed-network graph.

Our model for the proposed system is Markovian—the state the robot transitions to depends only on the current state and action. We model the navigation problem as a Markov decision process [3]. To compute the best action at a given vertex, we use the value iteration algorithm on the set of vertices $S - s_g$, where s_g is the goal. The general idea behind value iteration is to compute the utilities for every state, then pick the actions that yield a path toward the goal with the maximum expected utility. The utility is incrementally computed as follows:

$$U_{t+1}(s) = C(s, a) + \max_{a \in A(s)} \sum_{s' \in S - s_g} P(s'|s, a) U_t(s')$$

where $C(s, a)$ is the cost associated with moving to the next vertex. Initially, the utility of the goal state is set to one and that of the other states is set to zero. Given the utilities, an action policy is computed for every state s as follows:

$$\pi(s) = \underset{a \in A(s)}{\arg\max} \sum_{s' \in S - s_g} P(s'|s,a)U(s')$$

The robot maintains a probabilistic transition model for the deployed-network graph and can compute the action policy at each node for any destination point. In practice, however, this is limiting since it requires the robot to traverse the network many times over to learn the transition model. Practically, a better approach is for the robot to compute the action policy as above and, while traversing the network, record the optimal action for the current node as it passes by. Each node can store this action and can emit it as part of the message directed to passing robot traffic. An even more attractive solution is to compute the action policy distributively in the deployed network. The idea is that every node in the network updates its utility and computes the optimal navigation action (for a robot in its vicinity) on its own. Both these refinements and practical issues regarding signal-strength-based node switching are discussed further in [3].

This algorithm has been tested extensively in simulation and using physical hardware (a Pioneer 2DX mobile robot and a Mica2-mote-based sensor network). The map of the experimental environment and deployed sensor network of nine sensor nodes is shown in Figure 10.1. The task of the robot is to navigate from the "home base" (vicinity of node A) toward different goals. Representative trajectories from node A to nodes C, E, and F are shown in Figure 10.2. The navigation results were very encouraging; over a large number of trials (approximately 1-km aggregate), the robot was able to navigate to the correct goal node in all cases.

10.3.2 Task Allocation

Having described algorithms for the deployment and maintenance of a network by robots and network-mediated robot navigation, I now describe an algorithm for network-mediated multirobot task allocation (MRTA). In particular, I am interested in a variant of the MRTA problem, where tasks arrive asynchronously in time and are physically spread out. Further, each task can be performed by exactly one robot, and

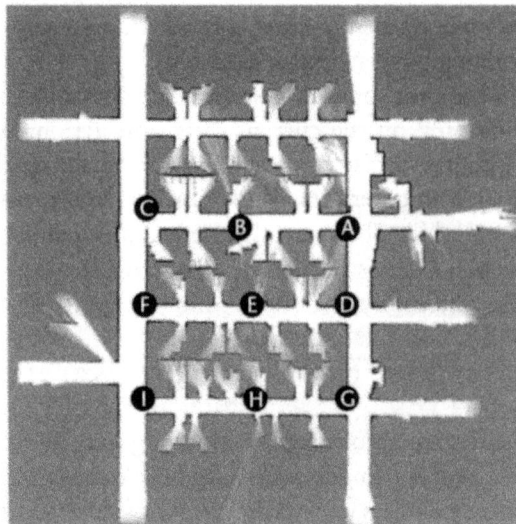

Figure 10.1 Map of the environment used in navigation experiments. Sensor nodes are marked A to I.

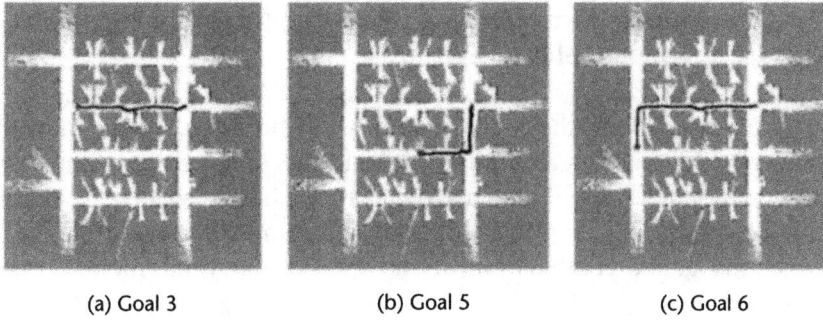

(a) Goal 3   (b) Goal 5   (c) Goal 6

Figure 10.2   Trajectories of a robot navigating to three different goals. The start location in each case is near node *A*. *(a) Goal 3, (b) Goal 5, and (c) Goal 6.*

one robot can only perform one task at a given time. Previous MRTA approaches in the robotics community [4] have focused on performing the task-allocation computation on the robots or at some centralized location external to the robots. All the sensing associated with tasks and robot localization is typically performed on the robots themselves. In contrast, my approach (described in detail in [5]) is based on the interaction between a sensor network and mobile robots. Tasks, upon arrival, are allocated to robots by a static sensor network. The basic idea is that as tasks are detected by the network, every node k in the network computes a suggested motion direction for a robot if it is in the vicinity of k. The ensemble of suggested directions computed over all nodes is called a *navigation field*. An adaptive distributed value-iteration algorithm is used to compute the navigation field as described in the previous section. I term this process Distributed In-Network Task Allocation (DINTA). DINTA-MF is a variant of DINTA, where multiple navigation fields (one for every task) are maintained in the network at a given time. Fields are assigned to robots using a greedy policy. The difference between the two approaches is that in DINTA-MF, every network node computes the direction that a robot should follow in its locality for every task in the environment. The performance of these algorithms is measured using the response time (i.e., the time elapsed between the task being spawned and its servicing by a robot). In my experiments I used the player/stage [6] simulation engine populated with simulated Pioneer 2DX mobile robots equipped with 180° field-of-view planar laser range finders (used for obstacle avoidance), wireless communication, and a mote base station (to communicate with the motes used as network nodes). A network of 25 motes was predeployed in a test environment. The communication range of motes and robots was set to approximately 4m. The robot team responded to tasks as they arrived by navigating to the site of the task as dictated by the network nodes. Both DINTA and DINTA-MF performed significantly better compared to the case where robots wander the environment in "explore-and-deploy" mode. DINTA-MF has also been statistically shown to outperform DINTA. The reason behind this is that every node in the network computes an assignment for every task; hence, all unassigned robots are assigned a task, which is not necessarily the case when only one assignment field is computed for every task. Note also that DINTA-MF does not "waste" resources (robots), whereas in DINTA, several robots can pursue the same task. The space and time requirements for DINTA-MF are linear in the number of tasks, which makes it realistic for implementation on our target node platform (the mote).

10.4   Coupling Local Sampling and Robotic Mobility to Detect Properties of a Scalar Field

Detecting and tracing the contours of a scalar field (e.g., temperature) in the environment is of interest to marine biologists, ecologists, and others studying the growth and life cycles of animals and plants. I have developed two approaches to contour detection. The first relies exclusively on sensor-based mobile robots, and the second uses a combination of static sensors and one mobile robot. I discuss these next.

10.4.1   Using a Large Number of Mobile Sensors Only

In this section, I briefly describe a distributed algorithm using multiple mobile sensors, local sensing, and local communication to detect a region of sharp temperature gradient in the environment, also known as a *thermocline*. The domain of interest is the marine underwater environment. Formally, a thermocline is a level set of the gradient of the temperature field with the property that no other level set has a greater value. The basic idea of my approach is sampling by divide and conquer. For simplicity, I address the problem in one dimension, namely depth. Suppose we have n nodes deployed in a vertical array, where the topmost node is connected to the external world (is an "edge" node). Each node has its own processor, memory, temperature sensor, and radio. However, the communication range is limited, and each node can only communicate with nearby nodes. Each node also has a calibrated pressure sensor allowing it to estimate its depth. We also assume that each node is able to change and regulate its depth (e.g., by using a simple buoyancy controller).

The search space is divided into regions. Every node uses its ability to move to explore one such region. The process is refined by splitting regions into halves (i.e., a binary search). Each node communicates with its neighbors and tries to persuade them that the thermocline lies within its search region as follows. At initialization, each node n_i collects temperature data at both end points of its search space. The temperature and the depth at each point, T_t, T_b, D_t, and D_b, are noted. Next, the node moves to $D_t + D_b/2$, where it collects a new temperature reading T and depth D. The new point divides the search space of node n_i into upper and lower parts. The differences between the new reading and the two previous readings are calculated. If $|T_t - T| > |T_b - T|$, D_b are replaced by T, D. Otherwise, the upper part is discarded, and T_t, D_t are replaced. The remaining part of the search region is designated as the new search region for node n_i. This process is repeated until termination conditions based on sensing resolution are satisfied (see Figure 10.3).

A process of data aggregation is enacted on the route from each node to the user to combine the conclusions (about the thermocline location) arrived at by the various nodes. For brevity, I do not discuss that process here; the reader is referred to [7] for details. A key advantage of in-network aggregation is the parsimonious use of energy. Not all nodes need to stay active in order for the process to converge, thereby saving energy overall. Further, it is interesting to note (see Figure 10.4) that after the first one or two steps of the distributed binary search, most nodes become inactive. However, the data-aggregation process results in some nodes that need to stay awake merely to forward messages from the active nodes to the network edge.

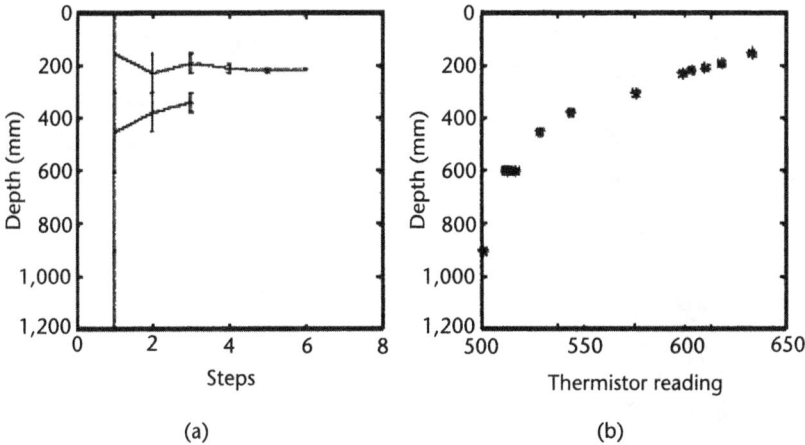

Figure 10.3   Results from a representative underwater experiment in a laboratory tank environment: (a) the output of the distributed binary-search algorithm, and (b) the corresponding temperature profile. Notice in (a) that two possible depths are viable candidates for the thermocline location until timestep 3, at which point there is only one survivor. By timestep 6, the system converges to one depth estimate for the thermocline (200 mm), which is reasonably accurate [compare with the depth-versus-temperature graph in (b)].

One way to alleviate this is to use a messenger, a robotic node that can move itself autonomously. This robot would move from the neighborhood of the edge node to the active node and forward the messages from one to another, thus acting as a data mule. The interested reader is referred to [7] for experimental results with a submarine data mule.

10.4.2   Using a Small Number of Mobile Sensors and a Large Number of Static Sensors

This section consider the case where the task is to detect and follow a specified level set of a scalar field, with the constraint that only one node in the network is mobile. Such a system may be used, for example, to detect and monitor the boundary of an oil spill in the ocean. One may imagine several static sensor buoys and a robot boat that uses information from these buoys to finds its way to the boundary (a level set of the oil concentration field). The basic idea behind the algorithm is simple. As the mobile node moves through the environment, it computes the local spatial gradient of the field by communicating with its immediate neighbors in the static sensor network. The algorithm causes the mobile node to perform gradient descent on the scalar field until it arrives at a location on the desired contour. From this point onward, the algorithm drives the mobile node to trace the desired contour without departing from it. Experiments in simulation indicate that the required contour is found with reasonable accuracy (between 80% and 90%) for networks with node degree of six or greater. Results also indicate that the paths generated by the algorithm are near optimal in terms of the distance traversed by the mobile node.

Denote the position of the mobile node by the (vector) x. Denote the field value at x to be $F(x)$. Denote the field value of the level set of interest to be F_d. The objective of the control law is to drive the node to locations that satisfy $G(x) = 0$, where $G(x) = F(x) - F_d$.

The mobile node collects the sensed values of the scalar field from all its neighbors within communication range, along with their location. Using this data, it identifies two neighbors (Figure 10.4), one with the highest gain-increase ratio and one with highest gain-decrease ratio (with respect to distance from itself). Having identified these two nodes, two unit gradient vectors $\nabla n_1(x)$ and $\nabla n_2(x)$ are defined in the

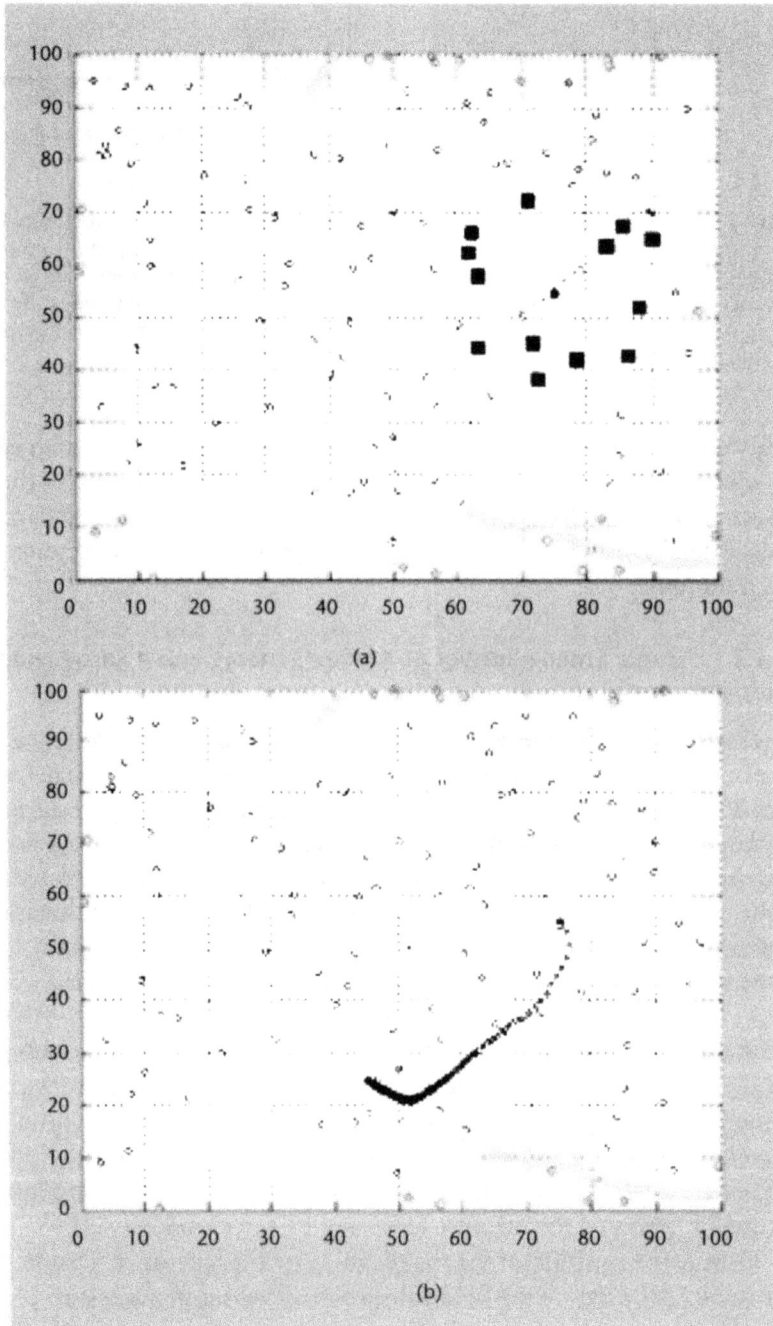

(a)

(b)

Figure 10.4 (a) A mobile node communicating with two neighbors with best gradients. The mobile node is the filled circle, the immediate neighbors are filled squares, and the circles are the stationary nodes. (b) The node trajectory to reach the contour of interest.

direction of each of the two chosen neighbors, and $\nabla G(x)$ is approximated by $\nabla G(x) = \nabla n_1(x) - n_2(x)$.

$G(x)$ and $\nabla G(x)$ are then used to compute a commanded velocity $\dot{x}$ for the mobile node, using the following control law:

$$\dot{x} = \alpha Null(\nabla G(x)) + \beta \nabla G(x)^+$$

where α, β are scalar gains, $Null(\nabla G(x))$ is the null space of $\nabla G(x)$, and $\nabla G(x)^+$ denotes the Penrose pseudoinverse of $\nabla G(x)$.

```
loop
    for i ∈{Current neighborlist of mobile node} do
            loc[i] ⇐ Location of i
            sense[i] ⇐ Sensor reading of i
    end for
    n1 ⇐ Node with best gain increase gradient {using loc[] and sense[]}
    n2 ⇐ Node with best gain decrease gradient {using loc[] and sense[]}
    Compute n₁(x) and n₂(x)
    Compute ∇G(x)(approximately) and G(x)(exactly)
    Compute x
    Command x to actuators
end loop
```

This control law has three interesting features: it is based on local sensor measurements and local communication, it is provably convergent (though I do not prove that here), and it results in a critically damped path for the mobile node (i.e., the node does not overshoot the contour or oscillate about it).

Based on an implementation of this control law, the following empirical conclusions about this algorithm, shown in Figure 10.6, suggest themselves. The percentage of successful contours found (as a fraction of all tests done) increases, then saturates, with increased network density irrespective of sensor noise. This saturation occurs at relatively low network densities (node degree approximately six to eight). While increased sensor noise adversely affects the percentage of successful completions (especially at short distances), the drop-off is gradual. The algorithm is independent of the type of decay experienced by the scalar field. For node degrees greater than six, the ratio of actual distance traveled by the mobile node to the straight-line distance (from the initial mobile-node location to the contour) saturates to approximately 1.1, which is very close to the best possible value, 1. Thus, this approach appears near optimal if the sensor network has a reasonable average node degree. As the percentage of sensor noise increases in the network, the ratio of distance traveled also increases. This is coupled with an increase in the variance of this ratio. This indicates that in high-noise scenarios, this algorithm might perform suboptimally, causing the mobile node to travel a longer distance than needed to reach the contour.

10.5 Related Work

The sensor-coordinated-actuation area is growing rapidly and has broad overlap with sensor networks and distributed robotics. This section briefly reviews the literature in areas most closely related to the themes discussed in this chapter. First, it is important to stress that there are fundamental problems in sensor-coordinated actuation that I simply have not addressed here. These include, among others, node localization [8], coverage with communication constraints [9, 10], and target tracking [11].

There is considerable work in network-mediated robot navigation. In addition to my previous work [3], there is a similar approach [12] that effectively causes the network to compute potential field-based solutions to effective and optimal navigation. In the area of network deployment and repair, my previous work (relying on LRV) is joined by aerial-robot-based deployment and repair [13]. In the mobile-robotics community, the problem of designing cooperative behavior for robot teams with communication constraints has been addressed in [14, 15] (based on designing communication-sensitive behaviors for individual robots) and [16] (a control-theoretic approach toward team behavior). The distributed estimation-and-control problem has been addressed for multirobot control applications. Recently, [17] has investigated the fundamental principles of adding small amounts of mobility to an otherwise static network. In the infrastructure-supported mobility community, the Networked Infomechanical Systems (NIMS) project has developed techniques for exploiting constrained mobililty (in combination with static sensors) for adaptive sampling and task allocation. As described in this chapter, infrastructure-supported mobility has also found application underwater, where a network of sensors capable of small amounts of motion can collectively determine the location of physical contours of interest.

10.6   Conclusion

I have described solutions to three problems: the automated deployment and maintenance of sensor networks, their use in mediating and coordinating mobile robots, and the detection of environmental phenomena using sensor-coordinated actuation. In each case, I presented and analyzed algorithms that exploit local interactions between the constituents of the system to meet global objectives.

References

[1]   Batalin, M., and Sukhatme G. S., "Coverage, Exploration and Deployment by a Mobile Robot and Communication Network," *Telecommunication Systems Journal, Special Issue on Wireless Sensor Networks*, Vol. 26, No. 2, 2004, pp. 181–196.

[2]   Batalin, M., and Sukhatme G. S., "The Analysis of an Effcient Algorithm for Robot Coverage and Exploration Based on Sensor Network Deployment," *Proc. IEEE International Conference on Robotics and Automation*, Barcelona, Spain, April 2005.

[3]   Batalin, M., Sukhatme G. S., and Hattig M., "Mobile Robot Navigation Using a Sensor Network," *Proc. IEEE International Conference on Robotics and Automation*, New Orleans, LA, April 2004, pp. 636–642.

[4]   Gerkey, B., and Matari´c M. J., "A Formal Framework for the Study of Task Allocation in Multi-robot Systems," *International Journal of Robotics Research*, Vol. 23, No. 9, 2004, pp. 939–954.

[5] Batalin, M., and Sukhatme G. S., "Sensor Network-based Multi-robot Task Allocation," *Proc. IEEE/RSJ International Conference on Intelligent Robots and Systems*, Las Vegas, NV, October 2003, pp. 1939–1944.

[6] Gerkey, B. P., et al., "Most Valuable Player: A Robot Device Server for Distributed Control," Proc. IEEE/RSJ International Conference on Intelligent Robots and Systems, Maui, HI, 2001, pp. 1226–1231.

[7] Zhang, B., Sukhatme G. S., and Requicha A. A., "Adaptive Sampling for Marine Microorganism Monitoring," *Proc. IEEE/RSJ International Conference on Intelligent Robots and Systems*, Sendai, Japan, 2004.

[8] Howard, A., M. J. Matari´c, and Sukhatme G. S., "Relaxation on a Mesh: A Formalism for Generalized Localization," *Proc. IEEE/RSJ International Conference on Intelligent Robots and Systems*, Wailea, HI, October 2001, pp. 1055–1060.

[9] Poduri, S., and Sukhatme G. S., "Constrained Coverage for Mobile Sensor Networks," *Proc. IEEE International Conference on Robotics and Automation*, New Orleans, LA, May 2004, pp. 165–172.

[10] Howard, A., Matari´c M. J., and Sukhatme G. S., "Mobile Sensor Network Deployment Using Potential Fields: A Distributed, Scalable Solution to the Area Coverage Problem," Proc. International Symposium on Distributed Autonomous Robotic Systems, Fukuota, Japan, 2002.

[11] Jung, B., and Sukhatme G. S., "Tracking Targets Using Multiple Robots: The Effect of Environment Occlusion," *Autonomous Robots*, Vol. 13, No. 3, 2002, pp. 191–205.

[12] Li, Q., Rosa M. D., and Rus D., "Distributed Algorithms for Guiding Navigation across a Sensor Network," *Proc. ACM MobiCom*, San Diego, CA, September 2003, pp. 313–325.

[13] Corke, P. I., et al., "Deployment and Connectivity Repair of a Sensor Net with a Flying Robot," *Proc. 9th International Symposium on Experimental Robotics (ISER)*, 2004.

[14] Wagner, A. R., and Arkin R. C., "Multi-robot Communication-Sensitive Reconnaissance," Proc. IEEE International Conference on Robotics and Automation, New Orleans, LA, May 2004.

[15] Ulam, P., and Arkin R. C., "When Good Comms Go Bad: Communications Recovery for Multirobot Teams," *Proc. IEEE International Conference on Robotics and Automation*, New Orleans, LA, May 2004.

[16] Fan, Z., Grocholsky B., and Kumar V., "Formations for Localization of Robot Networks," *Proc. IEEE International Conference on Robotics and Automation*, New Orleans, LA, May 2004.

[17] Kansal, A., et al., "Sensing Uncertainty Reduction Using Low Complexity Actuation," Proc. ACM IPSN, Berkeley, CA, 2004.

CHAPTER 11

Sensor-Network Tomography

Jerry Zhao and Ramesh Govindan

More than with most distributed systems, it is important to have a monitoring-and-management infrastructure for WSNs to indicate overall network status, to track node failures and resource depletion, and to assess system performance. In this chapter, we first review recent work in sensor-network monitoring and then describe *sensor-network tomography* (SNT), a monitoring architecture that addresses the unique challenges of sensor-network monitoring. We illustrate its elements using specific examples of sensor-network-monitoring tools that trade detailed network status for reduced energy expenditure.

11.1 Introduction

WSN systems consist of a large number of wireless sensors deployed in possibly harsh environments. With small, compact form factor and low cost, these nodes are expected to be powered by batteries and autonomously deployed in an ad hoc fashion. The working environment for these sensor nodes might be unpredictable and will thus affect the performance of the network dramatically. The high node-to-human ratio makes it infeasible to maintain individual sensors constantly. Given their unattended nature and their complexity, a monitoring-and-management infrastructure that can provide indications of network health is an important component of WSNs.

Such information is very useful for detecting and managing node failures, resource depletion, and other abnormalities in sensor networks. For example, knowing the remaining energy-resource distribution within a sensor field, a user may be able to determine whether any part of the network is about to fail in the near future due to depleted energy. Similarly, given the practical difficulties in precisely planning sensor field deployments, a network-monitoring system can guide incremental deployment of sensors. By examining the distribution of node density, communication quality, and other resources in the sensor field, additional sensors can be placed selectively in those regions that have fewer resources. Finally, information about the overall response of the sensors to some known stimulus can help test collaborative sensing algorithms. Such information is useful for validating expected sensing functionality or for fine-tuning detection algorithms.

Existing management protocols for the Internet, such as Simple Network Management Protocol (SNMP) [1], cannot be directly applied to sensor networks.

Unlike wired networks and infrastructure-based wireless networks, many sensor networks consist of a large number of nodes with limited energy and bandwidth resources. Thus, it is infeasible to maintain per-device databases for network management. Additionally, wireless sensors implement intricate distributed and collaborative protocols. A comprehensive monitoring system is necessary for users to understand the execution of the network as a whole. Finally, nodes detect physical events in the environment, and the variability in the physical phenomena they sense at least rivals the variability in wireless communication. Noisy network connectivity and sensor readings dictate new solutions for robust network management.

In this chapter, we first elaborate the challenges in collecting system-health indications from large-scale sensor systems with limited energy resources. After reviewing recent work in sensor-network monitoring and diagnosis, we describe in detail SNT, a monitoring architecture for WSN systems, that consists of three components: *digest*, *scan*, and *dump*. These components are invoked at different spatial and temporal scales to provide accurate yet low-overhead sensor-network-monitoring services. We further discuss in detail the design of residual-energy scan and sensor-network digest to illustrate these design principles by trading between the details of network status and associated energy expenditure.

11.2   Monitoring WSNs

11.2.1   Design Challenges

Network management is a challenging problem in most distributed systems, more so for sensor networks, where the ratio of the number of devices to the number of personnel to maintain the network is large. A suite of tools that simplifies maintenance is therefore a prerequisite in a deployed sensor network. Most typical systems, from the computer operating systems that we use everyday to servers and routers in the Internet, employ some kind of logging facility to track system state. For example, SNMP [1] provides per-device information databases for network monitoring and management, where node state and logs can be centrally collected and processed. However, energy constraints, thus the bandwidth constraints in sensor networks, preclude a centralized solution that communicates extensive logs from individual nodes. Rather, for such networks, we need methods that carefully select the data to be sent and use in-network aggregation techniques to reduce the volume of data.

Additionally, sensor networks implement intricate distributed and collaborative applications. In such applications, intermediate nodes might process and reduce the data. System performance and status cannot be measured from a few predefined servers or boundary routers, as in the Internet. Different from wired networks and ad hoc wireless networks, sensor networks need a comprehensive system for understanding the execution of the network *as a whole*. Collaboration between redundant sensor nodes also imposes challenges to the definition of system failure and status. One or a few node crashes do not necessarily undermine system functionality. A regional or global view of the system is helpful to identify the possibility of failures.

Finally, sensor nodes use wireless transceivers, the vagaries of which are well documented. Wireless communication is susceptible to environmental noise and interference between simultaneous transmitters. The propagation characteristics of

RF signals can result in highly variable wireless links. Nodes detect physical events in the environment, and the variability in the physical phenomena they sense at least rivals that of wireless communication. The dynamics of network connectivity and sensing activity can complicate the design of a network-management system since such a system needs to be highly robust.

The key problem for sensor-network management is to design *energy-efficient* and *robust* monitoring protocols that extract explicit network-status indications. However, such a system must support diverse management tasks, from distributed debugging to diagnosis of specific failures, and continuous performance monitoring. Analysis tools for collected data are necessary to identify, assess, and visualize known or unknown system abnormalities. Furthermore, such a network-monitoring solution also needs to tackle networks of different sizes. For example, *laboratory-scale* experiments typically use a handful of nodes and need very detailed status information from each node to support distributed debugging. *In-field* experiments are conducted at a slightly larger scale and need continuous but less detailed information about network operation to profile the performance of different design choices. For an *operational network* consisting of hundreds of nodes, low-overhead monitoring protocols are necessary to extract critical network-health indications continuously. For each of these scenarios, the trade-off between the system-status detail and corresponding energy expenditure has to be carefully examined.

11.2.2   Related Work

We start to tackle the problem of sensor-network monitoring and management by broadly reviewing state-of-the-art solutions. Diagnosis protocols for several specific failures in sensor networks have been well discussed. For example, it is important to access the coverage of a sensor networks after deployment. The coverage and exposure problem in WSNs is studied in [2, 3]. By exchanging sensor-coverage information within a local neighborhood, these techniques discover the maximal breach path (i.e., where sensor coverage is weakest) and the maximal support path (i.e., where sensor coverage is strongest) between a given source and destination positions. In [4, 5], node failures are detected using similar techniques. Limiting state exchange between nodes to their local vicinities, these protocols successfully detect specific system failures without communicating individual node state over long distance.

Often, the failure mode itself is not easily defined. Thus, retrieving a snapshot of system status is necessary to access the system performance. In Sensor Topology Extraction at Multiple Resolutions (STREAM) [6, 7], sensor-network topology maps are collected by in-network aggregation. Parameterized by virtual range and resolution factor, STREAM can support a wide range of topology queries from network backbone to complete connectivity graph. In Section 11.2.3, we will discuss in more detail a similar tool called eScan, which maps the remaining energy resources within the network. Another dimension of aggregation is to compress network states over time. For example, using a network-state model to represent node behavior and associated power consumption, the communication overhead of reporting energy levels can be reduced by probabilistically predicting the energy consumption [8].

There is another class of interesting approaches to achieve good scalability with randomized sampling techniques. Recent work on synopsis diffusion [9] and summation sketch [10] has been motivated by data-mining techniques in data streaming. Based on P. Flajolet and G. N. Martin's approximate counting algorithm [11], these two approaches can compute a wide range of aggregation functions with probabilistically bounded error, including duplicate sensitive functions such as *sum* and *median*. Similar techniques can be applied in digest diffusion for network monitoring.

11.2.3   Sensor-Network Tomography: A Monitoring Architecture

It is unlikely that one single mechanism or tool discussed in Section 11.2.2 could satisfy the diverse requirements of network monitoring. What, then, are the different kinds of tools one might use for sensor-network monitoring? How do they interact with each other to form a coherent architecture for sensor-network monitoring? We discuss these questions in this section by outlining SNT, a sensor-network-monitoring architecture.

Illustrated in Figure 11.1, SNT is distinguished by three levels of monitoring facilities, where each level consists of a class of tools and protocols that provide different spatiotemporal resolutions of monitoring data:

The first component consists of protocols such as *dump*, which are akin to the functionality provided by protocols like SNMP. Upon a user's request, dump collects detailed information from a small number of nodes. For example, the raw temperature readings from some sensors could be dumped to debug a collaborative event-detection algorithm. Dump can be implemented as a generic application over Directed Diffusion [12] or other data-dissemination mechanisms. Because the amount of data per node may be large, dump should be invoked only at small spatial scales (i.e., from a few nodes) and only when there is a reasonable certainty of a problem at those nodes.

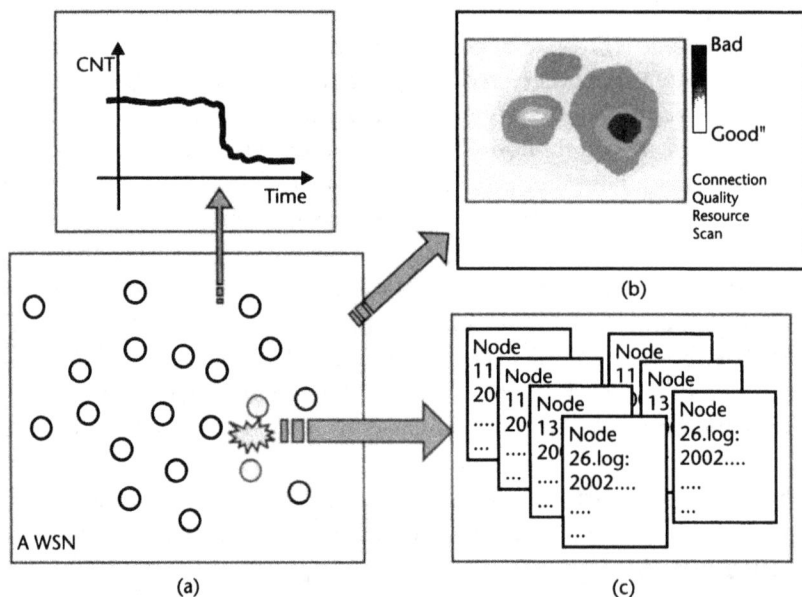

Figure 11.1   SNT: (a) aggregate network properties, (b) abstracted network scan, and (c) detailed network dump.

To guide system administrators to the location of problems, we envision a second class of tools that we call *scans*. Scans represent abstracted views of resource availability or network activity throughout the entire network or throughout a significant section of the network. Thus, this class of tools has a significantly greater spatial extent than dump. One example of a scan is the residual-energy scan *eScan* [13].To compute an eScan, a special user gateway node initiates collection of node state, for instance residual-energy-supply level, from every node in the system. Instead of delivering the raw data to the user node, eScan computation takes advantage of in-network aggregation. Residual-energy-level data from individual nodes is combined into a more compact form if, and only if, those nodes are nearby and have similar energy levels. By pushing the data processing into the network, eScan constructs an approximate, systemwide view of energy levels but incurs lower communication overhead compared to centralized data collection. From such a global view, users are able to isolate those nodes for which more detailed information is needed; a user can extract information from those nodes using tools such as dump.

Clearly, the energy cost of collecting a scan can be significant, and our third class of tools, *digests* [14], can help alert users to error conditions within the network. Such error conditions can trigger scans. A digest is defined as a compact (i.e., representable in a few bytes) aggregate of a system metric that indicates overall network state. For example, the size of network (i.e., the node count) can indicate several system-health conditions: a sudden drop in network size can indicate massive node failure or network partitioning; a significant oscillation in this number indicates frequent network connectivity changes. Digests, like scans, also span large spatial extent. However, unlike scans, they can be computed continuously because they are highly compact. Digests are not intended to isolate network problems, merely to indicate when a user might need to invoke networkwide scans.

11.3   Sensor-Network Scan and Digest

11.3.1   Design Principles

We discuss the design of *eScan* and *digest* in detail to illustrate the components of SNT. From a mechanistic aspect, these tools share several design principles that ensure energy efficiency, scalability, and robustness.

- *In-network aggregation.* Researchers have recognized that data-communication costs dominate energy consumption in WSNs. For example, empirical studies show the energy consumed in transmitting a 1-kb packet over 100m is approximately the same as performing 3 million CPU instructions on prototype wireless sensor nodes [15]. For this reason, sensors will prefer to perform significant, local, collaborative processing of data, rather than to transmit data over long distances. It is necessary to apply aggressive in-network aggregation for monitoring protocols since these protocols represent pure overhead, especially when they extract network status from the entire network or a large part thereof. Trading communication cost for computation cost, multiple data records can be selectively merged to reduce communication overhead.

- *Vertical integration.* Several data-centric communication paradigms, such as Directed Diffusion [12], have been proposed to support energy-efficient, large-scale data access from sensor networks. Monitoring protocols can be implemented as a special class of applications over these data-management frameworks. However, we propose to push monitoring services as low as possible into the network stack to reduce unnecessary messaging overhead. For example, small, periodic, monitoring data in digest can be embedded into MAC-layer beacons, even when the data logically belongs to a higher layer (e.g., a count of the number of packets seen at a node). Vertical integration of monitoring services over the entire network stack may introduce complexity in protocol design and implementation but can reduce energy consumption.

- *Localized signaling.* Energy efficiency is not the only reason to push monitoring services as low as possible into the network stack. Such a design can also improve the robustness of the monitoring protocols to failures on upper layers. More specifically, it is encouraged to use message exchange within a localized vicinity, such as one-hop broadcast or unicast. With little or no dependency on routing or upper layers, the monitoring service might perhaps be able to function well enough before or during a massive failure and to collect monitoring data as much as possible for postmortem analysis.

11.3.2 Sensor-Network Scan: Taking a Global Snapshot

A *residual-energy scan*, or *eScan*, depicts remaining energy supply in a sensor network. An eScan might look something like the image shown in Figure 11.1(b). Different regions of the sensor field are shaded depending on the average energy resources within that region. An eScan can help users to decide where new sensor nodes should be deployed to avoid energy depletion. It may also help verifying the behavior of energy-aware, adaptive routing protocols [16].

A naive approach to collecting an eScan is for an individual node to send its location and remaining energy level to a base station, which is obviously inefficient due to the massive messaging overhead incurred. As we briefly described in Section 11.2.3, eScans can be constructed by in-network aggregation on their path to the user gateway node. Here, we describe in detail of the representation and aggregation scheme for eScan.

A sensor-network scan represents an abstracted view of particular network characteristics. More precisely, we can define an eScan as a collection of (value, coverage) tuples:

- *Value* is the quantitative representation of the network state we are monitoring. It may have a more complex form than a single scalar value. In eScan, we use *value* = ⟨*min, max*⟩, where *min* and *max* are the minimum and maximal residual-energy level of the nodes, respectively. For example in Figure 11.2(a), the eScan value in scan *A* is (35%, 37%).

- *Coverage* denotes the region that value describes. In eScan, coverage of a scan is described by a polygon, which covers those nodes when energy levels fall within the range of *value.min* and *value.max*. The vertices of the coverage polygon are the locations of those boundary nodes. The polygon is not

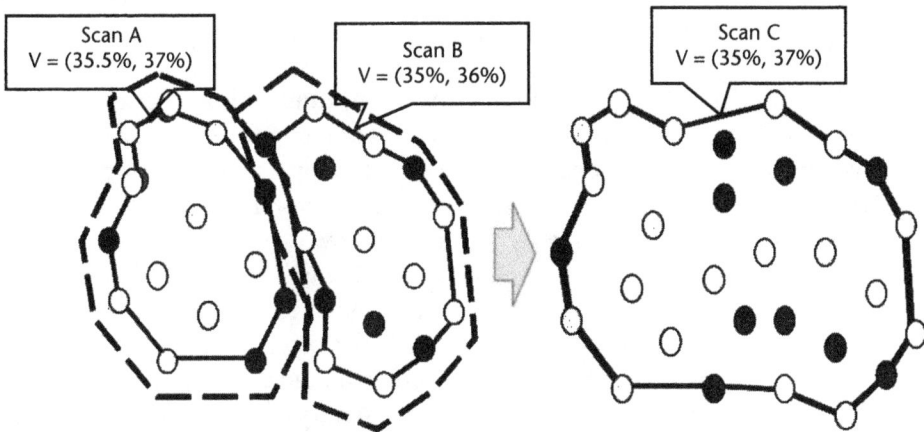

Figure 11.2 Representation and aggregation of sensor-network scans.

necessarily convex but is not self-overlapping. In Figure 11.2(a), the coverage polygon of the eScan is shown using a solid line.

This representation scheme leads to energy savings on messaging cost. Combining locations of the nodes with a region, the polygon representation is more compact. Intuitively, if all n nodes within a square region have similar values, instead of a list of n locations, the polygon representation uses a list proportional to $\sqrt{n}$. Information about residual energy at each node and the locations for the interior nodes will be lost. However, it is still helpful as an indication of overall energy-resource distribution. The eScans are enough in most cases to identify the near-depleted regions or discover energy-consumption patterns.

One of the most important characteristics of scans is that multiple scans can be combined. For example, eScan A and eScan B can be aggregated if $A.value$ and $B.value$ are similar,

$$\frac{max\{A.max, B.max\} - min\{A.min, B.min\}}{Avg\{A.min, A.max, B.min, B.max\}} \leq T$$

and $A.coverage$ and $B.coverage$ are adjacent,

$$Distance(A.coverage, B.coverage) < R$$

where tolerance T denotes the maximum allowed relative error of the residual-energy value by aggregation, resolution R decides when two regions are adjacent, and $Distance$ gives the minimum distance between any pair of points, with each point from one of the coverage sets. When both conditions are met, the aggregated eScan C can be obtained in the form of

$$C.min = min\{A.min, B.min\}$$

$$C.max = max\{A.max, B.max\}$$

$$C.coverage = Merge\{A.coverage, B.coverage, R\}$$

Where $Merge(P, Q, R)$ combines two polygons with resolution R. Obviously, $size(C) < size(A) + size(B)$, because $Merge$ only removes vertices from coverage and never adds new ones. An example of aggregation operation is shown in Figure 11.2, and the scan size is reduced by removing the location information of five nodes.

T and R provide the "knobs" to control how deeply we aggregate scans. They also decide the "fidelity" of outcome scans. Though aggregation consumes energy on local CPU processing, but such cost is much lower than delivering raw data across the network. Simulations shown in [13] indicate that in-network aggregation in eScan can achieve an order-of-magnitude reduction in communication compared to centralized collection. For a network consisting of 400 nodes, eScan can reduce messaging costs by a factor of 12.5 but only introduces 5% distortion.

11.3.3 Sensor-Network Digest: Collecting Vital System Metrics

As described in Section 11.2.3, *digest* is another component that leverages even more aggressive in-network aggregation to collect critical network performance metrics.

A *sensor-network digest* is a highly compact (only a few bytes) representation of an aggregated performance metric. Abstractly, a digest is defined by a digest function $f(v_1, v_2, ..., v_n)$, where v_i is the value contributed by each node i. We consider those v_i's that represent some aspect of network operation in compact format: node energy level, degree of connectivity, volume of traffic seen, and so forth. A key property of the class of aggregates we are interested in is that f is decomposable by function g:

$$f(v_1, v_2, ..., v_n) = g\big(f(v_1, v_2, ..., v_k)f\big)f(v_{k+1}, v_{k+2}, ... v_n)$$

Decomposable digest functions include *min, max, average,* and *count; median,* for example, is not decomposable.

The problem of digest computation is then that each node i provides a value v_i as its contribution to the digest function f, where v_i may change over time. The goal of the digest computation mechanism is for each node in the network to contain a continuous estimate for the current value of f.

The general application framework for aggregation in WSNs sensor networks has been well studied [12, 17, 18]. However, computing digests for sensor networks poses unique design challenges. Spanning the entire network and computed continuously, digest protocols must be aggressively energy efficient, far more so than other components of the system. Additionally, without a natural initiator for a digest (e.g., a user node), the routing structures for digest computations must be autonomously derived.

A standard way of computing these digest functions using in-network processing is to use a hierarchy and propagate the digest up to the root, computing partial values along the way. Instead of using more heavyweight hierarchy and clustering techniques, our approach uses a novel approach to set up a tree autonomously (i.e., without explicit initiation by a base station), using a technique we call *digest diffusion.* The tree built by digest diffusion can be used to compute other digests by propagating partial results up toward the root. We now describe these in some more detail.

In our description, we limit the digest functions we consider to $V_{MAX}, V_{SUM}, V_{AVG}$ and V_{CNT}, which respectively denote the maximum, sum, average of v_i, and number of the nodes in the network. V_{AVG} can be defined as V_{SUM}/V_{CNT}.

There is a specific rationale for our choice of digest functions since these functions are qualitatively different from each other. Using terminology from [17], a digest function is *monotonic* if, and only if, when two partial results r_1 and r_2 are combined by a function $r = g(r_1, r_2)$, the result r satisfies $\forall i = 1, 2; r \geq r_i$ for an ordering relationship $\geq$. It is *exemplary* if the final result can be determined from one single contribution value. In our set of digest functions, V_{MAX} is monotonic and exemplary, while V_{CNT} is monotonic but not exemplary. V_{SUM} and V_{AVG} may not be monotonic (if negative values are allowed) and are certainly not exemplary. Finally, as we shall argue later, the loss and duplicate sensitivity of V_{AVG} may be different from that of V_{MAX}.

We note that monotonic and exemplary digest functions can be computed efficiently by localized information exchanges between one-hop neighbors. We call this technique *digest diffusion*. We now describe digest diffusion for V_{MAX}. Initially, each node i sets its perceived maximum value $m_i = v_i$, source of maximum $s_i = i$, and hop distance $h_i = 0$ and periodically sends a tuple $M = (m_i, s_i, h_i)$ to its neighbors. Upon receiving a message (m_j, s_j, h_j) from neighboring node j with $m_j > m_i$, node i sets $m_i = m_j, s_i = s_j, h_i = h_j + 1$, and parent $p_i = j$. If $m_j = m_i$, it further checks if $s_j > s_i$, which guarantees strict monotonicity. Node i may switch its parent node from node j to k, when k provides the same maximum value but a shorter hop distance $h_k < h_j$. Gradually within $O(d/(1-p))$ steps (d is the diameter of the network; p is packet-loss rate per link), all nodes agree on a node s with the maximum value v.

This fusion-based approach is simple but efficient. It is fully distributed and requires no base station or user node to initiate the computation. The computation converges in a time proportional to the network diameter. It is energy efficient and scales well with network size since the overhead at each node is constant over time. The information exchanged between neighbors is small and can easily be piggybacked onto other neighbor-to-neighbor communication (e.g., beacons sent by MAC protocols or protocols for topology adaptation).

Digest diffusion *cannot be used to compute nonexemplary digests*, such as V_{AVG}. One of the fundamental reasons is that V_{AVG} is *duplicate sensitive*. When a node tries to aggregate the V_{AVG} partial results from its neighbors, it is difficult to determine if there are any overlaps between those results. Thus, merely merging partial results from neighbors will yield an erroneous result.

However, digest diffusion implicitly constructs a tree whose root is the node that contributes to the value of the exemplary digest (e.g., the node that has the maximum value in a V_{MAX} digest). Digest also computes a parent p_i for each node i. We call this tree the *digest tree*. Other digest functions can be computed easily on this tree. For example, with the aggregation tree from V_{max} computation, it is straightforward to calculate V_{MAX}: node i periodically calculates a partial result from most recent reports from its children $c_1, c_2, ..., c_k$ for node count

$$n_i = \sum_{j=1}^{k} n_{c_j} + 1$$

and average value

$$a_i = \frac{\sum_{j=1}^{k} a_c . n_{c_j} = v_i}{n_i}$$

It then sends out $< a_i, n_i >$ to its parent π along the tree. Hop by hop, the partial results are propagated up to the root, where the final result of V_{AVG} is calculated. It takes $O(d / (1 - p))$ time to converge on the correct result.

The digest-computation process is fully distributed and robust. The digest tree migrates adaptively when the current root fails since digest diffusion will try to find the new value for V_{MAX}. There are many factors that need to be considered in a practical implementation of digests [14]. For example, maintenance of the digest tree against topology changes, such as node failure and addition, is necessary due to the high dynamics in wireless links. We may also further reduce communication cost by *incrementally updating* the partial digests. Only those subtrees that have nodes whose values have changed beyond a certain threshold need to send their partial results.

11.4   Conclusion

In this chapter, we have describe SNT, an architecture for sensor-network monitoring. We have illustrated the design principles of in-network aggregation to push data processing into the network, vertical integration to reduce unnecessary messaging overhead, and localized signaling to increase tolerance to routing failures. Network management is still the subject of ongoing research. We use SNT as a starting point to address the challenges in managing large-scale sensor networks with limited resources. A variety of protocols and tools to support debugging, diagnosis, monitoring, and management are required to foster experimentation and deployment of emerging WSNs.

References

[1]   Case, J., et al., "Simple Network Management Protocol," Internet Request for Comments 1908, 1996.

[2]   Meguerdichian, S., et al., "Coverage Problems in Wireless Ad Hoc Sensor Networks," *Proc. IEEE Infocom*, Anchorage, AK, 2001.

[3]   Meguerdichian, S., et al., "Exposure in Wireless Ad Hoc Sensor Networks," *Proc. ACM/IEEE International Conference on Mobile Computing and Networking*, Rome, Italy, 2001, pp. 139–150.

[4]   Chessa, S., and Santi P., "Comparison-based System-Level Fault Diagnosis in Ad Hoc Networks," *Proc. 20th IEEE Symposium on Reliable Distributed Systems*, New Orleans, LA, 2001.

[5] Staddon, J., Balfanz D., and Durfee G., "Efficient Tracing of Failed Nodes in Sensor Networks," *Proc. 1st ACM International Workshop on Wireless Sensor Networks and Applications*, Atlanta, GA, 2002, pp. 122–130.

[6] Deb, B., Bhatangar S., and Nath B., "A Topology Discovery Algorithm for Sensor Networks with Applications to Network Management," *Proc. IEEE CAS Workshop on Wireless Communications and Networking*, Orlando, FL, 2002.

[7] Deb, B., S. Bhatnagar, and B. Nath, "Multi-resolution State Retrieval in Sensor Networks," *Proc. IEEE ICC Workshop on Sensor Network Protocols and Applications*, 2003.

[8] Mini, R. A. F., Nath B., and Loureiro A. A. F., "Prediction-based Approaches to Construct the Energy Map for Wireless Sensor Networks," *Proc. 21st Simpsio Brasileiro de Redes de Computadores*, 2003.

[9] Nath, S., et al., "Synopsis Diffusion for Robust Aggregation in Sensor Networks," *Proc. ACM Conference on Embedded Networked Sensor Systems*, Baltimore, MD, 2004

[10] Kollios, G., et al., "Approximate Aggregation Techniques for Sensor Databases," *Proc. International Conference on Data Engineering*, Seattle, WA, 2004.

[11] Flajolet, P., and Martin G. N., "Probabilistic Counting Algorithms for Database Applications," *Journal of Computer and System Sciences*, Vol. 31, 1985, pp. 182–209.

[12] Intanagonwiwat, C., Govindan R., and Estrin D., "Directed Diffusion: A Scalable and Robust Communication Paradigm for Sensor Networks," *Proc. ACM/IEEE International Conference on Mobile Computing and Networking*, Boston, MA, 2000, pp. 56–67.

[13] Zhao, J., Govindan R., and Estrin D., "Residual Energy Scans for Monitoring Wireless Sensor Networks," *Proc. IEEE Wireless Communications and Networking Conference*, Orlando, FL, 2002.

[14] Zhao, J., Govindan R., and Estrin D., "Computing Aggregates for Monitoring Wireless Sensor Networks," *Proc. IEEE ICC Workshop on Sensor Network Protocols and Applications*, Anchorage, AK, 2003.

[15] Pottie, G. J., and Kaiser W. J., "Embedding the Internet: Wireless Integrated Network Sensors," *Communications of the ACM*, Vol. 43, No. 5, 2000, pp. 51–58.

[16] Xu, Y., Heidemann J., and Estrin D., "Geography Informed Energy Conservation for Ad Hoc Routing," *Proc. ACM/IEEE International Conference on Mobile Computing and Networking*, Rome, Italy, 2001, pp. 70–84.

[17] Madden, S. R., et al., "Supporting Aggregate Queries over Ad Hoc Wireless Sensor Networks," *Proc. Workshop on Mobile Computing Systems and Applications*, Callicoon, NY, 2002.

[18] Madden, S., et al., "TAG: A Tiny Aggregation Service for Ad Hoc Sensor Networks," *Proc. USENIX Symposium on Operating Systems Design and Implementation*, Boston, MA, 2002.

CHAPTER 12

Enabling Data- and Event-centric Communications

Wendi B. Heinzelman, Amy L. Murphy, and Mark A. Perillo

12.1 Introduction

Sensor networks represent a new way to link the physical world with the virtual world, providing end users with data describing phenomena in the environment in which they are deployed. While sensor networks have the potential to revolutionize the way that we interact with the world, the environments in which they operate are dynamic and complex, making the design of reliable, robust, and long-lived sensor networks difficult. Specifically, sensor networks operate under the following constraints:

- *Inherent distribution.* The sensors are distributed throughout a physical space and are primarily connected wirelessly. This introduces delays, communication costs, and data-throughput limitations based on available bandwidth.
- *Dynamic availability of data sources.* Mobility, addition of new sensors, and loss of existing sensors can cause the set of available sensors and the network topology to change over time. The network must detect and adapt to these changes.
- *Resource limitations.* As many sensor networks use battery-operated sensors and wireless communication, network bandwidth and sensor energy are constrained and must be carefully managed.

From the application's perspective, sensor networks themselves are a resource that must operate at a minimum quality of service (QoS) for an extended time. There may be many ways to achieve the desired QoS (e.g., different sensors may offer data or services that meet the application's requirements). Furthermore, the required QoS or means of meeting this QoS can change over time, based on events detected by the sensors or the introduction of new sensors. At the same time, sensor networks may support collaborative applications, where sensor nodes cooperatively share available network resources (e.g., channel bandwidth) and sacrifice local commodities (e.g., battery energy, memory) for the good of the network. This opens opportunities for cooperation within and across applications for both data and resource usage, enabling a set of applications to operate more energy-efficiently than if they were to manage the sensor and network resources independently.

In order to address these issues, the network must be managed on multiple levels. First, to meet data-fidelity requirements (one component of QoS) at a minimum cost, application queries and data flows must be managed efficiently. In other words, it must be determined which sensors provide data of interest to the application and the properties of their data generation and flows. Second, network-level protocols must be handled properly. Many protocols specifically designed for sensor networks expose certain tunable parameters to allow applications to balance various trade-offs. These features should be exploited for maximal energy efficiency.

However, it is often overly complex for the application to manipulate these network details directly. Instead, mechanisms to provide QoS support and resource management may be elevated above the protocol levels via middleware encapsulations, allowing applications to interact with the sensor network as a whole rather than deal with individual sensors. For example, typical applications may make high-level requests of the sensor network, such as

- Reproduce a data field with a given signal-to-noise ratio.
- Provide k-coverage of a certain region in space (e.g., ensure that every point in the space is within the sensing range of at least k sensors).
- Ensure that data is received within a given delay after being produced by the sensor.
- Provide data from a camera near a specific location.
- Guarantee with a certain probability that an intrusion will be detected.
- Ensure that the network lasts for a given lifetime, while providing maximal coverage during that period.

Different middleware services can be used to ease the application's burden of managing the data flow and system resources for these types of requests. For example, some middleware products provide services such as enabling direct access to data and events of interest, while others provide cross-layer management to interface between the application and the tunable low-layer protocols. The strength of exploiting middleware as part of the application design is the ability of middleware to abstract the sensor-network capabilities in a clean, simple-to-use API, whose functionality is sufficient to meet application requirements. The trade-off in incorporating middleware, however, is that fine-grained control of low-level network details is lost, as this control is handed off to the middleware.

In this chapter, we describe data- and event-centric sensor-network applications and discuss different software architectures for managing data flows, as well as the sensor and network resources, that benefit these applications, using examples from the literature to highlight these different architectures. We review low-layer protocols for resource management and then describe middleware and cross-layer approaches for providing abstractions to allow the application to interact with the network. All of these approaches provide different services to support the data- and event-centric communications that characterize many WSN applications. As an illustration of the benefit of designing applications using these software architectures, we provide details of the implementation of an example application, using one of the cross-layer management schemes. Finally, we provide directions for future research to help ease the burden of programming long-lifetime data- and event-centric applications for WSNs.

12.2 Example Data- and Event-centric Applications

Sensor-network applications represent a new class of applications that are either or both of the following:

- *Data-centric*, meaning that the applications collect and analyze data from the environment, and depending on redundancy, noise, and the properties of the sensors themselves, the data has a certain "value" to the application;
- *Event-centric*, meaning that the applications may be interested in specific events of interest, and the sensor-network goal is to detect and identify such events.

Often for such data- and event-centric applications, system tasks such as routing and storage should be modified to take into account the actual content and structure of the data rather than treating the data as random sequences of bits. Furthermore, the notion of QoS is quite different for data- and event-centric applications, where the actual data rather than the amount of data transported on the network is important. As battery-operated sensors have a limited lifetime during which they provide data to the application, a challenge of the design of sensor networks is to maximize network lifetime (which requires that the sensors and the network be energy efficient), while meeting these unique application QoS requirements, such as data quality (measured in terms of error tolerance, signal-to-noise ratio, resolution, or other application-specific metrics) and latency.

There is often a trade-off among these parameters: energy efficiency may be obtained at the cost of an increase in latency (e.g., sensors operate at a low duty cycle to save energy, but this delays event notification) or a reduction in data quality (e.g., fewer sensors send data to save energy, but this reduces the aggregate data's signal-to-noise ratio). For data- and event-centric applications, the needs of the application (which may vary over time based on the state of the system being monitored) should dictate the appropriate trade-off between energy efficiency, data quality, and latency. Oftentimes, this trade-off can be directly mapped to how the network and the sensors themselves are configured, managed, and operated. However, to reduce the burden of such management on the application, a middleware layer may be used to encapsulate the low-level configuration decisions. To further illustrate this point, we discuss some specific sensor-network applications and how they can benefit from this form of interaction.

12.2.1 Environmental Surveillance

Consider an environment where multiple sensors (e.g., acoustic, seismic, video) are distributed throughout a field, supporting a surveillance application that provides information to an end user about events in the environment [see Figure 12.1(a)]. Suppose that the sensor network consists of sensors with overlapping coverage areas that provide redundant information. To achieve maximum QoS, the

Figure 12.1 Example data- and event-centric applications for WSNs: (a) a video-based surveillance network, and (b) a personal-health-monitoring system.

application may consider it ideal for the network to transport all of the data in real time. However, this will come at the expense of a short network lifetime. Furthermore, this may be infeasible if the bandwidth of the channel is not high enough to support real-time transmission of the data from each sensor. Alternatively, the application may only require that data from active sensors provide a certain spatial resolution and that notification of interesting events reach the end user after some tolerable delay.

Balancing this trade-off between QoS and network lifetime and energy efficiency may require that the application interact with the low levels of the sensor network's protocol stack to manage the sensors over time. Such management can be as simple as turning sensors on and off or as complex as selecting the routes for data to take from each sensor to the collection point in a multihop network and tuning the MAC protocol so that the delay constraints can be met. Furthermore, the needs of the surveillance application may change as a result of changes in phenomena or events detected in the network. For example, if the application detects an intrusion into the area where the network is deployed, the application may require more data to classify the intruder more accurately. The implementation of these tasks can be complex, and they are difficult to incorporate directly into applications. Rather than controlling the sensors directly, the application may simply periodically specify to the middleware in simple terms its requirements, such as the required data resolution and the tolerable delay.

12.2.2 Medical Monitoring

As another example, consider a personal-health-monitoring application running on a PDA that receives and analyzes data from a number of body-worn sensors (e.g., Electrocardiogram (ECG), Electromyogram (EMG), blood pressure, blood flow, pulse oxymeter) and external sensors (e.g., video cameras), as shown in Figure 12.1(b). The monitor reacts to potential health risks and records health information in a local database or alerts the patient with audio and written advice. Considering that most sensors used by the personal health monitor are battery-operated and use wireless communication, it is clear that this application can benefit from intelligent sensor management that provides a way to trade off between energy efficiency, data quality, and latency. Furthermore, the desired trade-off may change over time with changes in the patient's state. For example, higher fidelity might be required for data used to determine certain health-related variables during high-stress situations, such as exercise, and lower fidelity might be required during low-stress activities, such as sleep. Balancing such trade-offs and controlling the sensors and network functionality can be done via low-level protocols or through high-layer abstractions that ease the programming burden of the application by taking over the management task.

12.3 Architectures to Support QoS and Resource Management

One of the distinguishing characteristics of sensor networks is their reliance on nonrenewable batteries, making their simultaneous need to remain active as long as possible and maintain application-specific QoS difficult to achieve. Therefore, using a traditional OSI protocol stack [see Figure 12.2(a)] with standard cumbersome protocols (e.g., TCP/IP) is generally not considered a good approach for sensor networks. To better meet the goals of sensor-network applications, new protocols have been developed that provide the functionality of the Open System Interconnection (OSI) stack layers, sometimes crossing layer boundaries to improve energy efficiency. However, for an application to make the best use of these protocols, it may be required to control the low-level features of the protocols directly, which is not

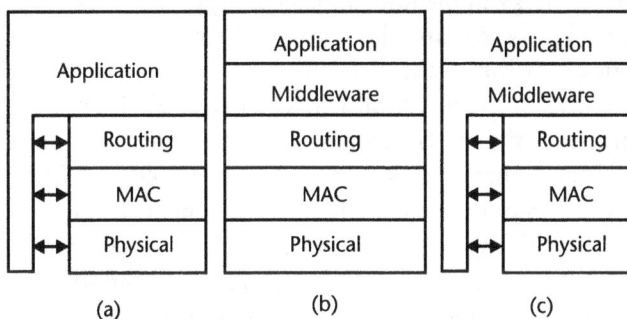

Figure 12.2 Different architectures to provide QoS support to the application while managing sensor and network resources. The arrows represent the flow of control information, while the interfaces between two layers represent the flow of data. (a) An architecture that provides low-layer support for the application. (b) An architecture that provides middleware support for the application. (c) An architecture that provides cross-layer support by managing the functions provided by the low-layer protocols.

desirable for many applications whose focus is on application concerns such as medical evaluation rather than network management.

Middleware has often been useful in traditional networked systems for bridging the gap between the operating system (a low-level component) and the application, easing the development of distributed applications. Because WSNs share many properties with traditional distributed systems, it is natural to consider using middleware abstractions as a means of bridging the gap between the application and the network [see Figure 12.2(b)]. Middleware is used to present a well-defined API that allows the application to specify QoS goals at a high level. The middleware is then responsible for managing the data flows and available resources, while meeting the application's QoS goals and enabling the type of data- and event-centric communications needed by many sensor-network applications.

Rather than designing middleware that accesses only the top level of the protocol stack, an alternative is to expose the deeper layers, allowing the middleware to apply user-defined QoS goals to set protocol parameters [see Figure 12.2(c)]. Using this approach, the middleware can exploit any tunable parameters provided by the network protocols to provide the best service to the application. Each of these approaches can be used to manage data flows to provide the application's required QoS, while simultaneously managing the sensor and network resources to provide long lifetime for data- and event-centric sensor applications. We will describe these different approaches and discuss the advantages and disadvantages of each.

12.3.1 Low-Layer Management

In this section, we describe some representative protocols that highlight the way that the parameters of some low-layer protocols can be configured to manage sensor and network resources, while meeting application QoS.

12.3.1.1 Medium Access Control Protocols

Several MAC protocols aim to reduce energy dissipation at the sensors by setting them into a sleep state as often as possible. As idle power can often be a significant factor in energy dissipation, this approach can greatly extend application lifetime. For example, S-MAC [1] is a CSMA protocol that periodically turns nodes off to avoid wasting energy in the idle mode. This approach trades latency in packet delivery for energy efficiency, and the sleep/awake ratio can be set to make this trade-off most beneficial for the application.

Berkeley-Media Access Control (B-MAC) [2] is another low-power, CSMA MAC protocol designed with many configuration options to trade-off throughput, latency, and energy dissipation. For example, B-MAC utilizes an adaptive thresholding technique for clear channel assessment (CCA) to better detect when the channel is busy given varying noise conditions. Use of CCA can be disabled if desired, for example, when a TDMA approach is found to be more appropriate than a CSMA approach. In addition, the back-off time, which represents how long a node must wait before reassessing the channel after the channel is deemed busy, is configurable. B-MAC also supports the optional use of link-layer acknowledgment messages for guaranteed packet delivery. Finally, B-MAC employs a configurable preamble-sampling technique to reduce idle listening time. The preamble length and interval between consecutive checks of the channel may be adapted depending on the application's desires.

The IEEE 802.15.4 standard [3] enables a similar trade-off between energy consumption and latency by adjusting the superframe parameters Superframe Order (S)) and Beacon Order (BO). As the BO becomes short, nodes waste more energy because they receive the beacon messages every $15.36 * 2^{BO}$ ms; however, as the BO becomes longer and the SO becomes shorter, latencies may be introduced, as data can only be transferred in the first $15.36 * 2^{SO}$ ms of a superframe.

12.3.1.2   Routing Protocols

Some routing protocols are also capable of balancing a trade-off between energy efficiency and latency. In [4] J. Chang and L. Tassiulas propose a minimum-cost forwarding approach using an energy-aware routing cost. The cost of sending a packet over a link is a tunable combination of the energy cost of transmission across that link and the residual and initial energy of the receiving node. By tuning the cost appropriately, either energy efficiency or short paths (i.e., low latency) can be emphasized. Similar trade-offs can be performed in many sensor-network routing protocols, including GEAR [5] and the protocol presented in [6].

12.3.1.3   Cross-Layer Protocols

Some protocols make use of node collaboration to reduce the energy cost of data transfer by aggregating data locally rather than sending all raw data to the application. For example, in LEACH [7], nodes form local clusters, and all data within a cluster are aggregated by the cluster head node before being transmitted to the base station. Providing means for in-network data aggregation is very helpful for many applications, but the application must specify its requirements for the data aggregation, including the type of data aggregation required (e.g., beamforming, averaging) and the maximum allowable distance between sources for the data to be aggregated.

The Bluetooth [8] protocol includes several low-power modes, which other nodes can utilize in order to save energy. In Bluetooth, a master node polls slaves for data, and either the master or the slaves can put the slaves into one of the low-power modes to save energy. How often this is done and for how long is again a trade-off between latency in data delivery and energy efficiency that should be set to meet the needs of the application. Bluetooth also allows the topology of the network to be chosen by the user. Certain roles (masters, bridges) are more power intensive than others, and nodes with the most energy resources should be chosen most frequently to serve in these roles.

12.3.2   Middleware Management

While the low-layer protocols discussed in Section 12.3.1 provide means to manage the limited sensor-network resources, oftentimes they cannot be effectively managed by the application. For example, suppose that an application wants to increase the resolution of data being collected. There are many ways to achieve this goal (e.g., perform less in-network aggregation, reduce sleep cycles for the active nodes and require them to transmit more data). As another example, suppose that the application is interested in only a certain type of data, such as any data that indicates an intrusion into the region of interest. Using low-layer management of the network, the application cannot easily manage its resources to ensure that only data of interest is transmitted.

For complex data- and event-centric applications, it is unreasonable to expect applications to manage the data flow and network resources themselves. Middleware that provides a well-defined API can be used to provide services to manage the data flow and network resources on behalf of the application. In this section, we discuss several different types of services that can be provided by middleware to manage the data flows and network resources, and we will see how each of these services eases the design of data- and event-centric applications.

12.3.2.1 Exposing Events

Many sensor-network applications can be designed around an application paradigm that comes from the realization that not all data from a sensor network need be collected and analyzed; instead, specific event values can be distinguished as important. When these values are sensed, the application is notified, and appropriate actions are taken. In this event-driven programming paradigm, depicted in Figure 12.3, the network is required only to describe the events of interest to the application; how these events are identified and delivered remains a detail of the implementation.

In Directed Diffusion [9], nodes collaborate to set up routes as *interests*, or queries for particular types of data or events (e.g., temperature readings over a given threshold), are disseminated through the network. Here, all data items generated by sensors are tagged with attribute-value pairs. Thus, an application's "interest query" needs only to specify the attributes of interest. These queries are diffused throughout the network and gradients are set up to allow data that matches the interest specified in the query to be sent back to the application. This allows the application to interact with the sensor network at a much higher level than specifying individual sensors from which it would like to receive data. Thus, Directed Diffusion effectively manages the data resources of the network on behalf of the application.

Adaptive publish-subscribe similarly supports event-driven applications. However, it assumes a more stable infrastructure over which routing optimizations can

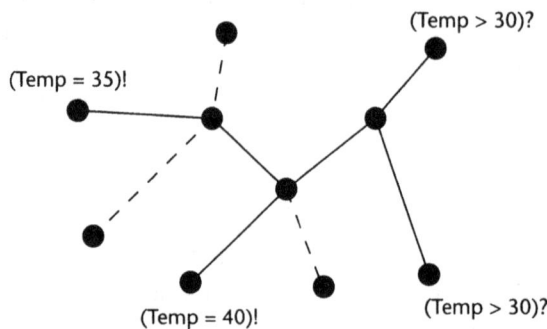

Figure 12.3 Event-centered systems allow subscribers to request (? in the figure) and publishers to announce (! in the figure) events and provide efficient routing of events to subscribers. In the figure, a tree-based scheme is shown with solid lines indicating the path of the events to the subscribers.

be performed. For example, in subscription forwarding, a single tree is formed to connect the event publishers (the sensors) with the subscribers (the applications). An application registers to receive events that match a specific pattern (e.g., temperature greater than 50°) by sending a subscription along all edges of the tree. The effect of this message is to establish routes for events to follow toward the application. Adaptation can be introduced [10], providing the ability to modify the overlay tree in order to optimize communication at any time. In practice, this publish-subscribe approach is most useful when multiple applications are exploiting the same sensors, and benefit is gained by sharing the expense of event propagation (an advantage similar to using multicast instead of multiple unicasts).

Another approach to supporting event-centric communication is employed by EnviroTrack [11], a middleware for environmental-tracking applications. EnviroTrack abstracts the low-level collaboration of sensors around a certain event of interest by allowing the application to view the event as a logical object. When an event is detected, sensors surrounding the event form a cluster, to which a context label is associated. A tracking object is attached to each cluster, and the cluster performs local operations on the event of interest. Using EnviroTrack, the burden on the application is eased, as for each context label (i.e., event to be sensed), the application only needs to provide the conditions for an event to be sensed, a definition of the aggregate state, and any object code to be attached with the context label.

At a higher level, a sensor-network middleware in development [12] allows applications to express the mapping between application quality and data quality. This middleware translates these data-quality values into thresholds on sensor readings that specify when sensors should transition from a monitoring state into an active state (i.e., specifying the parameters that signify events of interest to the application, for which the application would like to receive data). Additionally, it is assumed that sensors can transition themselves to a low-power-consumption state when sensed data follows predefined prediction models to within certain error bounds.

12.3.2.2   Supporting Direct Data Access

Another approach to interacting with a sensor network is to access the data according to a traditional database model (see Figure 12.4). The main challenges, however, are that both the organization of the data and the type of query are different than in traditional databases. From the organizational perspective, traditional distributed databases have few data locations, each with either the entire dataset replicated or with a large portion of a nonreplicated data store. In sensor networks, each individual sensor is a source for its own data only, yielding a large number of data sources, each providing minimal information. Further, queries for sensor networks tend to be long-lived (e.g., For the time interval dawn to dusk, return the temperature of an area every 10 minutes). In contrast, traditional queries are most often performed to retrieve current information a single time. Many approaches have been taken to adapt sensor-network queries to fit this paradigm.

TinyDB [13], described in Chapter 14, is a processing engine that provides a generic, easy-to-use interface to the network through an enhanced Structured Query Language (SQL)–like API. Enhancements to the SQL interface include hooks that allow a certain required lifetime to be specified. TinyDB can subsequently adapt

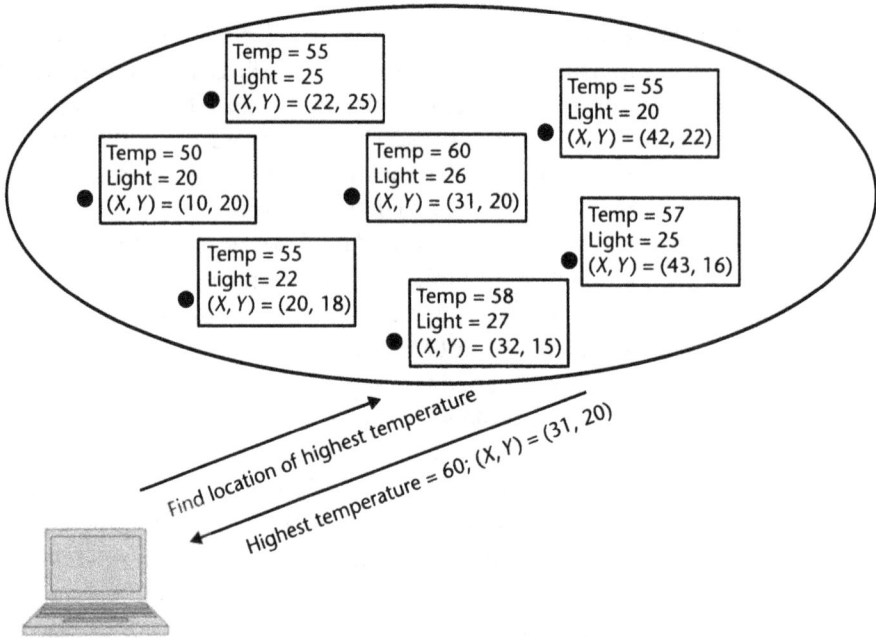

Figure 12.4   Database model for querying a sensor network.

sensor settings such as the sensing rate so that this lifetime is met. Several features also allow queries to be optimized at several levels within the network. *Storage points* containing windows of sensor data can be created so that queries over the data streams can be executed more easily. Such storage points may be useful, for example, in sliding-window-type queries (e.g., Find the average temperature in a room over the previous hour once per minute). Queries can also be performed upon the occurrence of specific events. Optimization of the scheduling of sensing tasks is performed for complex queries. This optimization considers sensing costs and the expected selectivity of the query such that the expected power consumption during a query is minimized. The network topology is optimized by considering the query in the formation of aggregating trees. Rather than requiring sensors to choose an upstream parent node solely based on link quality, the predicates of the query (i.e., the conditions that should be met for inclusion in the query) are also considered.

Both the Cougar [14] and Sensor Information Networking Architecture (SINA) [15] systems provide a distributed database-query interface with the information from a sensor network that emphasizes power management. Cougar proactively manages power resources by distributing the query among the sensor nodes to minimize the energy consumed to collect the data and calculate the query result. SINA, on the other hand, incorporates low-level mechanisms for hierarchical clustering of sensors for efficient data aggregation, as well as mechanisms that limit the retransmission of similar information from geographically proximate sensor nodes.

To address the redundant nature of sensor networks, DSWare [16] supports aggregation of information collected by geographically proximate sensors. This allows applications to interact with many individual sensors as a single entity, simplifying programming and yielding more accurate readings as a result of redundancy and aggregation. An added advantage is the ability to mask the failure or loss of one or more sensors in a region as long as enough sensors remain active to provide a valid measurement.

Quality-Aware Sensing Architecture (QUASAR) [17] provides the application with a means to query a sensor database with a quality-aware query (QaQ). QaQs can express quality requirements as either set based (e.g., Find at least 90% of the sensors with temperature greater than 50°) or value based (e.g., Estimate the average temperature within 1°). After the queries are expressed by the application, they are executed to meet the quality requirements at minimum cost.

12.3.2.3 Controlling Spatial Resolution

In contrast to event subscription and notification or direct querying for data of interest, another approach is for applications to express their requirements explicitly in terms of sensor-network resolution. Applications that fall into this category are those that wish to replicate a data image of some environmental phenomenon and those that require a certain level of coverage of the environment.

Probing environment and adaptive sleeping protocol (PEAS) [18] and coverage configuration protocol (CCP) [19] are coverage-preserving protocols that aim to turn nodes off as much as possible, while ensuring that coverage is maintained over the entire sensing field. These approaches assume a given sensing range, within which a sensor is capable of providing information about events of interest. A sensor's coverage area may be expressed as a probability of detection, where the probability of detection of events decreases the further the event is from the sensor. For a given required detection probability P_d, the sensing range can be found by determining the maximum distance from the sensor at which events are detected with probability greater than P_d and essentially represents the desired spatial resolution of the active sensors. Using other protocols such as those presented in [20, 21], the required spatial resolution can be expressed by the application more directly.

By specifying the desired spatial resolution, the application can effectively specify its QoS requirements. However, as stated previously, applications' data-quality or -latency requirements may change drastically as a result of events in the environment. In such cases, it might be appropriate to utilize a cross-layer approach, such as those described in Section 12.3.3 to change the specification of P_d or the coverage requirements (e.g., switching from 1-coverage to 4-coverage), using these protocols for spatial-resolution management.

12.3.3 Cross-Layer Management

The previously mentioned middleware approaches were designed to provide different services for the application, largely ignoring the services already provided by the low-layer protocols for sensor networks described in Section 12.3.1. At the same time, as discussed previously, the low-layer protocols often cannot be effectively managed directly by the applications. Thus, middleware that provides cross-layer management by converting high-level application QoS requirements, specified through a well-defined API, to parameters for the low-layer protocols is needed. This approach allows the middleware to adaptively set the actual parameters of the network over time to manage the network resources efficiently, while simultaneously meeting the requirements of the application.

One such middleware product that provides cross-layer management is Middleware Linking Applications and Networks (MiLAN) [22]. MiLAN receives a description of QoS requirements from the application, monitors the current network conditions, and optimizes sensor and network configurations by adjusting the parameters of the low-layer protocols to maximize application lifetime. To perform these functions, applications represent their requirements to MiLAN through specialized graphs that incorporate changes in application needs due to event detection. Based on this information, MiLAN makes decisions about how to control the network as well as the sensors themselves to balance application QoS and energy efficiency.

To enable MiLAN to configure any low-layer network protocol, MiLAN has an architecture that extends into the network protocol stack, as shown in Figure 12.2(c). As MiLAN is intended to sit on top of multiple physical networks, an abstraction layer is provided that allows network-specific plug-ins to convert MiLAN commands to protocol-specific commands. Therefore, MiLAN can continuously adapt to the specific features of whichever protocols are being used for communication (e.g., determining sleep/awake cycles in S-MAC [1], reporting rate specified in the interests of Diffusion [9], sensing range in PEAS [18]) in order to best manage the resources, while meeting the application's QoS requirements over time.

While MiLAN abstracts sensor-network QoS goals and events at a high-level, simplifying network design, abstraction at a lower levels for managing a sensor node's neighborhood may be more appropriate in some networks, as neighborhood-based algorithms are typically simpler to design. Several programming primitives, such as Hood [23] and Abstract Regions [24] (described in Chapter 14), allow this level of abstraction and provide the means necessary for establishing neighborhood membership, sharing data, caching data, and messaging.

12.4   Example Application Design Using a Cross-Layer Architecture

To further illustrate how data- and event-centric applications can be more easily designed using the software architectures described in this chapter, here we provide details of the implementation of an example application (the health monitor described in Section 12.2.2) using one of the cross-layer management schemes, MiLAN [22].

For data- and event-centric applications, application performance can be characterized by the QoS of different variables of interest to the application, where the QoS of the different variables depends on which sensors provide data to the application. For example, in the personal health monitor, variables such as blood pressure, respiratory rate, and heart rate may be determined based on measurements obtained from any of several sensors [25]. Each sensor has a certain QoS in characterizing each of the application's variables. For example, a blood-pressure sensor directly measures blood pressure, so it provides a quality of 1.0^1 in determining this variable. In addition, the blood-pressure sensor can indirectly measure other variables, such as heart rate, so it provides some quality, although less than 1.0, in determining these variables. The quality of the heart-rate measurement would be improved through high-level fusion of the blood-pressure measurements with data from additional sensors, such as a blood-flow sensor.

In order to determine how to best serve the application, MiLAN must know (1) the variables of interest to the application, (2) the required QoS for each variable, and (3) the level of QoS that data from each sensor or set of sensors can provide for each variable. Note that all of these may change based on the application's current state. During initialization of the application, this information is conveyed from the application to MiLAN via state-based-variable-requirements and sensor-QoS graphs. Examples of these graphs are shown in Figures 12.5 and 12.6, respectively. Figure 12.5(a), an abstract state-based-variable-requirements graph, shows the required QoS for each variable of interest based on the current state of the system and the variables of interest to the application where these states are based on the application's analysis of previously received data. Figure 12.5(b) shows the state-based-variable-requirements graph for the personal health monitor. This application has a system state that includes the patient's overall stress level and multiple states for each variable that can be monitored. The state-based-variable-requirements graph specifies to MiLAN the application's minimum acceptable QoS for each variable (e.g., blood pressure, respiratory rate) based on the current state of the patient. For example, the figure shows that when a patient is in a medium-stress state and the blood pressure is low, the blood-oxygen level must be monitored with a quality level of 0.7, and the blood pressure must be monitored with a quality level of 0.8.

For a given application, the QoS for each variable can be satisfied using data from one or more sensors. The application specifies this information to MiLAN through the sensor-QoS graph [see Figure 12.6(a)]. Figure 12.6(b) shows the sensor-QoS graph for the personal health monitor. This graph illustrates the important variables to monitor when determining a patient's condition and indicates the sensors that can provide at least some quality to the measurement of these variables. Each line between a sensor (or virtual sensor) and a variable is labeled with the quality that the sensor (or virtual sensor) can provide to the measurement of that variable. For example, using data from a blood-pressure sensor, the heart rate can be determined with a 0.7 quality level, but combining this with data from a blood-flow sensor increases the quality level to 1.0.

Given the information from these graphs, as well as the current application state, MiLAN can determine which sets of sensors satisfy all of the application's QoS requirements for each variable. These sets of sensors define the application feasible set $\mathcal{F}_A$, where each element in $\mathcal{F}_A$ is a set of sensors that provides QoS greater than or equal to the application-specified minimum-acceptable QoS for each specified variable. For example, in the personal health monitor, for a patient in medium stress with a high heart rate, normal respiratory rate, and low blood pressure, the application feasible sets in $\mathcal{F}_A$ that MiLAN should choose to meet the specified application QoS are shown in Table 12.1. MiLAN must choose which element of $\mathcal{F}_A$ should be provided to the application. This decision depends on network-level information.

The properties of specific network types, as well as the current condition of the network, can constrain the set of feasible sets to a subset of those in $\mathcal{F}_A$. It is the network plug-in's job to determine which sets of nodes (sensors) can be supported by the network, as well as other protocol-specific information, such as what role each node must play. The power cost of using a node is a combination of the power to run the device, the power to transmit its data, the power to forward the data of other nodes in the set,

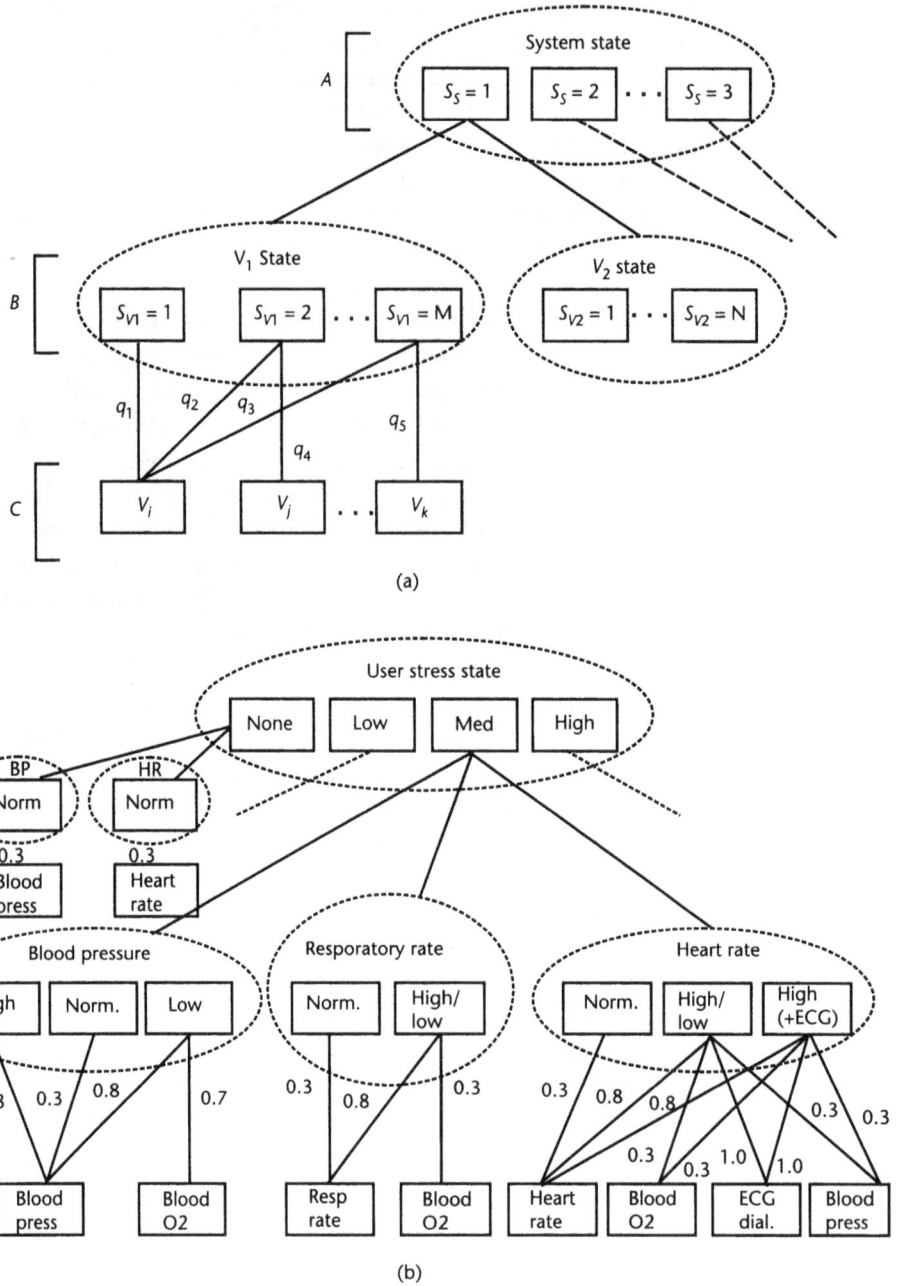

Figure 12.5 State-based-variable-requirements graph of MiLAN for specifying the variables and the required QoS when the application is in various states: (a) abstract example, and (b) example for the personal-health-monitoring application. This graph illustrates only a subset of the application's possible states. (*Source:* © 2004 IEEE.)

and the overhead of maintaining its role in the network. These costs can be influenced by MiLAN through setting the roles and protocol parameters appropriately. For example, MiLAN should determine roles like master or bridge node in Bluetooth scatternets

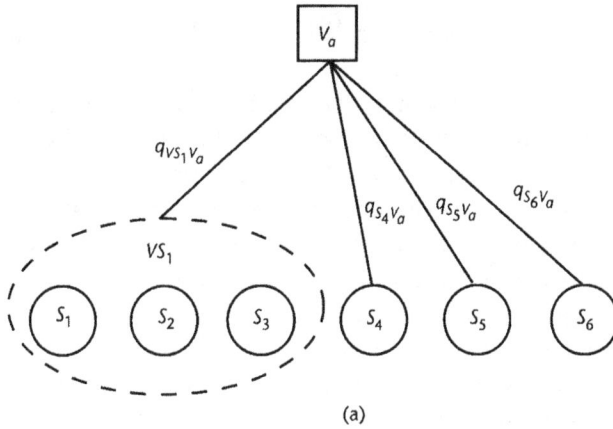

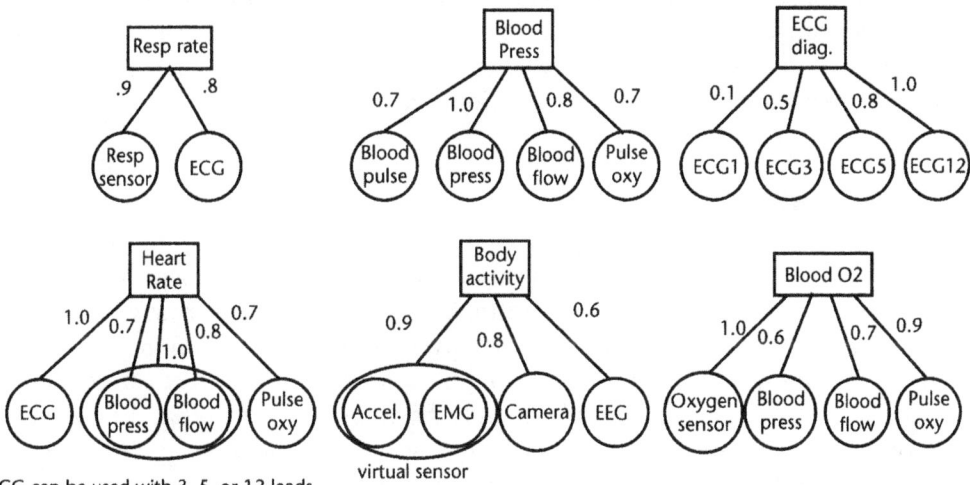

Figure 12.6   Sensor-QoS graph of MiLAN for specifying which sensors, or sets of sensors, can provide what level of QoS for each variable: (a) abstract example, and (b) example for the personal-health-monitoring application. This graph illustrates only a subset of the variables that should be considered by the application. (*Source:* © 2004 IEEE).

Table 12.1   Feasible Sets $\mathcal{F}_A$ Calculated within MiLAN for the Personal Health-Monitoring Application for a Patient in Medium Stress with High Heart Rate, Normal Respiratory Rate, and Low Blood Pressure

Set #	Sensors
1	Flood flow, resp. rate
2	Blood flow, ECG (3 leads)
3	Pulse oxymeter, blood pressure, ECG (1 lead), resp. rate
4	Pulse oxymeter, blood pressure, ECG (3 leads)
5	Oxygen measurement, blood pressure, ECG (1 lead), resp. rate
6	Oxygen measurement, blood pressure, ECG (3 leads)

and aggregation type in LEACH. Furthermore, MiLAN should set the parameters of the particular protocols, such as the sleep-awake ratio in S-MAC, back-off and contention timers in B-MAC, SO and BO in 802.15.4, and route-cost tuning to empha-size either energy efficiency or short paths.

MiLAN must choose an element of $\mathcal{F}_A$ that can be supported by the network and represents the best performance-cost trade-off. The best trade-off depends on the application—the MiLAN framework supports any method of deciding how to select a feasible set. In many sensor-network applications, we want to allow the application to last as long as possible using the limited energy of each of the sensors. Thus, the goal may be to schedule feasible sets so as to maximize network lifetime. For other applications, the goal may be to maximize some quality for a given lifetime. MiLAN is flexible enough to incorporate any optimization criteria.

Once a feasible set and all the protocol parameters have been selected, the sensors are configured appropriately and start sending data to the application. The network operates in this state until there is some change, such as a change in application state, a change in the environment being monitored, or a change in the network resources (e.g., a sensor dies), at which point MiLAN selects new sensors and protocol configurations to meet the application's goals.

12.5 Discussion

In this chapter, we have described a variety of approaches to manage the complexity of designing data- and event-centric applications for the sensor-network environment. The choice to adopt one approach over another depends on the properties of the application and which aspects of the sensor-network architecture must be controlled. An additional consideration is the amount of effort that the application programmer can invest into designing management of the sensor network directly into the application. For simple applications, effort can be invested in explicitly tailoring low-level protocols to extract the maximum benefit from the sensor network. If, instead, the application itself is complex, a designer may choose to delegate the details of sensor-network management to middleware and, instead, use the abstraction it provides. The network management provided by the middleware may consist of query handling and data-flow setup and even extend to low-level protocol handling, offering the best of both worlds: high-level programming abstraction and low-level control. Summaries of the features of the tunable low-layer protocols and the middleware solutions described in this chapter are given in Tables 12.2 and 12.3.

As sensor networks continue to grow as a viable technology, the demands on approaches to support application development will continue to grow. To give an idea of future directions, we outline some of the ongoing areas of research that are necessary for easing the programming of sensor-network applications.

One area for improvement is to continue to narrow the performance gap created by the introduction of abstractions to ease programming. In standard middleware, this approach requires applications to provide additional information; however, this usually means breaking the abstraction in order to control low-level features. Techniques must be designed that allow the natural expression of quality needs for sensor networks as well as how these needs change over time as events occur and new data is collected.

Table 12.2 Summary of Low-Layer Protocols for WSNs That Can Be Tuned by the Application for Data- and Event-centric Communication

Protocol	Tunable Parameters
S-MAC [1]	Sleep/awake cycle time
B-MAC [2]	CCA use, backoff timers, use of ACKs, preamble settings
IEEE 802.15.4 [3]	Superframe parameters SO and BO
Chang and Tassiulas [4]	Latency-energy routing-cost parameter
GEAR [5]	Latency-energy routing-cost parameter
Shah et al. [6]	Latency-energy routing-cost parameter
LEACH [7]	Aggregation type, amount
Bluetooth [8]	Power-save mode intervals, topology

Table 12.3 Summary of Middleware Solutions for Easing Data- and Event-centric Communication for WSN Applications

Middleware	Services
Directed Diffusion [9]	Query handling, data aggregation
Adaptive publish-subscribe [10]	Event registration, multiple application support
EnviroTrack [11]	Event abstraction
Yu et al. [12]	QoS abstraction
TinyDB [13]	SQL-based query handling
Cougar [14]	Query handling, power management, data aggregation
SINA [15]	Query handling, power management, data aggregation
DSWare [16]	Event registration, data aggregation
QUASAR [17]	Query handling, power management
PEAS [18]	Coverage preservation
CCP [19]	Coverage preservation
MiLAN [22]	QoS abstraction, support for existing network protocols and middleware services
Hood [23]	Neighborhood formation
Abstract Regions [24]	Neighborhood formation

Another issue is flexibility. We have outlined several systems that each address the different issues of sensor-network applications. However, many applications face multiple issues that are not supported within a single infrastructure. One challenge is to develop systems that incorporate all of the needs of sensor-network applications without requiring bloated pieces of software that are no longer suitable for the sensor-network environment.

A related issue is the ability of a single sensor-network to support multiple applications simultaneously. Many sensor-network systems use a single base station to collect data and optimize for a single application. However, there is no inherent reason that sensor networks cannot be shared among multiple cooperative applications. This involves basic research on two fronts. First, traditional sensor-network approaches must be modified to allow multiple points of data collection. Second, cooperative applications must be built to express their needs yet yield to one another when appropriate. While some techniques for access control exist in traditional distributed networks, they are not directly applicable to the complex environments of sensor networks.

One key difference between the approaches to easing data- and event-centric communication described in this chapter and traditional approaches to distributed system design is the reliance on a centralized component for optimization and processing. Traditional distributed computing is steadily moving away from all centralization to eliminate communication bottlenecks and single points of failure. On the other hand, sensor networks are inherently composed of heterogeneous components, some of which are clearly more powerful than others in terms of energy reserve, computational capabilities, and even storage space. These more powerful nodes can legitimately act as control points in the sensor-network environment. Nonetheless, sensor-network middleware should not rely exclusively on powerful nodes to perform all computation. Algorithms and techniques must be designed to distribute the computation of the sensor-network operations, as well as the middleware control operations, to the nodes of the network according to each node's capabilities. This requires that the distributed software be explicitly built to have a small footprint in order to execute on simple devices, and it requires using only limited communication in order to limit energy consumption.

Finally, as with all distributed systems, robustness and security must be addressed in sensor networks. To some degree, robustness to sensor loss is already supported by most of the systems presented here. However, few systems characterize exactly how much loss can be tolerated, a requirement for critical software systems. Further, security remains an issue from the perspective of tampering with the physical devices as well as the transfer of information. As sensor networks are used for various home- and environmental-monitoring applications, the task of ensuring that the information is restricted to the intended recipient(s) and that the information itself can be trusted becomes increasingly important.

References

[1]   Wei, Y., Heidemann J., and Estrin D., "An Energy-Efficient MAC Protocol for Wireless Sensor Networks," *Proc. 21st International Annual Joint Conference of the IEEE Computer and Communications Societies (INFOCOM)*, New York, NY, 2002.

[2]   Polastre, J., Hill J., and Culler D., "Versatile Low Power Media Access for Wireless Sensor Networks," *Proc. Second ACM Conference on Embedded Networked Sensor Systems (SenSys)*, Baltimore, MD, 2004.

[3]   Gutierrez, J. A., Callaway E. H., and Barrett R., *Low-Rate Wireless Personal Area Networks: Enabling Wireless Sensors with IEEE 802.15.4*, IEEE Press, New York, 2003.

[4]   Chang, J., and Tassiulas L., "Energy Conserving Routing in Wireless Ad Hoc Networks," *Proc. 19th International Annual Joint Conference of the IEEE Computer and Communications Societies (INFOCOM)*, Tel Aviv, Israel, 2000.

[5]   Yu, Y., Govindan R., and Estrin D., "Geographical and Energy Aware Routing: A Recursive Data Dissemination Protocol for Wireless Sensor Networks," Technical Report UCLA/CSD-TR-01-0023, UCLA Computer Science Department, May 2001.

[6]   Shah, R., and Rabaey J., "Energy Aware Routing for Low Energy Ad Hoc Sensor Networks," *Proc. IEEE Wireless Communications and Networking Conference (WCNC)*, Orlando, FL, 2002.

[7]   Heinzelman, W., Chandrakasan A., and Balakrishnan H., "Energy-Efficient Communication Protocol for Wireless Microsensor Networks," *IEEE Transactions on Wireless Communication*, Vol. 1, No. 4, 2002, pp. 660–670.

[8]   Bluetooth protocol, at http://www.bluetooth.org.

[9]   Intanagonwiwat, C., Govindan R., and Estrin D., "Directed Diffusion: A Scalable and Robust Communication Paradigm for Sensor Networks," *Proc. 6th Annual International Conference on Mobile Computing and Networks (MobiCom)*, Boston, MA, 2000.

[10] Picco, G. P., Cugola G., and Murphy A. L., "Efficient Content-based Event Dispatching in Presence of Topological Reconfiguration," *Proc. 23rd International Conference on Distributed Computing Systems*, Providence, RI, May 2003, pp. 234–243.

[11] Abdelzaher, T., et al., "EnviroTrack: Towards an Environmental Computing Paradigm for Distributed Sensor Networks," *Proc. 24th International Conference on Distributed Computing Systems*, Tokyo, Japan, 2004.

[12] Yu, X., et al., "Adaptive Middleware for Distributed Sensor Environments," *Distributed Systems Online*, May 2003, Vol. 4, No. 5.

[13] Madden, S. R., et al., "The Design of an Acquisitional Query Processor for Sensor Networks," *Proc. 2003 ACM SIGMOD International Conference on Management of Data*, San Diego, CA, 2003.

[14] Bonnet, P., Gehrke J., and Seshadri P., "Querying the Physical World," *IEEE Personal Communication*, Vol. 7, No. 5, 2000, pp. 10–15.

[15] Shen, C.C., Srisathapornphat C., and Jaikaeo C., "Sensor Information Networking Architecture and Applications," *IEEE Personal Communication*, Vol. 8, No. 4, 2001, pp. 52–59.

[16] Li, S., Son S., and Stankovic J., "Event Detection Services Using Data Service Middleware in Distributed Sensor Networks," *Proc. 2nd International Workshop on Information Processing in Sensor Networks*, Palo Alto, CA, April 2003.

[17] Lazaridis, I., et al., "QUASAR: Quality-Aware Sensing Architecture," *SIGMOD Record*, Vol. 33, No. 1, 2004, pp. 26–31.

[18] Ye, F., et al., "PEAS: A Robust Energy Conserving Protocol for Long-lived Sensor Networks," *Proc. 23rd International Conference on Distributed Computing Systems*, Providence, RI, 2003.

[19] Wang, X., et al., "Integrated Coverage and Connectivity Configuration in Wireless Sensor Networks," *Proc. Second ACM Conference on Embedded Networked Sensor Systems (SenSys)*, Los Angeles, CA, 2003.

[20] Deb, B., Bhatnagar S., and Nath B., "Multi-resolution State Retrieval in Sensor Networks," *Proc. 1st IEEE Workshop on Sensor Network Protocols and Applications (SNPA)*, Anchorage, AK, 2003.

[21] Perillo, M., Ignjatovic Z., and Heinzelman W., "An Energy Conservation Method for Wireless Sensor Networks Employing a Blue Noise Spatial Sampling Technique," *Proc. 3rd International Symposium on Information Processing in Sensor Networks (IPSN)*, Berkeley, CA, 2004.

[22] Heinzelman, W., et al., "Middleware to Support Sensor Network Applications," *IEEE Network Magazine*, Special Issue, Vol. 18, No. 1, 2004, pp. 6–14.

[23] Whitehouse, K., et al., "Hood: A Neighborhood Abstraction for Sensor Networks," *Proc. ACM International Conference on Mobile Systems, Applications, and Services (MobiSYS)*, Boston, MA, 2004.

[24] Welsh, M., and Mainland G., "Programming Sensor Networks Using Abstract Regions," *Proc. 1st USENIX/ACM Symposium on Networked Systems Design and Implementation (NSDI)*, San Francisco, CA, 2004.

[25] Conway, J. C. D., et al., "Wearable Computer as a Multi-parametric Monitor for Physiological Signals," *Proc. IEEE International Symposium on Bioinformatics and Bioengineering (BIBE)*, Washington, D.C., 2000, pp. 236–242.

Storage Issues in Sensor Networks

Deepak Ganesan and Ben Greenstein

13.1 Introduction

One of the key challenges in WSNs is the storage and querying of useful sensor data, broadly referred to as *data management*. The term *useful sensor data* is application-specific and has different meanings in different application scenarios. For instance, in a target-tracking application, users are interested in detecting vehicles and tracking their movement. Here, useful sensor data comprises target detections (timestamp and location) as well as their tracks. In a structure-monitoring application like [1], scientists are interested in spatiotemporal analysis of sensor data, such as vibrations measured at different points of a building in response to an excitation over a period of time. This requires access to vibration-event data corresponding to periods when the building is excited. Data management captures the broad spectrum of how useful data is handled in a sensor network and addresses three key questions:

- Where is data stored in a network? Is it stored locally at each sensor node (local storage), in a distributed manner within the network (distributed storage), or at the edge of the network at the base station (centralized storage)?
- How are queries routed to stored data? Can we use attributes of the search to make it efficient?
- How can the network deal with the storage limitations of individual sensor nodes?

Many schemes have been proposed for data management in sensor networks. The main ideas are summarized in Figure 13.1 along the axes of communication required for data storage and communication required for query processing. The figure depicts a fundamental trade-off between centralized storage and local storage. At one extreme is a conventional approach of centralized data management where all useful sensor data is transmitted from sensors to a central repository that has ample power and storage resources. This task is communication intensive but facilitates querying. Queries over this data are handled at the central location and incur no energy cost to the sensor network. Centralized storage is appropriate for low-data-rate, small-scale sensor networks, where there are infrequent events. As

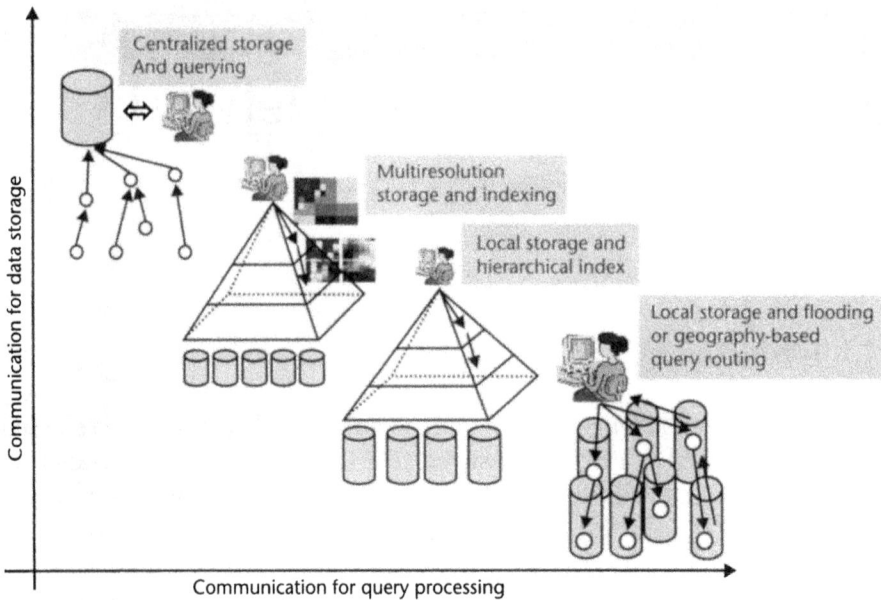

Figure 13.1 Taxonomy of data-storage solutions.

the useful-data rate increases and the network scale grows, centralized storage is not feasible in sensor networks due the power constraints on sensor nodes. The aggregate costs of transmitting the data to a base station can quickly drain power on all nodes in the network. Nodes that are closer to the base station are especially impacted due the need to relay data from many other nodes in the network. Thus, dealing with power constraints requires alternate approaches that involve storing sensor data within the sensor network and performing query processing on this distributed data store.

Many parameters are at work in determining which data-management scheme best suits an application. Is the deployment short term or long term? What are the energy, storage, and processing resources on sensor nodes? What kinds of queries are posed on the data and with what frequency? How much useful sensor data does the network generate? What in-network data processing is appropriate for the sensor data? Dealing with the wide range of requirements and constraints in sensor networks requires a family of solutions. These solutions are different from the distributed storage systems used in wide-area networks in two respects. First, energy and storage limitations introduce more stringent communication constraints. Second, data in sensor networks exhibits spatiotemporal correlations, which can be exploited in the storage, processing, and querying of data. This chapter will first survey various storage and indexing solutions that have been proposed for data management in sensor networks, focusing on one such solution, DIMENSIONS, a multiresolution in-network storage and search system where data is stored at different spatiotemporal resolutions within the network, and queries are processed on these views. The rest of this chapter is structured as follows. In Section 13.2, we describe the key building blocks for a distributed storage and search system. Then, we provide an overview of the salient features of different kinds of data-management schemes shown in Figure 13.1. In Section 13.4, we provide an overview of one instance of a distributed storage system, DIMENSIONS, and usage models. we conclude in Section 13.5.

13.2   Key Systems Building Blocks

There are three key systems building blocks that in-network storage and search systems are based on: geographic routing, geographic hashing for rendezvous, and hierarchical multiscale processing.

13.2.1   Geographic Routing

A key component of in-network storage and querying is the ability to route queries to nodes that have appropriate data. An important attribute in categorizing sensor data is the notion of location or geography. Most applications that involve sensing care about where an event or piece of data of interest was sensed. Therefore, many queries can be expected to be associated with a geographic attribute. An instance of such a query in the target-tracking scenario is, How many vehicles passed through a region R in the last day? For energy efficiency, such geographically scoped queries should be directed solely to the region of interest rather than the entire network. Geographic routing schemes provide a framework to enable such scoped querying. GPSR [2] is a popular geographic forwarding scheme proposed for sensor networks and offers many useful features. GPSR uses greedy forwarding when possible (i.e., a node forwards a packet to the closest of its all neighbors to the destination). Greedy forwarding fails when a node is closer than any of its neighbors to the destination. This case results from there being a void (an area with no nodes) between a node and the destination. To route around such voids, GPSR constructs a planar subgraph of the network. It then forwards the packet around the perimeter of the void on the edges of the planar subgraph.

13.2.2   Geographic Hashing for Rendezvous

Geographic routing can be used to locate data when the query explicitly specifies where the data is located. However, queries are often unaware of this information; instead, they seek to find where an event occurred. For instance, a typical target-tracking query is, *Report all trucks that were detected in the last hour*. Geographic queries cannot deal with such queries. A geographic hash table (GHT) [3] proposes an approach to create in-network rendezvous points where queries and data meet. In geographic hashing, data names (e.g., target of type TRUCK) are hashed to rendezvous points (locations) in the geographic space occupied by the sensor network. All event detections that match the type and all queries that request data about a particular type of event are routed to this rendezvous location; hence, search complexity is reduced. Data is then stored at the node closest to the location obtained by hashing its associated name. This is achieved by using a clever modification of GPSR to reach the node that is closest to a target location. Structured replication [3] uses a hierarchical decomposition of the geographic space into nested grids that is useful for aggregation and lends locality in storage, as shown in Figure 13.2. Such a feature is extremely useful in an in-network rendezvous scheme, where queries and data meet at a certain location in the network.

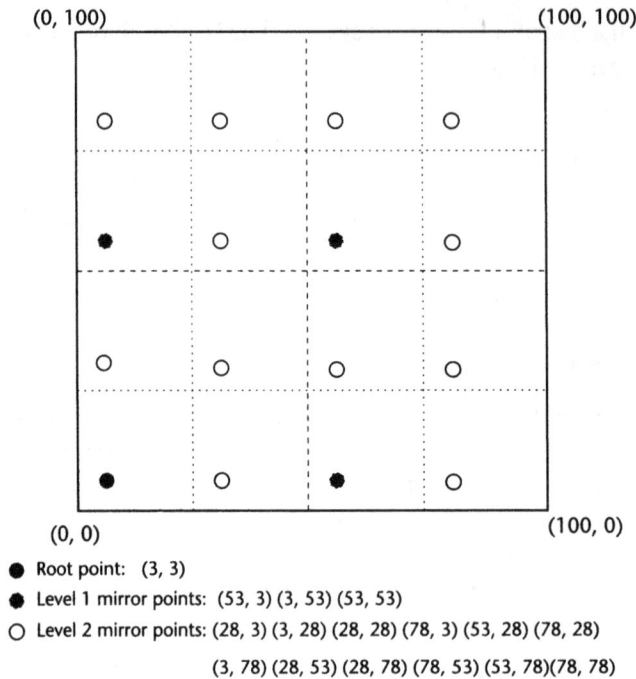

Figure 13.2 Structured Replication.

13.2.3 Hierarchical Indexing and Multiresolution Processing

Both geographic routing and hash tables provide distributed routing infrastructure over which index structures can be built. The third core building block concerns the different kinds of indexes and multiresolution processing approaches that can be built over such a hierarchical framework. Geographic hash tables provide a geography-based hierarchical structure that can be used to construct different kinds of index structures. For instance, hierarchical search data structures such as quad trees [4] and KD trees [5] can be constructed to enable a search based on specific attributes. An instance of such a hierarchical index for the Distributed Index of Features in Sensornets (DIFS) is shown in Figure 13.3. In addition to indexes, multiresolution views can be stored on a distributed hierarchy. Multiresolution views of data exploit inherent spatiotemporal correlations in the data to compress them and store summaries of the data rather than all of the raw data. DIMENSIONS [4] proposes such an approach.

Having described the building blocks for in-network storage and search, we briefly review how local storage and distributed indexing schemes can be composed from these components.

13.3   Taxonomy of Data Storage and Indexing Solutions

Figure 13.1 shows three solutions for distributed storage and search that address limitations of the centralized approach. This section addresses under which circumstances each of these solutions is applicable, as well as their strengths and weaknesses.

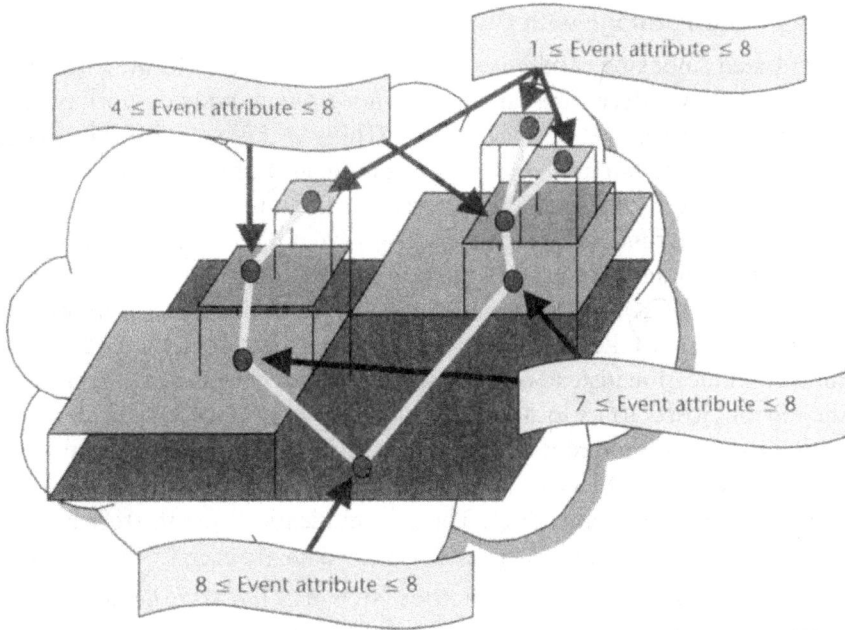

Figure 13.3 Hierarchical indexing in DIFS.

13.3.1 Local Storage and Geographical Search

At the lower right of the spectrum shown in Figure 13.1 is a fully local storage scheme where all useful sensor data is stored locally on each node and queries are routed to locations where the data is stored. Since data is stored locally at the sensing nodes that produced them, there is no communication cost involved in constructing the distributed store. However, in-network search and query processing incurs high overhead. Because data can reside anywhere in the network, a query that does not explicitly constrain the physical search space must be flooded to all storage nodes in the network. This costs $O(n)$, where n is the number of nodes in the network. Responses are sent back to the source of the query at a cost of $O(\sqrt{n})$ (since the network diameter is approximately $\sqrt{n}$). If only a few queries are issued during the lifetime of a network, answering these queries might involve little communication cost. A large number of flooded queries that each involve significant communication, however, can drain a network's energy reserves. Many of the initial ideas pursued within the context of Directed Diffusion [6] followed such a paradigm. Sources publish the events that they detect, and sinks with interest in specific events can subscribe to these events. The Directed Diffusion substrate routes queries to specific locations if the query has geographic information embedded in it (e.g., Find temperature in the southwest quadrant); if it does not, the query is flooded throughout the network. There are three drawbacks to such a scheme. First, for queries that are not geographically scoped, search cost ($O(n)$) might be prohibitive for large networks or frequent queries. Second, queries that process spatiotemporal data (e.g., edges) need to perform significant distributed data processing each time a query is posed, which can be expensive. Third, these techniques need to be enhanced to deal with storage limitations on sensor nodes.

13.3.2 Local Storage with Distributed Indexing

Distributed indexing addresses search issues in the local storage mechanism described above. Recent research has seen a growing body of work on data indexing schemes for sensor networks [3, 5, 7]. These techniques differ in the aggregation mechanisms used but are loosely based on the idea of geographic hashing and structured replication. One such indexing scheme is data-centric storage (DCS) [3], which provides a hash function for mapping from event name to location. DCS constructs a distributed storage structure that groups events together spatially by their named type (e.g., TRUCK). In traditional databases, a table is indexed in order to speed up common queries. Likewise, DCS indexes its data but does so in a manner optimized for communication instead of latency. A node that detects an event stores the event at the mirror closest to its location. A search using structured replication would begin with the root, descend to its four children, descend to each of the children's four children, and so forth. DCS uses structured replication to register the existence of events at replicated rendezvous nodes. The communication cost to store a piece of data is $O(\sqrt{n})$; the costs to send a query and retrieve data are each $O(\sqrt{n})$. DIFS [7] and Multidimensional Range Queries in Sensor Networks (DIM) [5] extend the data-centric storage approach to provide spatially distributed hierarchies of indexes to data. In these two techniques, the storage atom is a high-level event described with attributes that each have associated numerical values. For example, if a network detects vehicles, then these schemes might subcategorize the vehicle detections by their associated velocities. In these schemes, subcategorization of the name space is transformed into a hierarchical partitioning of the physical space of the sensor network. General names such as "vehicle" are stored near the root of a hierarchy. Specific names such as "truck with velocity of 35 mpg" are stored near the bottom.

13.3.3 Multiresolution Storage and Indexing

While indexing addresses the question of in-network search for event-based sensor networks, two other constraints need to be addressed. Multiresolution storage is intended for in-network storage and search in systems where (1) storage constraints on individual sensor nodes necessitate aging of old data, and (2) queries are complex and perform spatiotemporal data analysis. A simple example of an application that involves these constraints is one that looks for spatiotemporal weather patterns in data and involves deployments of small devices, such as Berkeley motes. Multiresolution storage is intended for applications where finding patterns in the sensor data is of primary interest. Thus, indexes to search through the data would themselves have to be summaries of the sensor data rather than simpler numerical indexes. Such an approach would be appropriate for applications that want to process queries on data at different timescales (past data). A system that constructs wavelet-based summaries is DIMENSIONS [4], which we discuss in greater detail in Section 13.4. Having described the various kinds of distributed storage and search schemes, we now provide a detailed discussion of one instance of a distributed storage system, DIMENSIONS [4].

13.4 Distributed Multiresolution Storage and Search

The DIMENSIONS system is motivated by the myriad applications of multiresolution processing and storage. First, sensor platforms are often storage constrained. Small sensor platforms such as Mica motes are currently equipped with 4 MB of Flash storage, whereas larger platforms, such as Intel Stargates, can be equipped with 1 GB or more of storage. A number of emerging scientific applications involve storing or transmitting considerable amounts of data. A simple distributed storage scheme would discard old data to make space for fresher data. In DIMENSIONS, a highly compressed, low-resolution view of the data can be stored in place of the older data, enabling a long-term but lossy storage model. Besides long-term storage, multiresolution processing and storage is also useful for complex spatiotemporal query processing for patterns in data. The design goals of DIMENSIONS are fourfold. First, the system should be energy efficient and incur low communication overhead. Conserving energy is fundamental to sensor networks. Second, the system should provide a distributed, load-balanced, long-term storage capability. Scientific applications of sensor networks are particularly demanding since scientists analyze data over long timescales. Third, in order to enable long-term storage, given limited storage capacity, the system should to provide a mechanism to age data. This aging procedure should be utility based (i.e., it should age data less useful while retaining data that is potentially useful for querying). The aging procedure should be graceful, and data quality should degrade slowly rather than abruptly over time. Finally, the system should exploit correlations in sensor data. Correlations in sensor data can be expected along multiple axes: temporal, spatial, and between multiple sensor modalities. The system should exploit these correlations in order to reduce communication and storage requirements.

13.4.1   Architecture

The DIMENSIONS architecture has three parts: (1) hierarchical multiresolution processing over a distributed hierarchy that constructs lossy multiresolution summaries, (2) hierarchical drill-down queries, and (3) a progressive data-aging scheme that determines how summaries should be discarded, given node storage requirements. Summarization and data aging are periodic processes that repeat every epoch. The choice of an epoch is application specific; for instance, if users of a microclimate-monitoring network [8] would like to query data at the end of every week, then a reasonable epoch would be a week. In practice, an epoch should at least be long enough to provide enough time for local raw data to accumulate for efficient summarization.

13.4.1.1   Multiresolution Summarization

The goal of this system is to support a wide range of queries for patterns in data. Since the goal is to be generally applicable rather than optimizing for a specific query, the system uses wavelets as the summarization mechanism. Wavelets have well-understood properties for data compression and feature extraction and offer good data reduction while preserving dominant features in data for typical spatiotemporal datasets [9, 10]. Hierarchical construction [shown in Figure 13.4(a)] using wavelets involves two phases. The first phase, *temporal summarization*, is cheap energywise since it involves only computation at a single sensor and incurs no communication overhead. This step consists of each node compressing the

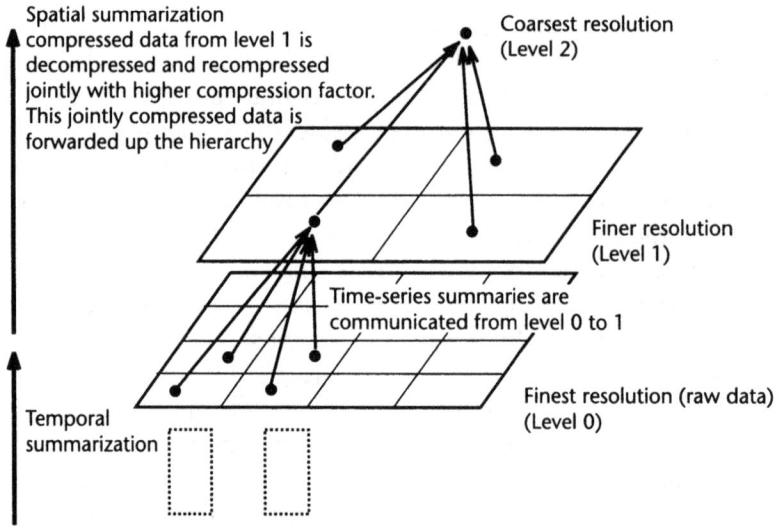

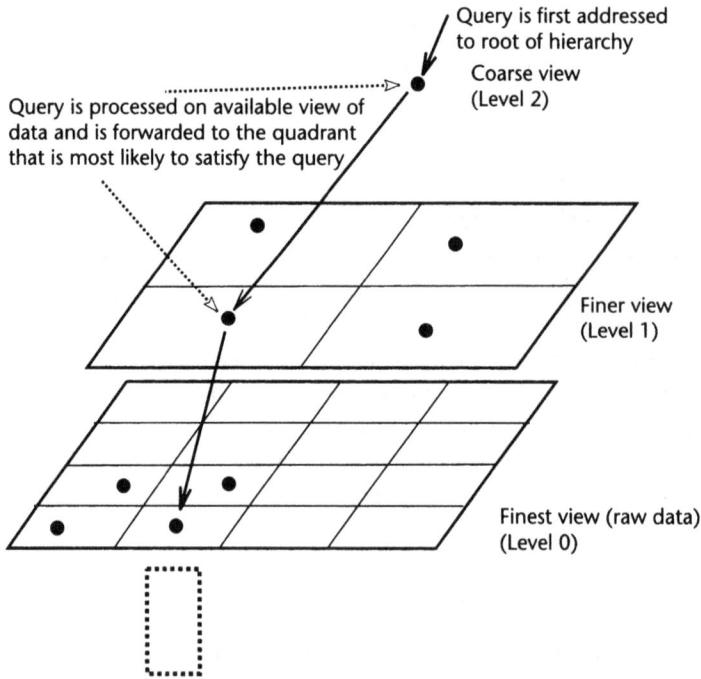

Figure 13.4   (a) Temporal and spatial summarization: (a) hierarchy construction. Temporal and spatial summarization: (b) drill-down querie.

time-series data by exploiting temporal redundancy in the signal. Significant benefit can be expected from merely temporal processing since a lot of sensor data is locally

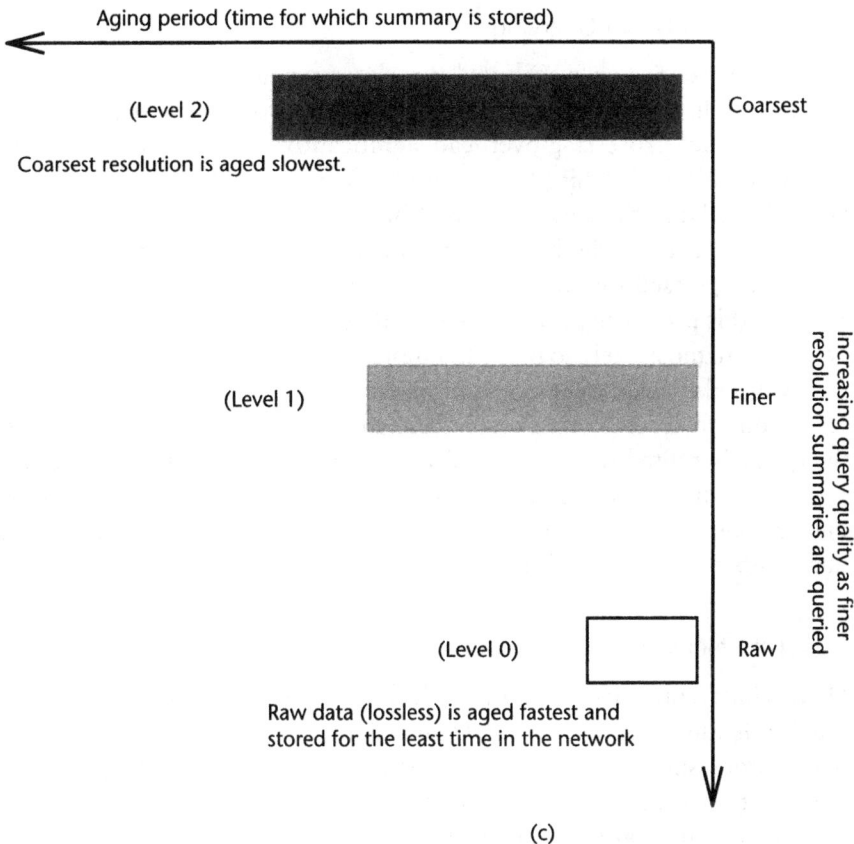

Aging period (time for which summary is stored)

(Level 2)   Coarsest

Coarsest resolution is aged slowest.

(Level 1)   Finer

Increasing query quality as finer
resolution summaries are queried

(Level 0)   Raw

Raw data (lossless) is aged fastest and
stored for the least time in the network

(c)

Figure 13.4 (continued)   Temporal and spatial summarization: (c) long-term storage.

generated at each node. The *spatial-summarization* phase constructs a hierarchical, grid-based overlay and uses spatiotemporal wavelet compression to resummarize data at each level. Figure 13.4(a) illustrates its construction: at each higher level of the hierarchy, summaries encompass larger spatial scales but are compressed more and are therefore lossier. At the highest level (level 2), one or a few nodes contain a very lossy summary for all data in the network.

13.4.1.2   Distributed Quad Tree

Distributed Quad Tree (DQT) is loosely based on the notion of structured replication introduced in DCS, which we discussed in Section 13.2. DQT adds load balancing to such a hashing and clustering scheme. Such a load-balancing scheme is essential when individual storage capacity on nodes is not substantial. Clearly, a simple hierarchical arrangement, like that shown in Figure 13.4(a), distributes load quite unevenly. DQT deals with this problem by adding a simple, probabilistic load-balancing mechanism, whereby each node assumes the role of a cluster head at any level for a limited time frame. After each time frame, a different node is probabilistically chosen to perform the role. As a result of such a load-balancing procedure, the responsibility for being a cluster head is shared among different nodes. The performance of such a scheme depends on the node distribution with uniform distribution of load in a regular setting.

13.4.1.3   Drill-Down Querying

Drill-down queries on distributed wavelet summaries can dramatically reduce the cost of search. By restricting search to a small portion of a large data store, such queries can reduce processing overhead significantly. These queries operate by using a coarse summary as a hint to decide which finer summaries to process further. In DIMENSIONS, drill-downs are used in a distributed context. Queries are injected into the network at the highest level of the hierarchy and processed on a coarse, highly compressed summary corresponding to a large spatiotemporal volume. The result of this processing is an approximate result that indicates which regions in the network are most likely to provide a more accurate response to the query. The query is forwarded to nodes that store summaries for these regions of the network and processed on more detailed views of these subregions. This procedure continues until the query is routed to a few nodes at the lowest level of the hierarchy or until an accurate enough result is found at some interior node. This procedure is shown in Figure 13.4(b), where a drill-down query is forwarded over a logical hierarchy to the quadrants that are most likely to satisfy it.

13.4.1.4   Networked Data Aging

Hierarchical summarization and drill-down querying address challenges in *searching* for features in distributed sensor data. Providing a long-term storage-and-query-processing capability requires storing summaries for long deployment periods. In storage-constrained networks, however, resources have to be allocated for storing new summaries by discarding older ones. The goal of networked data aging is to discard summaries such that network storage resources are efficiently utilized and graceful quality degradation over time is achieved. In other words, this work addresses the problem of apportioning the limited storage capacity in the network between different summaries. Each summary represents a view of a spatial area for an epoch, and its aging renders such a view unavailable for query processing. For instance, storing only the highest level (level 2) summary in Figure 13.4(c) provides a condensed data representation of the entire network and consequently low storage overhead compared to finer summaries but may not offer sufficiently accurate query responses. Storing a level 1 summary (finer) in addition to the level 2 one enables another level of drill-down and offers better accuracy but incurs more storage overhead. Figure 13.4(c) shows a typical instance of gracefully degrading storage, the coarsest summary being stored for the longest time and subsequent lower level summaries being stored for progressively shorter periods. To enable networked data aging, [4] proposes a training-based constraint-optimization scheme, where the storage partitioning between summaries is performed during a training period.

13.4.2   Usage Models

A distributed multiresolution storage infrastructure with system components as described above benefits search and retrieval of datasets that exhibit spatial correlation and applications that use such data. This section briefly describes some of the different applications that this system can be used for.

- *Long-term storage.* DIMENSIONS provides long-term storage to applications that are willing to sacrifice data fidelity for the ability to provide long-term storage. The rationale in balancing the need to retain detailed datasets for multiresolution data collection and to provide long-term storage is that if scientists were interested in detailed datasets, they would extract them within a reasonable interval (weeks). Long-term storage is primarily intended to enable querying of long-term spatiotemporal patterns, for which it is sufficient to store summaries that retain key features of data. Thus, the wavelet compression threshold is aged progressively, lending older data to progressively better compression but retaining key features of the data [9, 10].

- *Querying for spatiotemporal features.* The hierarchical organization of data can be used to search for spatiotemporal patterns efficiently by reducing the search space. Spatial features such as edges or discontinuities are important for many applications as well as for systems components. Detecting edges is important for applications like geographic routing, localization, and beacon placement. By progressively querying for the specific features, the communication overhead of searching for features can be restricted to only a few nodes in the network. Temporal patterns can be efficiently queried by drilling down the wavelet hierarchy by eliminating branches whose wavelet coefficients do not partially match the pattern, thereby reducing the number of nodes queried.

- *Multiresolution data collection.* During real-time monitoring of events in a bandwidth-constrained network, all data cannot be extracted from the network in real time. In such an instance, DIMENSIONS can create a compressed summary of the data that can be communicated in real time. The rest of the data can be collected in nonreal time when the bandwidth utilization in the network is less ([1]). I refer to such a facility as multiresolution data collection.

- *Approximate querying of wavelet coefficients.* Summarized coefficients that result from wavelet decomposition have been found to be excellent for approximate querying [11, 12] and to obtain statistical estimates from large bodies of data. Often, good estimates for counting queries [13, 14] can be obtained from higher-level wavelet coefficients (range sum queries [12]). Coefficients at higher levels of the decomposition are often effective in capturing many interesting features of the original data. The hierarchy is aged progressively, with more compressed coefficients being stored for longer periods of time. Queries that mine patterns over long timescales are executed on the compressed coefficients rather than the original dataset.

- *Network monitoring.* Network-monitoring data is another class of datasets that exhibits high correlation. Consider wireless packet throughput data: throughput from a specific transmitter to two receivers that are spatially proximate is closely correlated; similarly, throughput from two proximate transmitters to a specific receiver is closely correlated. DIMENSIONS serves two purposes for these datasets: (1) they can be used to extract aggregate statistics with low communication cost, and (2) discontinuities represent network hotspots, deep fades, or the effects of interference, which are important protocol parameters and can be easily queried.

13.5   Conclusion

In-network storage and search in sensor networks is one of the main aspects of data management and poses considerable challenges. In-network storage is necessary for sensor networks because in power-limited systems, it is more efficient to store data locally than to transmit it to a central location. Significant research challenges emerge due to the need to optimize for resources, power, and the types of queries that are posed on the data. This chapter has reviewed techniques for in-network indexing and storage of sensor data and has considered systems approaches that address two key problems: reducing the cost of flooding in search and addressing storage constraints in sensor data. Three building blocks are important for building distributed storage and indexing systems: geographic routing, hierarchical structures such as geographic hash tables, and hierarchical indexing and multiresolution processing. We have described how different distributed storage and search schemes can be composed of these building blocks. Finally, we described a storage architecture, DIMENSIONS, that uses these building blocks to provide a wavelet-based summarization for multiresolution processing of sensor data. This system provides a graceful data-degradation architecture for storage-constrained networks.

While some of the research challenges in sensor-network data management have been studied, many issues remain. Current research has not addressed power requirements for storage and how they can be minimized. Storage in Flash memory consumes power and, in current platforms, is only an order of magnitude less than the energy for communication. Thus, an interesting question is how to build systems such that storage and communication costs can be jointly reduced. The problem of irregular topologies poses considerable challenges as well. Many of the distributed storage schemes are designed for regular settings. Under highly irregular settings, both the processing techniques, as well as routing and distributed storage frameworks, will need to be improved [15]. For instance, wavelet compression can easily be adapted to regular sensor networks, but applying the same techniques to highly irregular settings is a harder problem.

As sensor networks start being deployed, the question of data storage and querying will become increasingly important. A closely related technological trend that demonstrates this importance is Radio Frequency Identification Tags (RFIDs). Data management in RFIDs is quickly becoming a critical problem as massive amounts of information are being generated by these systems. Similarly, sensing the physical world makes it essential to deal with the large volumes of data generated by sensor networks. The need for storage and querying in sensor networks will spur significant research in the coming years.

References

[1]   Xu, N., et al., "A Wireless Sensor Network for Structural Monitoring," *Proc. Second ACM Conference on Embedded Networked Sensor Systems (SenSys)*, Baltimore, MD, November 2004.

[2]   Karp, B., and Kung H. T., "GPSR: Greedy Perimeter Stateless Routing for Wireless Networks," In *Mobile Computing and Networking*, 2000, Boston, MA, pp. 243–254.

[3]   Ratnasamy, S., et al., "Ght a Geographic Hash Table for Data-centric Storage," *Proc. 1st ACM International Workshop on Wireless Sensor Networks and Their Applications,* Atlanta, GA, 2002.

[4]   Ganesan, D., et al., "Multi-resolution Storage in Sensor Networks," *Proc. 1st ACM Conference on Embedded Networked Sensor Systems (SenSys),* Los Angeles, CA, 2003.

[5]   Li, X., et al., "Multi-dimensional Range Queries in Sensor Networks," *Proc. 1st ACM Conference on Embedded Networked Sensor Systems (SenSys),* Los Angeles, CA, 2003.

[6]   Intanagonwiwat, C., Govindan R., and Estrin D., "Directed Diffusion: A Scalable and Robust Communication Paradigm for Sensor Networks," *Proc. Sixth Annual International Conference on Mobile Computing and Networking,* Boston, MA: ACM Press, 2000, pp. 56–67.

[7]   Greenstein, B., et al., "Difs: A Distributed Index for Features in Sensor Networks," *Elsevier Journal of Ad Hoc Networks,* Vol. 1, No. 2–3, Sept. 2003, pp. 333–349.

[8]   Hamilton, M., James San Jacinto Mountains Reserve, http://www.jamesreserve.edu.

[9]   Rao, R. M., and Bopardikar A. S.. *Wavelet Transforms: Introduction to Theory and Applications,* Boston, MA: Addison-Wesley, 1998.

[10] Vetterli, M., and Kovacevic J., *Wavelets and Subband Coding,* Upper Saddle River, NJ: Prentice Hall, 1995.

[11] Chakrabarti, K., et al., "Approximate Query Processing Using Wavelets," *VLDB Journal: Very Large Data Bases,* Vol. 10, No. 3, 2001, pp. 199–223.

[12] Vitter, J. S., Wang M., and Iyer B., "Data Cube Approximation and Histograms via Wavelets." In D. Lomet, (ed.), *Proc. CIKM '98,* Washington, D.C., November 1998, pp. 69–84.

[13] Hellerstein, J., et al., "Beyond Average: Towards Sophisticated Sensing with Queries," *Proc. IPSN '03,* Vol. 1, Palo Alto, CA, 2003.

[14] Zhao, Y., Govindan R., and Estrin D., "Residual Energy Scans for Monitoring Wireless Sensor Networks," *Proc. IEEE Wireless Communications and Networking Conference,* Orlando, FL, March 2002.

[15] Ganesan, D., et al., "Coping with Irregular Spatio-temporal Sampling in Sensor Networks," *Proc. 2nd Workshop on Hot Topics in Networks (Hotnets),* Cambridge, MA, 2003.

CHAPTER 14

A Whole-Network Approach to Sensor-Network Programming

Matt Welsh and Sam Madden

Programming sensor networks can be incredibly difficult for several reasons, including the limited capabilities and energy resources of sensor nodes, as well as the inherent unreliability communications, uncalibrated sensor data, and possibility of node failure. In this chapter, we describe several approaches to these problems that take the view of programming the sensor network as a whole rather than as individual sensor nodes. The idea behind *macroprogramming* is to write sensor-network applications in a high-level language that allows the programmer to describe the global behavior of the network. This high-level program is then compiled down to the low-level code that executes on sensor nodes, base stations, or gateways.

A number of approaches to macroprogramming have been explored by us as well as by other research groups. These fall into three broad categories:

1. *Underlying runtime and communication models that support the development of higher-level programming interfaces.* We describe *abstract regions*, a suite of general-purpose communication primitives for sensor networks that provide addressing, data sharing, and reduction within local regions of the network.
2. *High-level programming and query languages that allow application developers to take a network-centric view.* As an example, we describe TinyDB, a continuous query-processing engine for sensor networks based on an SQL-like query language.
3. *Tools and frameworks to support sensor-network deployment and management.* Tiny Application Sensor Kit (TASK) is a set of tools, based on TinyDB, that greatly simplifies tasking and control of sensor-network deployments.

We begin with a discussion of the overall goals and challenges for sensor-network macroprogramming. We then describe the three systems above in detail and conclude with a discussion of future research directions and highlights on other work in this area.

14.1   Introduction: Goals and Challenges

The core difficulty with building sensor-network applications is that high-level, global behavior must be expressed in terms of complex, local actions taken at each node. Because of limited energy, communication, and computational resources, sensor-network programs tend to be extremely low-level and fairly complex. High-level programming interfaces are typically absent from current sensor-network runtime environments; programmers must implement communication protocols, concurrency support, and power-management policies directly.

A more serious concern for application designers is the development of efficient *distributed* programs that coordinate the operation of multiple nodes. A key goal is to save energy and increase the lifetime of the system by trading increased computation for reduced radio communication (which is relatively expensive in terms of energy cost). Rather than collect samples centrally, it is generally desirable to perform local compression, aggregation, or summarization within the sensor network to reduce overall communication overheads.

As a simple example, consider determining the boundary of group of nodes in the network sharing some local property (e.g., the set of nodes measuring a temperature above some threshold). It is not possible to collect readings from all nodes centrally and simply run standard edge-detection algorithms on the complete set of data. Rather, it is necessary to implement distributed programs that rely chiefly on local data and limited communication between nearby nodes.

This challenge is exacerbated by the volatility of the sensor network itself. Radio communication is unreliable, and the limited buffer space on individual nodes precludes the use of expensive protocols such as TCP. The quality of the radio channel between any pair of nodes fluctuates over time and may exhibit sudden dropouts. In addition, the network should continue to operate even if nodes fail. Ideally, programmers would like to abstract away these details and implement distributed algorithms according to an idealized interface that masks these complications.

Apart from simplifying program development, automatic global-to-local compilation has the potential to achieve greater efficiency than manual node programming. Taking a global view of the network allows the compiler to task individual sensors, optimize collaborative activities across groups of nodes, and schedule node operations according to high-level program semantics. These kinds of optimizations are difficult to implement at the level of individual nodes since they require an understanding of networkwide behavior. As an example, a macroprogramming compiler might determine that only a fraction of the nodes in the network need to be active to implement a particular aggregation operation, whereas a programmer would have to build up a great deal of machinery to implement this behavior at the node level.

14.2   Proposed Approach: Macroprogramming

Our proposed approach to addressing these challenges is *macroprogramming*, or "programming in the large." A macroprogramming environment consists of either a language or programming interfaces for an existing language that simplify the development of distributed sensor-network applications. The macroprogram is automatically compiled or transformed into a program that runs on individual sensor nodes.

The nature of the macroprogramming model, interfaces, and language depends to a large extent on the goals of the underlying application. Many sensor networks are focused on *data collection*, which involves collecting data from sensor nodes at one or more wired base stations, possibly aggregating data along the way. For these kinds of applications, a query-language approach (Section 14.4) is often appropriate. Other applications require more sophisticated *in-network processing*, such as distributed event detection or filtering and compression of sensor signals. Programming these algorithms can make use of runtime support for distributed coordination (Section 14.3). Another style of programming involves *distributed actuation and control*, causing sensor nodes to take actions or initiate new computations when local criteria are met. We describe several programming models supporting such in-network triggers (Section 14.6).

14.3   Abstract Regions: Energy-Aware Collective Communications for Macroprogramming

The first step in designing any macroprogramming environment is to define the runtime environment and interfaces that the macroprogram will employ when running on individual nodes. To support coordination across nodes in the network, communication abstractions are needed that provide services common across a range of applications. Such interfaces are useful even without macrolanguage mapping since they allow sensor developers to program at the node level but take advantage of the facilities provided by the higher-level runtime environment.

Given such extreme resource limitations, it is critical that any communication abstraction for sensor networks allow the application some measure of control over resource consumption. Managing communication overheads is essential for meeting energy and bandwidth goals. Moreover, given the reactive nature of sensor networks, there is the opportunity to design applications that can adapt to changing network or environmental conditions, tuning the bandwidth or energy usage of the communication model to achieve given targets of latency, accuracy, or lifetime.

In this section, we we describe *abstract regions* [1, 2], a family of spatial operators that capture local communication within regions of the network, which may be defined in terms of radio connectivity, geographic location, or other properties of nodes. In addition to providing a flexible means of node addressing, abstract regions support data sharing, using a tuple-space-like programming model, as well as efficient aggregation operations. Applications can adapt to changing network conditions by tuning the energy and bandwidth usage of the underlying communication substrate. Abstract regions provide feedback on the accuracy of collective operations, as well as an interface for controlling resource usage. They are general enough to support a wide range of sensor-network applications and form the basis for other higher-level communication models.

14.3.1 Abstract Regions Programming Model

Sensor-network applications are often expressed in terms of groups of nodes over which local sampling, computation, and communication occur. For example, tracking a moving object involves aggregating sensor readings from nodes near the object. Abstract regions are a communication abstraction intended to simplify application development by providing a region-based collective communication interface. Abstract regions capture the inherent locality of communication and hide the details of data dissemination and aggregation within regions.

An abstract region defines a *neighborhood relationship* between a particular node and other nodes in the network. Examples of neighborhood predicates include "the set of nodes within N radio hops" and "the set of nodes within distance d." Local, spatial operations in the sensor network are performed by sharing data and coordinating activity among nodes in this neighborhood. In general, each node in the sensor network defines multiple abstract regions that it wishes to operate over, depending on application requirements. A node may also be a member of multiple regions at once: for example, a node will belong to multiple single-hop radio regions, one for each of its own set of single-hop neighbors.

Regions capture a range of common idioms in sensor-network programming. These include identification of neighboring nodes, data sharing, and data reduction within local neighborhoods. These operators allow nodes to query the state of neighboring nodes and implement efficient aggregation, compression, and summarization of local data. Abstract regions support the following set of operators:

- *Neighbor discovery.* Before performing other operations on a region, each node initiates the process of discovering neighboring nodes. Depending on the type of region, this may require broadcasting messages, collecting information on node locations, or estimating radio link quality. This is a continuous process, and each node is informed of changes in the region membership set, due to nodes joining, leaving, or moving within the network. A node may terminate this process at any time to avoid additional discovery messages. When terminated, the neighbor-discovery operator returns a quality metric that measures the accuracy of the region formation, such as the percentage of candidate nodes that responded to the discovery request.

- *Enumeration.* The enumeration operator returns the set of nodes participating in the region, allowing them to be addressed, for example, for direct message communication. Supplemental information, such as the location of each node, may be returned as well.

- *Data sharing.* The data-sharing operator allows variables, represented as key/value pairs, to be shared among nodes in the region. A shared variable supports two operations: *get* and *put*. The operation *get(v, n)* retrieves the value of variable v from node n, and *put(v, l)* stores the value $<i>l<i>$ in variable v at the local node. A simple implementation of *get(v, n)* sends a message to node n requesting the value of v. In this case, *put(v, l)* simply stores the value of v locally. An alternate implementation might broadcast or gossip the values of shared variables among nodes in the region, allowing *get(v, n)* to fetch locally cached values.

- *Reduction.* The reduction operator takes a shared-variable key and an associative operator (such as *sum*, *max*, or *min*) and reduces the shared variable across nodes in the region, storing the result in a shared variable. Reduction may be implemented in a number of ways, such as by collecting all values locally or forming a spanning tree, propagating values up the branches of the tree, and performing reduction at each level. As with shared variables, the region hides the details of reductions from the programmer.

Abstract regions simplify application design by shielding developers from the complexity of routing, data dissemination, and state management. Additionally, regions provide a unified interface regardless of the particular definition of the region membership; that is, one can readily interchange the underlying region implementation without necessarily affecting higher-level application logic. For example, an application that makes use of an N-radio-hop region can be readily modified to use a geographic neighborhood region in its place.

14.3.2   Abstract-Region Implementations

Given the diverse needs of sensor-network applications, we expect a range of abstract-region definitions will be useful to programmers. We have completed several abstract-region implementations, several others are underway. Three examples are shown in Figure 14.1. They include the following:

- *N radio hop:* Nodes within N radio hops;
- *N radio hop with geographic filter:* Nodes within N radio hops and distance d;

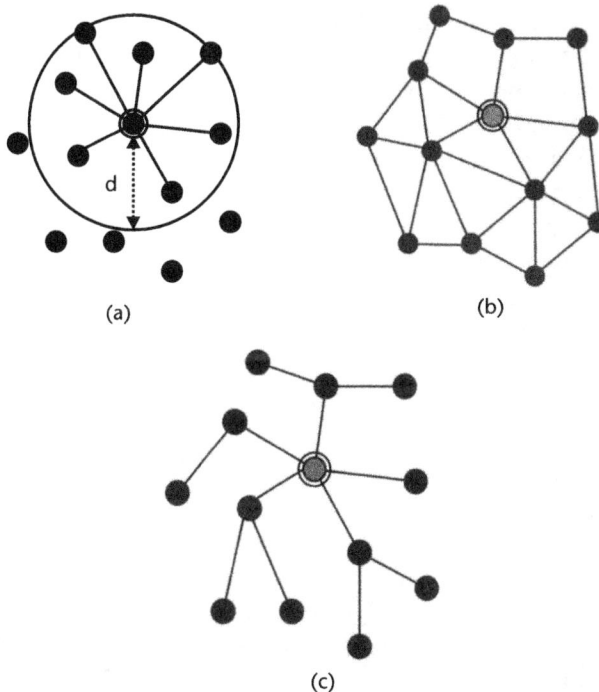

Figure 14.1   Examples of abstract regions: (a) geographic, (b) planar mesh, and (c) spanning tree.

- *k nearest neighbor:* k nearest nodes within N radio hops;
- *k best neighbor:* k nodes within N radio hops with the highest link quality, measured as the fraction of packets successfully transmitted during some measurement interval;
- *Approximate planar mesh:* A mesh with a small number (possibly zero) crossing edges;
- *Spanning tree:* A spanning tree rooted at a single node used for aggregating values over the entire network.

14.3.3 Quality Feedback and Tuning Interface

The collective operations provided by abstract regions are inherently unreliable. The quality of region discovery, the fraction of nodes contacted during a reduction operation, and the reliability of shared-variable operations all depend on the number of messages used to perform those operations. Generally, increasing the number of messages (hence, the energy usage) improves communication reliability and the accuracy of collective operations. However, given a limited energy or bandwidth budget, the application may wish to perform region operations with reduced fidelity.

Abstract regions expose this trade-off between resource consumption and the accuracy or yield of collective operations. This is in contrast to most existing approaches to sensor-network communication, which hide these resource trade-offs from the application. As a concrete example of such a trade-off, the communication layer may tune the number and frequency of message retransmissions to obtain a given degree of reliability from the underlying radio channel. Similarly, nodes may vary the rate at which they broadcast location advertisements to neighboring nodes, affecting the quality of region discovery.

Abstract regions provide feedback to the application in the form of a *quality measure* that represents the completeness or accuracy of a given operation. For member discovery, the quality measure represents the fraction of candidate nodes that responded to the discovery request. For example, when discovering one-hop neighbors within distance d, the quality measure represents the fraction of one-hop nodes that responded to the region-formation request. For reduction, the quality measure represents the fraction of nodes in the region that participated in the reduction. In addition, each operation supports a time-out mechanism, which causes the operation to fail if it has not completed within a given time interval. For example, when performing a shared-variable *get* operation, a time-out indicates that the data could not be retrieved from the requested node.

Applications can use this quality feedback to affect resource consumption of collective operations through a *tuning interface.* The tuning interface allows the application to specify low-level parameters of the region implementation, such as the number of message broadcasts, amount of time, or number of candidate nodes to consider when forming a region. Likewise, region operations perform message retransmission and acknowledgment to increase the reliability of communication; the depth of the transmit queue and number of retransmission attempts can be tuned by the application.

14.3.4 Application Examples

To demonstrate the use of abstract regions for sensor-network programming, we briefly describe two applications built using this interface: tracking moving vehicles and finding spatial contours in a sensor field. Each of these applications requires extensive coordination across sensor nodes and exploits the use of local communication patterns provided by regions.

14.3.4.1 Vehicle Tracking

Object tracking is an oft-cited application for sensor networks [3–5]. In its simplest form, tracking involves determining the location of a moving object by detecting changes in relevant sensor readings, such as magnetic field. In our vehicle-tracking system, each node takes periodic magnetometer readings and compares them to a threshold value. Nodes above the threshold communicate with their neighbors and elect a leader node, which is the node with the largest magnetometer reading. The leader computes the centroid of its neighbors' sensor readings and transmits the result to a base station.

The following pseudocode shows this application expressed in terms of abstract regions:

```
location = get_location();
/* Get 8 nearest neighbors */
region = k_nearest_region.create(8);
while (true) {
    reading = get_sensor_reading();
    /* Store local data as shared variables */
    region.putvar(reading_key, reading);
    region.putvar(reg_x_key, reading * location.x);
    region.putvar(reg_y_key, reading * location.y);
    if (reading > threshold) {
            /* ID of the node with the max value */
            max_id = region.reduce(OP_MAXID, reading_key);
            /* If I am the leader node ... */
            if (max_id == my_id) {
            /* Perform reductions and compute centroid */
            sum = region.reduce(OP_SUM, reading_key);
            sum_x = region.reduce(OP_SUM, reg_x_key);
            sum_y = region.reduce(OP_SUM, reg_y_key);
            centroid.x = sum_x / sum;
            centroid.y = sum_y / sum;
            send_to_basestation(centroid);
        }
    }
sleep(periodic_delay);
}
```

The program performs essentially all communication through the abstract-regions interface, in this case the k-nearest-neighbor region. Nodes store their local sensor reading and the reading scaled by the x and y dimensions of their location as shared variables. Nodes above the threshold perform a reduction to determine the node with the maximum sensor reading, which is responsible for calculating the centroid of its neighbors' readings. A series of sum-reductions is performed over the shared variables, which are used to compute the centroid.

The use of regions greatly simplifies application design. The programmer need not be concerned with the details of routing, data sharing, or identifying the appropriate set of neighbor nodes for performing reductions. As a result, the application code is very concise. Our actual implementation is just 134 lines of code, of which 70 lines make up the main application loop above. The rest consists of variable and constant declarations as well as initialization code.

14.3.4.2 Contour Finding

Contour finding involves determining a set of points in space that lie along, or close to, an isoline in the gradient of sensor readings. Contours might represent the boundary of a region of interest, such as a group of sensors with an interesting range of readings. Contour finding is a valuable spatial operation as it compresses the per-node sensor data into a low-dimensional surface. This primitive could be used for detecting thermoclines or tracking the flow of contaminants through soil [6].

An implementation of contour finding using abstract regions is depicted in Figure 14.2 and is expressed in pseudocode as

```
location = get_location();
/* Form approximate planar mesh */
region = apmesh_region.create();
region.putvar(loc_key, location);
```

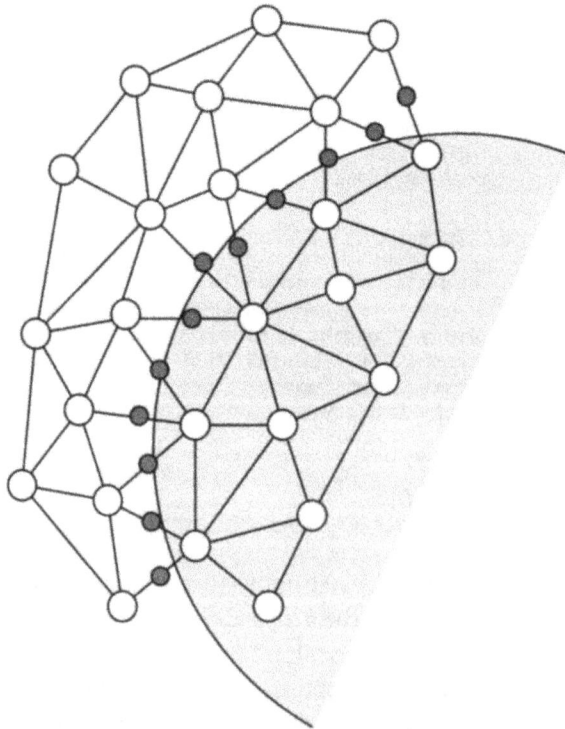

Figure 14.2 Contour-finding application. The shaded region represents an area where sensor readings fall above a threshold. Nodes (unfilled circles) are connected to an approximate planar mesh. Contour points (filled circles) are chosen as midpoints between nodes above the threshold and nodes below the threshold.

```
while (true) {
    reading = get_sensor_reading();
    region.putvar(reading_key, reading);

    if (reading > threshold) {
            foreach (nbr in region.get_neighbors()) {
                    /* Fetch neighbor s reading */
                    rem_reading = region.getvar(reading_key, nbr);
                    if (rem_reading <= threshold) {
                            rem_loc = region.getvar(loc_key, nbr);
                            contour_point = midpoint(location, rem_loc);
                            send_to_basestation(contour_point);
                    }
            }
    }
sleep(periodic_delay);|
}
```

Nodes first form an approximate-planar-mesh region, and store their location and sensor reading as a shared variable. Nodes above the sensor threshold of interest fetch the readings and locations of their neighbors. For each neighbor below the threshold, the node computes a contour point as the midpoint between itself and its neighbor and sends the result to the base station.

The use of the approximate-planar-mesh region ensures that few edges will cross and that nodes will generally select neighbors that are geographically near. These properties are vital for our contour-finding algorithm as it is based on pairwise comparisons of sensor readings, and values are computed along edges in the mesh. As with the object-tracking application, abstract regions shield the programmer from the details of mesh formation and data sharing; the application code is very straightforward. It is only 118 lines of nesC code, 56 lines of which are devoted to the main application loop.

14.4 TinyDB

The TinyDB system provides a complete macroprogramming environment, replete with a simple declarative programming language based on queries and a runtime environment for executing those queries. This runtime environment uses a region- based spanning-tree abstraction for data collection.

In TinyDB, queries are input at the user's PC in a simple SQL-like language that describes the data the user wishes to collect and the ways in which he or she would like to combine, transform, and summarize it. The most significant way in which the variant of SQL we have developed differs from traditional SQL is that queries are *continuous* and *periodic*. That is, users register an interest in certain kinds of sensor readings (e.g., temperatures from sensors on the fourth floor every 5 seconds), and the system streams these results out to the user. We call each period in which a result is produced an *epoch*. The *epoch duration* or *sample period* of a query refers to the amount of time between successive samples; for this example, the sample period would be 5 seconds. As we discuss various aspects of our system, we will show some examples of our language syntax and discuss its other features (both new and common to traditional SQL) in more detail.

14.4.1   Query Language

Queries in TinyDB, as in SQL, consist of SELECT-FROM-WHERE-GROUPBY-HAVING blocks supporting selection, joining, projection, aggregation, and grouping. TinyDB also includes explicit support for windowing and subqueries via *materialization points*. In queries, we view sensor data as a single virtual table with one column per sensor type. Tuples are appended to this table periodically at well-defined intervals that are a parameter of the query. This period between each sample interval is the epoch, as described above. Epochs provide a convenient mechanism for structuring computation to minimize power consumption. As an example, consider the following query:

```
SELECT nodeid, light, temp
FROM sensors
SAMPLE PERIOD 1s
FOR 10s
```

This query specifies that each sensor should report its own ID, light, and temperature readings once per second for 10 seconds. The virtual table sensors contains one column for every attribute available in the catalog and one row for every possible instant in time. The term *virtual* means that these rows and columns are not materialized—only the attributes and rows referenced in active queries are actually generated.

Results of this query stream out of the network where they may be logged, output to the user, or fed into another database system. The output consists of an ever-growing sequence of tuples, clustered into 1 second time intervals. Each tuple includes a timestamp corresponding to the time it was produced.

Note that the sensors table is (conceptually) an unbounded, continuous stream of data values; as is the case in other streaming and on-line systems, certain blocking operations (such as sort and symmetric join) are not allowed over such streams unless a bounded subset of the stream, or *window*, is specified. Windows in TinyDB are defined as fixed-size materialization points over the sensor streams. Such materialization points accumulate a small buffer of data that may be used in other queries. Consider, as an example, the following query:

```
CREATE STORAGE POINT recentlight SIZE 8 seconds
AS (SELECT nodeid, light FROM sensors
SAMPLE PERIOD 1s)
```

This statement provides a shared, local (i.e., single-node) location to store a streaming view of recent data similar to materialization points in other streaming systems, like Aurora or STREAM [7, 8], or to materialized views in conventional databases.

Joins are allowed between two storage points on the same node or between a storage point and the sensors relation, in which case sensors is used as the outer relation in a nested-loops join; that is, when a sensors tuple arrives, it is joined with tuples in the storage point at its time of arrival. This is effectively a *landmark query* [9], common in streaming systems. Consider the following as an example:

```
SELECT COUNT(*)
FROM sensors AS s, recentLight AS rl
WHERE rl.nodeid = s.nodeid
```

```
AND s.light < rl.light
SAMPLE PERIOD 10s
```

This query outputs a stream of counts indicating the number of recent light readings (from zero to eight samples in the past) that were brighter than the current reading.

14.4.2 Aggregation Queries

TinyDB also includes support for *grouped aggregation* queries. Aggregation has the attractive property that it reduces the quantity of data that must be transmitted through the network; other sensor-network research has noted that aggregation is perhaps the most common operation in the domain (see, e.g., [10, 11]).

Aggregate query syntax and semantics.   Consider a user who wishes to monitor the occupancy of the conference rooms on a particular floor of a building. She chooses to do this by using microphone sensors attached to nodes and looking for rooms where the average volume is over some threshold (assuming that rooms can have multiple sensors). Her query could be expressed as follows:

```
SELECT AVG(volume),room FROM sensors
WHERE floor = 6
GROUP BY room
HAVING AVG(volume) > threshold
SAMPLE PERIOD 30s
```

This query partitions nodes on the sixth floor according to the room where they are located (which may be a hard-coded constant in each device or determined via some localization component available to the devices.) The query then reports all rooms where the average volume is over a specified threshold. Updates are delivered every 30 seconds. The query runs until it is deregistered.

Recall that the primary semantic difference between TinyDB queries and SQL queries is that the output of a TinyDB query is a stream of values, rather than a single aggregate value (or batched result). For these streaming queries, each aggregate record consists of one *<group id, aggregate value>* pair per group. Each group is time-stamped with an epoch number, and the readings used to compute an aggregate record all belong to the same epoch.

Structure of aggregates.   TinyDB structures aggregates similarly to shared-nothing parallel database engines (e.g., [12]). The approach used in such systems (and followed in TinyDB) is to implement *agg* via three functions: a merging function f, an initializer i, and an evaluator e. In general, f has the following structure:

$$< z > = f(< x >, < y >)$$

where x and y are multivalued *partial-state records,* computed over one or more sensor values, representing the intermediate state over those values that will be required to compute an aggregate. The partial-state record $<z>$ results from the application of function f to $<x>$ and $<y>$. For example, if f is the merging function for AVERAGE, each partial-state record might consist of a pair of values: SUM and COUNT, with f specified as follows, given two state records $<S_1, C_1>$ and $<S_2, C_2>$:

$$f(< S_1, C_1 >, < S_2, C_2 >) = < S_1 + S_2, C_1 + C_2 >$$

The initializer i is needed to specify how to instantiate a state record for a single sensor value; for our AVERAGE aggregate over a sensor value of x, the initializer $i(x)$ would return the tuple $<x, 1>$. Finally, evaluator e takes a partial-state record and computes the actual value of the aggregate. For AVERAGE, the evaluator $e(<S, C>)$ would simply return S/C.

These three functions can easily be derived for the basic SQL aggregates; in general, any operation that can be expressed as a commutative application of a binary function is expressible. TinyDB includes a simple facility for allowing programmers to extend the system with new aggregates by authoring software modules that implement these three functions.

In addition to aggregates over values produced during the same sample interval, users need to be able to perform temporal operations. For example, in a building-monitoring system for conference rooms, users may detect occupancy by measuring maximum sound volume over time and reporting that volume periodically:

```
SELECT WINAVG(volume, 30s, 5s)
FROM sensors
SAMPLE PERIOD 1s
```

This query will report the average volume over the last 30 seconds once every 5 seconds, acquiring a sample once per second. This is an example of a *sliding-window* query common in many streaming systems [8, 9].

When a query is issued in TinyDB, it is assigned an identifier (id) that is returned to the issuer. This identifier can be used explicitly to stop a query via a STOP QUERY id command. Alternatively, queries can be limited to run for a specific period via a FOR clause or can include a stopping condition as a *triggering condition* or *event*. Our recent SIGMOD paper on *acquisitional query processing* [13] provides more detail regarding these language constructs.

14.4.3   Query Dissemination and Result Collection

Once a query has been input, it is parsed and converted to a compact binary representation. It is then *disseminated* into the network. We discuss one basic communication primitive, a *routing tree*. A routing tree is rooted at a base station; it allows the root of the network to disseminate a query and to collect query results. This routing tree is formed by forwarding the query from every node in the network: the root initially transmits the query; all children that hear it process it and forward it on to their children, and so on, until the entire network has heard about the query.

Each radio message contains a hop count, or *level*, indicating the distance from the broadcaster to the root. To determine their own level, nodes pick a *parent* node that is (by definition) one level closer to the root than they are. This parent will be responsible for forwarding the node's (and its children's) query results to the base station.

Figure 14.3 shows an example sensor-network topology and routing tree. Solid arrows indicate parent nodes, while dotted lines indicate nodes that can hear each other but do no use each other for routing. In general, a node may have several possible choices of parent; a simple approach is to chose as the parent the ancestor node with the lowest level. In practice, it turns out that making a proper choice of parent is quite important in terms of communication and data-collection efficiency and that network topologies are much less regular and more complex than one might expect [14]. Unfortunately, the details of the best-known techniques forming trees in real networks are quite complicated and outside the scope of our discussion in this chapter. For a more complete discussion of these and other issues, see, for example, recent work from the TinyOS group at the University of California, Berkeley [15]. We note that there has been a plethora of work on routing in ad hoc and sensor networks [10, 16–18], including energy-aware routing [19, 20] and special MAC protocols [21]. The techniques used to build this routing tree are essentially the same as techniques used by the spanning region discussed in Section 14.3, suggesting a software layering in which TinyDB is in fact a region-based program. At the time of this writing, such a layering has not been attempted due to the parallel development of the two approaches.

Once a routing tree has been constructed, each node has a connection to the root of the tree, which is just a few radio hops long. We can then use this tree to collect data from sensors by having them forward query results up this path. In TinyDB, the routing tree evolves over time as new nodes come on-line, interference patterns change, or nodes run out of power. Tree maintenance is done locally, at every node, by keeping a set of candidate parents and an estimate of the quality of the communication link with each of them; when the quality of the link to the current parent is sufficiently worse than the quality to another candidate parent, a switch is made.

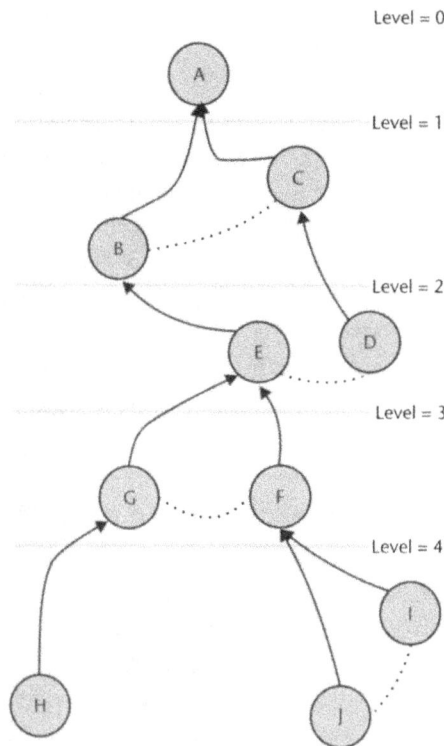

Figure 14.3   A sensor-network topology, with routing tree overlay.

A simple routing structure, such as a routing tree, is well suited to our scenario: Sensor-network query processors impose communication workloads on the multi-hop communication network that are very different from those of traditional ad hoc networks with mobile nodes. Since the sensor network is programmed only through queries, there are very regular communication patterns, mainly consisting of the collection of sensor readings from a region at a single node or the base station.

A second advantage of the stylized tree-based communication pattern used to execute in TinyDB is that this makes it possible to impose a power-efficient communication schedule on the nodes; we discuss this possibility in more detail in Section 14.4.5. First, we turn our attention to the basics of query processing.

14.4.4   Query Processing

Once a query has been disseminated, each node begins processing it. Processing is a simple loop: once per epoch, readings, or *samples*, are acquired from sensors corresponding to the fields, or *attributes*, referenced in the query. This acquisition is done by a special *acquisition operator*. This set of readings, or *tuple*, is routed through the query plan built in the optimization phase. The plan consists of a number of operators that are applied in a fixed order; each operator may pass the tuple on to the next operator, reject it, or combine it with one or more other tuples from the local node or received from neighboring nodes. Any tuple that successfully passes the plan is transmitted up the routing tree to the node's parent, which may in turn forward the result on or combine it with its own data or data collected from its other children. In the current version of TinyDB, every node in the network runs an identical version of the query plan.

The acquisition operator uses a *catalog* of available attributes to map names referenced in queries into low-level operating system functions that can be invoked to provide their values. This catalog abstraction allows sophisticated users to extend the sensor network with new kinds of sensors and also provides support for sensors that are accessed via different software interfaces. For example, in the TinyDB system, users can not only run queries over sensor attributes like light and temperature but can also query attributes that reflect the state of the device or operating system, such as the free RAM in the dynamic-memory allocator.

Once a node has applied its local query-processing operators to its tuples, it forwards them on up the routing tree. Tuples may be rejected if they do not pass a selection predicate (i.e., a statement in the WHERE clause of the query). Tuples that are forwarded to parents may need to be joined or aggregated with tuples from other nodes but do not need to have selection operators applied again.

14.4.5   Communication Scheduling and Aggregate Queries

When processing aggregate queries, some care must be taken to coordinate the times when parents and children are awake so that parents nodes have access to their children's readings before aggregating. The basic idea is to subdivide the epoch into a number intervals and assign nodes to intervals based on their position in the routing tree. Because this mechanism makes relatively efficient use of the radio channel and has good power-consumption characteristics, TinyDB uses this scheduling approach for all queries (not just aggregates).

In this *slotted approach*, each epoch is divided into a number of fixed-length time intervals. These intervals are numbered in reverse order such that interval 1 is the last interval in the epoch. Then, each node is assigned to the interval equal to its level, or number of hops from the root, in the routing tree. In the interval preceding their own, nodes listen to their radios, collecting results from any child nodes (which are one level below them in the tree, and thus communicating in this interval.) During a node's interval, it then computes the partial-state record consisting of the combination of any child values it heard with its own local sensor readings. After this computation, the node transmits its partial-state record up the network. In this way, information travels up the tree in a staggered fashion, eventually reaching the root of the network during interval 1.

Figure 14.4 illustrates this in-network aggregation scheme for a simple COUNT query that reports the number of nodes in the network. In the figure, time advances from left to right, and different nodes in the communication topology are shown along the y-axis. Nodes transmit during the interval corresponding to their depth in the tree, so H, I, and J transmit first, during interval 4, because they are at level 4.

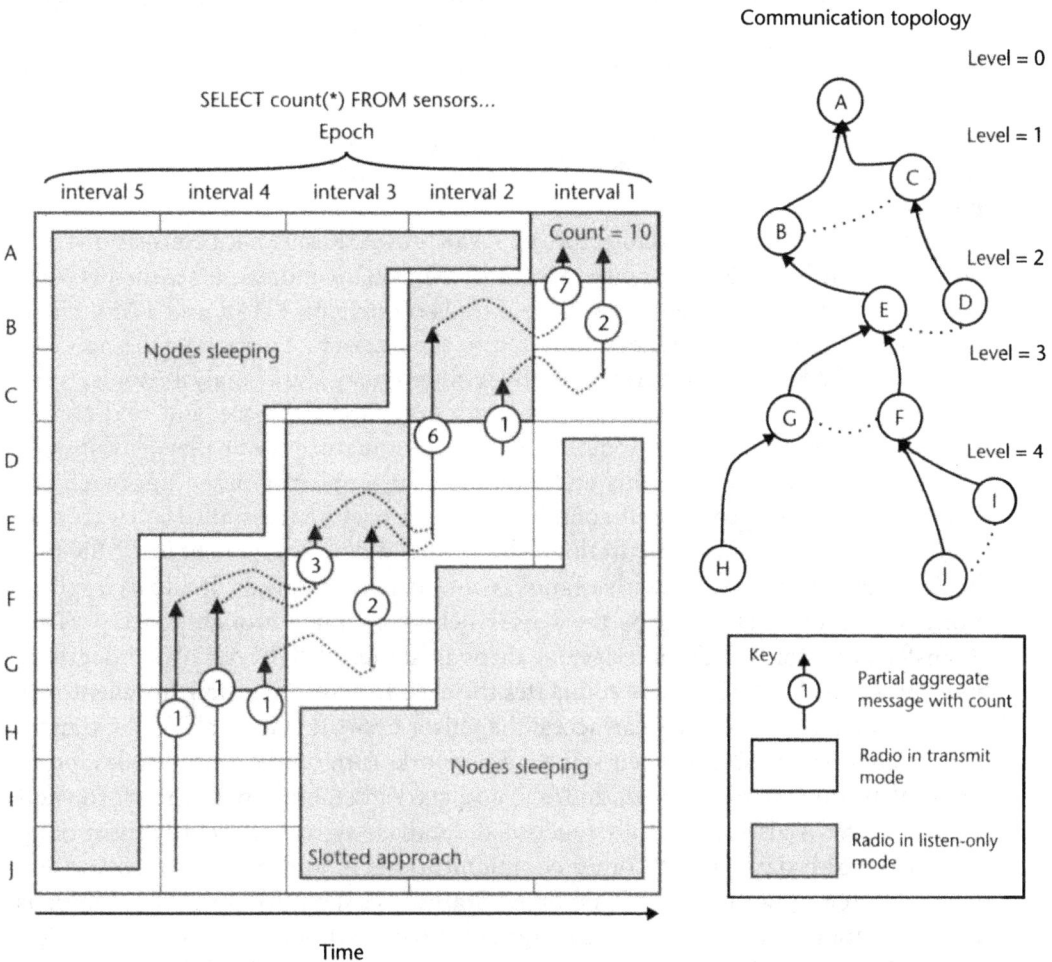

Figure 14.4 In-network aggregation via scheduled communication. Partial-state records flowing up the tree during an epoch using the interval-based communication approach.

Transmissions are indicated by arrows from sender to receiver, and the numbers in circles on the arrows represent COUNTs contained within each partial-state record. Readings from these three nodes are combined via the COUNT merging function, at nodes G and F, both of which transmit new partial-state records during interval 3. Readings flow up the tree in this manner until they reach node A, which then computes the final count of 10. Notice that nodes are idle for a significant portion of each epoch so that they can enter a low-power sleeping state. A detailed analysis of the accuracy and benefit of this approach in TinyDB can be found in [22].

This scheduled-communication approach makes it possible to execute queries in a power-efficient manner.

TinyDB includes a number of other features designed to promote efficient execution of queries via static and runtime optimization; the reader is referred to [22] for a detailed discussion.

14.5   TASK

TASK is a high-level toolkit designed to ease the deployment of sensor networks for remote monitoring. It uses TinyDB for the core data-collection features, although it includes a number of other components designed to assist with deployments. The principal goal is to eliminate the need for users to do any programming or installation of code onto sensor devices—they simply obtain a TASK kit, place the sensors, configure them through a graphical user interface (GUI), and begin collecting data.

Figure 14.5 shows the architecture of TASK. A TASK sensor kit consists of a collection of sensor nodes (currently Mica2 or Mica2Dot motes), a sensor-network appliance (SNA), a number of TASK field tools running on PDAs, and TASK client tools running on any computer connected to the Internet. To support the needs of end users, TASK also integrates easily with popular data analysis tools, such as MATLAB, Excel, and Labview, through standard database and text-export interfaces. A principal design requirement was to minimize the challenge of installing sensornet software. To this end, we chose an appliance-based approach for TASK, in which we provide self-contained hardware with preinstalled software that requires a minimum of configuration. The SNA is a core component in TASK that bridges the sensor network with a familiar Internet environment. It allows applications on the Internet to access the sensor network either through an HyperText Transfer Protocol (HTTP) interface or through standard JDBC/ODBC™ database interfaces. The HTTP interface eliminates the need for users to install any client-side software on a PC before they can access the sensor network. Given a TASK kit, users simply connect the SNA to their standard network, turn on the sensor nodes, point their Web browser to the SNA address, and start interacting with the sensor network. The SNA also runs a fully functional database system to record sensor data, network-health data, and all other communications in and out of the sensor network. Currently, we use the xScale-based Stargate [23] platform running Linux as our SNA. It includes a connector to interface directly with a mote, as well as a Personal Computer Memory Card International Association (PCMCIA) connector suitable for bridging to a wireless or wired Ethernet.

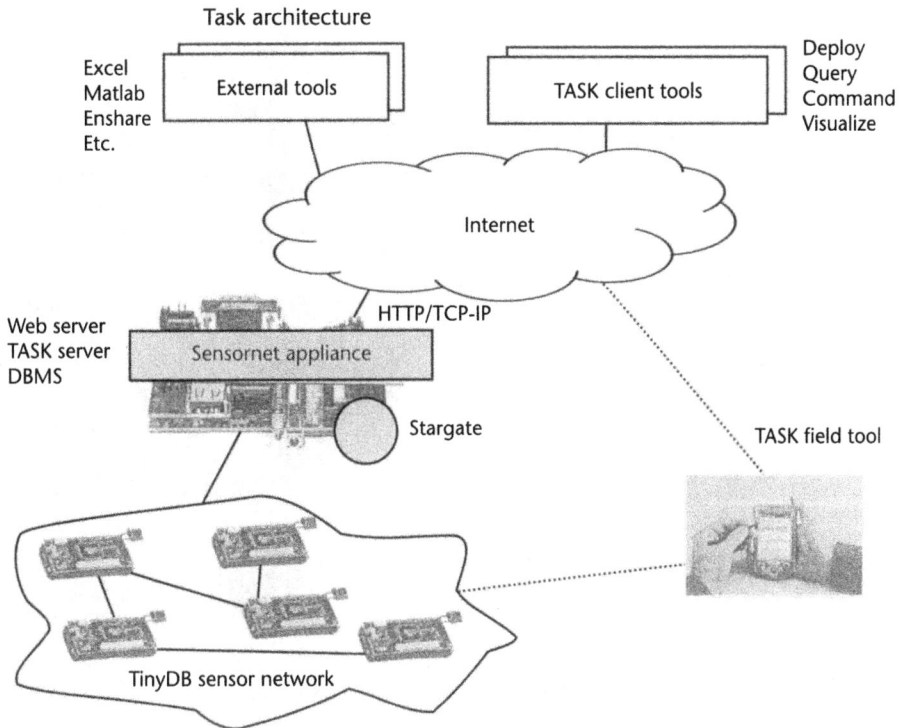

Figure 14.5   The TASK architecture.

On the sensor-network side, we installed TinyDB on each mote. TinyDB's query interface makes it easy for end users to reconfigure and extend the network via SQL queries. TASK's TinyDB installation is configured with a variety of introspective attributes for network topology and system parameters, as well as support for the standard sensor boards that are commonly shipped with motes.

The field tool provides a simple network diagnostics GUI on a PDA for users to interact with sensor nodes in close range and diagnose problems. It is particularly useful when some nodes become unreachable from the SNA.

Finally, TASK comes with two versions of the client tools: a simple Web-based tool for interacting with the sensor network through the SNA and a full-featured Visual Basic–based tool called TASKView that provides more advanced features such as node-deployment bookkeeping, network visualization, and near-real-time continuous monitoring. Figure 14.6 is a sample screen shot of TASKView. The lower left window shows the deployment-and-configuration screen, while the two other windows show a tabular and graphical view of data from a set of motes.

14.5.1   The Garden Deployment

To illustrates the utility of the TASK toolkit, we discuss a real-world deployment that was undertaken in a redwood grove in the Berkeley Botanical Garden. The majority of the deployment was done over two days, once the needed hardware components had been obtained. It took the majority of one day to get the motes programmed with TASK/TinyDB and the appropriate software installed on our SNA. It took another day to place the devices in the trees (we had a team of biology

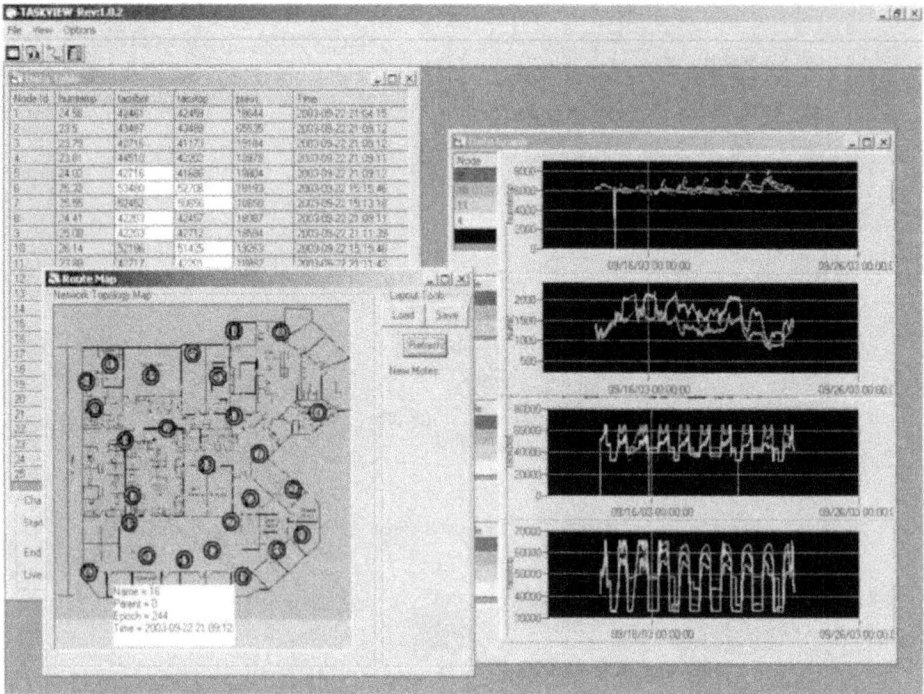

Figure 14.6   The TASKView client tool.

graduate students who were skilled at tree climbing), set up our data-collection hardware, and connect to the Internet from a building near the garden.

The garden deployment consisted of 23 motes running for about 20 days (480 hours) during the summer of 2003 at a sample period of 30 seconds. In this case, motes were deployed in special weatherproof packages with 800-mAh lithium-ion batteries. A picture of the mote and packaging are shown in Figure 14.7. Motes sensed photosynthetically active radiation (PAR), humidity, temperature, and barometric pressure.

Devices were placed in a 34m redwood tree at four different altitudes. The goal of the deployment was to measure how environmental parameters varied throughout the day at different heights in the forest canopy. Such variances can have a significant effect on the models used to predict the growth of trees [24]; for example, variations of just 20% in humidity (either up or down) can reduce tree growth from a normal rate to almost nothing. Such models play a critical role in forestry management because incorrect predictions can have a significant, negative economic impact on the logging industry.

Motes radioed data from the tree, which was in the middle of a small grove, across about 100m to a building where we had installed a large antenna and the TASK base station, which collected results on a local database. In many cases, the network was single hop, although for motes on the far side of the tree, several hops were needed.

Figure 14.8 shows data from five of the sensors collected during the first few days of August 2003. The periodic bumps in the graph correspond to daytime readings; at night, the temperature drops significantly, and humidity becomes very

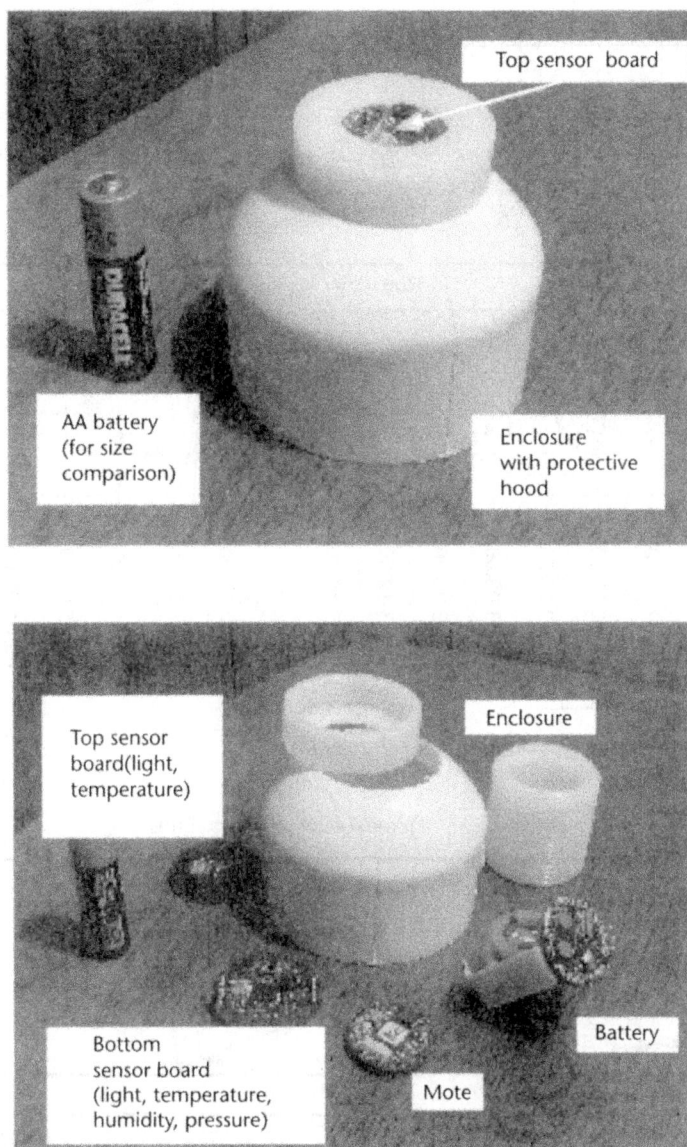

Figure 14.7 The Mica2Dot mote and packaging used in the redwood garden deployment: (a) assembled mote, and (b) disassembled mote.

high as fog rolls in. August 2 and 3 were relatively cool (below 21°C) and likely overcast as many summer days at the botanical garden are; August 4 was sunny and somewhat warmer, reaching as high as 24°C at the tops of the tree. Note that during the daytime, it can be as much as 8° cooler and 25% more humid at the bottom of the canopy (sensor 1) than at the top (sensors 15 and 16). On hot days, this effect is particularly pronounced; compare, for example, sensors 1 and 15 just after 10 a.m. on August 4. Anyone who has ever been walking in a redwood forest has felt this cool dampness under the cover of these trees.

This increase in humidity and decrease in temperature is one of the effects the biologists involved in the project were interested in studying. In this case, the trees act as a buffer, releasing water into the air on hot days and absorbing it on foggy

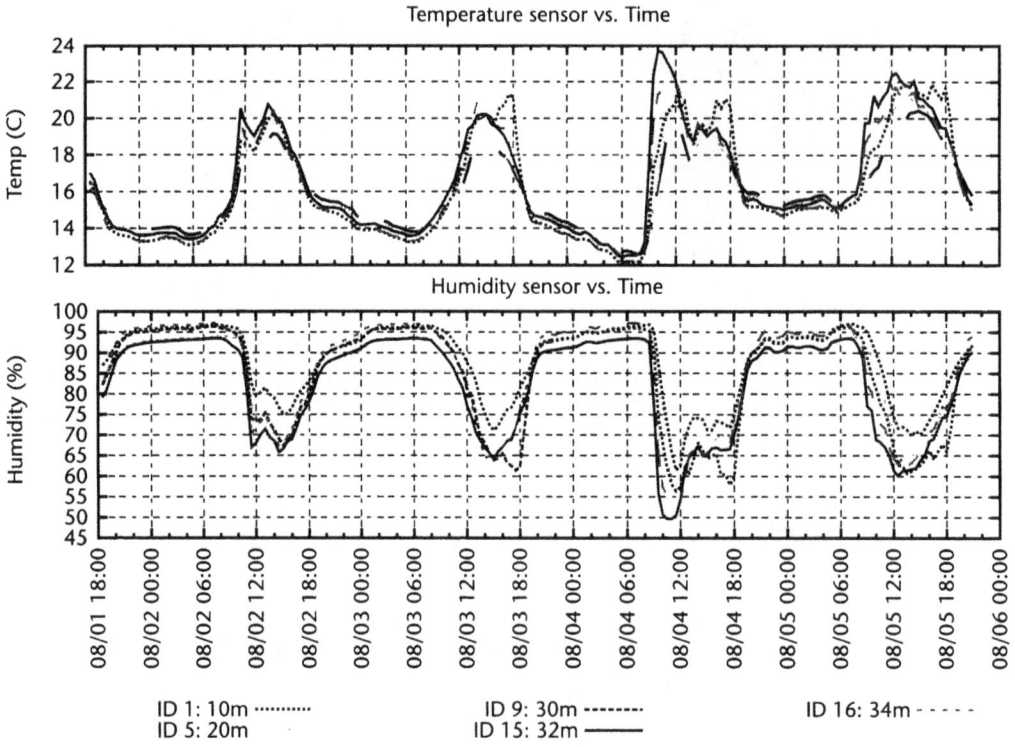

(a)

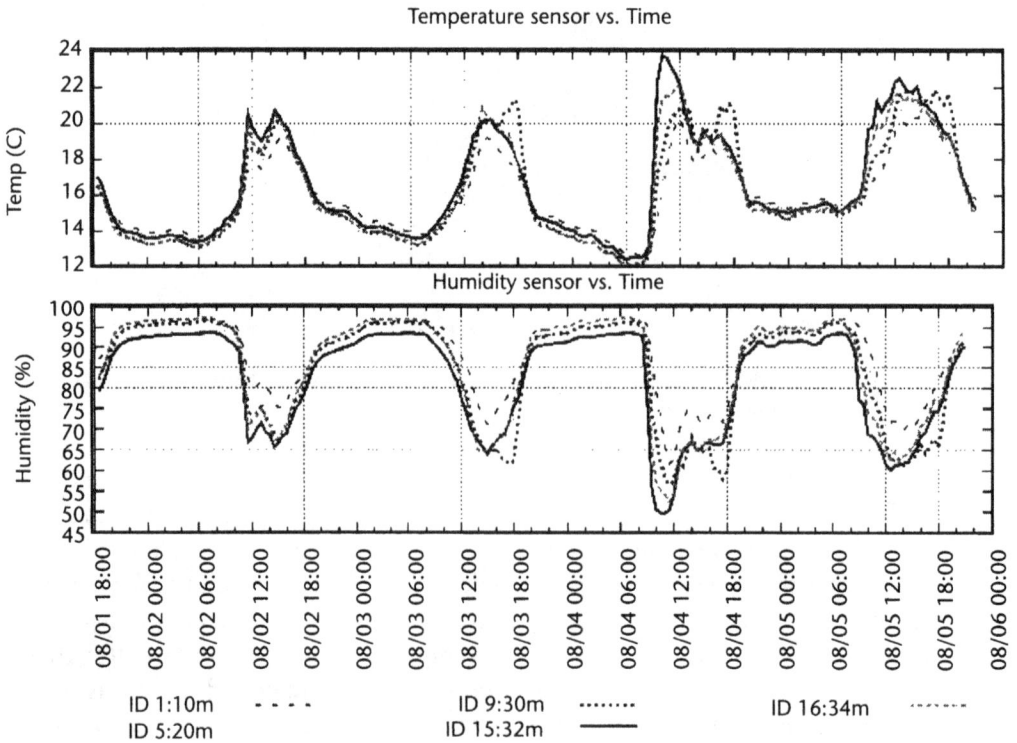

(b)

Figure 14.8   (a) Temperature and (b) humidity readings from five sensors in the botanical garden.

days. Looking closely at the line for sensor 1, it is apparent that not only is the mote cooler and wetter during the day, but at night it is somewhat warmer and less humid than the exposed sensors near the treetop. Furthermore, the rate of change of humidity and temperature at the bottom of the tree is substantially less than at the top: sensor 1 is the last sensor to begin to warm up and the last sensor to cool off.

14.6   Related Work

Apart from the systems we have focused on in this chapter, a number of other approaches to high-level sensor-network programming have been proposed. Cougar [11] is a system similar to TinyDB that provides an SQL-like query interface to real-time sensor data; the motivation between the two systems is quite similar, although each has focused on a different set of optimizations and power-efficient routing techniques. Cougar has not been evaluated through a complete implementation or any real-world deployments.

One of the first proposed communication models for sensor networks was Directed Diffusion [10], which provides a framework for distributed event detection and propagation. The network is seeded with a set of *interests* generated by one or more data *sinks*, and matching sensor data is routed to the appropriate sink. GHT [25] provides a hash table–like storage mechanism within sensor networks, using a variant of GPSR [26] to route data geographically to the node responsible for its storage. Other communication abstractions include Spatial Programming [27], DIFS [28], SPIN [29], and DIMENSIONS [30]. These systems are generally focused on developing a communication or aggregation model for specific applications.

Maté [31] is a tiny virtual machine for sensor networks, providing an alternative to programming in a low-level language. Maté programs, or *capsules*, are compact enough to fit into a single radio message, allowing the network to be rapidly reprogrammed by disseminating programs over the radio. In addition, nodes can emit new program fragments that "infect" other nodes in a viral fashion, triggering new behaviors.

IrisNet [32], Medusa [33], and Hourglass [34] are three systems that provide remote access to multiple geographically distributed sensor networks or other sources of streaming data. These systems focus on discovery of remote sensors, mapping data types and schemas across multiple services, and dynamic optimization of service and query operator placement across hosts on the network. They are not strictly focused on programming an individual sensor network but rather on providing a framework for collecting data from multiple sensor networks.

14.7   Future Directions

Macroprogramming raises a number of new opportunities for language, algorithm, and protocol design. One of our goals is to produce an extremely expressive, complete programming language that captures the operation of the sensor network at a global level. TinyDB is a step in this direction but is focused on data collection; a more general language is needed to support a wider range of applications. We are developing a functional macroprogramming language called *Regiment* that is based on a data model called *region streams*. A region stream represents a spatially distributed, time-varying collection of node states, for example, the set of sensor values across all nodes within an abstract region. Regiment is a purely functional language, which gives the compiler considerable leeway in terms of realizing region-stream operations across sensor nodes and exploiting redundancy within the network.

Another open problem is that of global resource allocation across a sensor network. Existing programming models often require that sensor nodes operate on a fixed schedule; for example, TinyDB nodes collect, aggregate, and transmit sensor data at a periodic rate that is identical across the network. However, in many cases it is desirable to tune the behavior of individual nodes based on their location or proximity to physical phenomena of interest. One direction we are exploring is that of using *virtual markets* to allocate sensor resources dynamically [35]. In this approach, individual sensor nodes are modeled as self-interested agents that operate in a virtual market and receive profit for performing local actions in response to globally advertised price information. Sensor nodes run a very simple cost-evaluation function, and global behavior is induced through- out the network by advertising price information that drives nodes to react. Nodes adapt their behavior according to changes in prices and network conditions. The prices can be dynamically tuned by the centralized market maker to meet systemwide goals of lifetime, accuracy, or latency based on the needs of the sensor-network programmer. The "macroprogram" is therefore encoded in the process to update price information in response to changing network conditions.

A third open problem has to do with the management of loss, uncertainty, and imprecision in macroprogramming systems. Though systems like TinyDB provide an attractive abstraction for data collection, they largely ignore any of the sources of uncertainty that are endemic to sensor networks: readings may be lost due to network contention; sensors themselves are imprecise; readings may not be available from a particular location or time.

We have begun to explore issues related to these concerns by building statistical and probabilistic models of uncertainty in sensor networks. The Barbie-Q: A Tiny Model Query System (BBQ) [36] captures correlations between sensors using time-varying multidimensional models, which allows confidence (ε, δ) bounds to be formulated for all sensors given readings from just a few locations or times. By quantifying uncertainty in this way, it is possible to make much more concrete statements about the effect of missing values or imprecise sensor readings. Similarly, such models can be used to detect unusual or faulty sensor readings that lie far outside the predicted range of sensor values.

Finally, we have are exploring extensions to the tuple-based model presented in TinyDB to deal with queries over *sequences*. Such extensions will make it possible to execute queries that extract trends and patterns from groups of readings; we expect to use well-known signal-processing techniques (like wavelets or Fourier transforms) to facilitate these extensions.

References

[1]   Welsh, M., "Exposing Resource Trade-offs in Region-based Communication Abstractions for Sensor Networks," *Proc. 2nd ACM Workshop on Hot Topics in Networks (HotNets-II)*, Cambridge, MA, November 2003.

[2]   Welsh, M., and Mainland G., "Programming Sensor Networks Using Abstract Regions," *Proc. 1st USENIX/ACM Symposium on Networked Systems Design and Implementation (NSDI '04)*, San Francisco, CA, March 2004.

[3]   Brooks, R., Ramanathan P., and Sayeed A., "Distributed Target Classification and Tracking in Sensor Networks," *Proc. IEEE*, November 2003, Vol. 91, No. 8, pp. 1163–1171.

[4]   Li, D., et al., "Detection, Classification and Tracking of Targets in Distributed Sensor Networks," *IEEE Signal Processing Magazine*, Vol. 19, No. 2, March 2002.

[5]   Xu, Y., and Lee W. C., "On Localized Prediction for Power Efficient Object Tracking in Sensor Networks," *Proc. 1st International Workshop on Mobile Distributed Computing*, Providence, RI, May 2003.

[6]   Center for Embedded Network Sensing, contaminant transport monitoring, at http://cens.ucla.edu/Research/Applications/ctm.htm.

[7]   Carney, D., et al., "Monitoring Streams—A New Class of Data Management Applications," *Proc. VLDB*, Hong Kong, China, 2002.

[8]   Motwani, R., et al., "Query Processing, Approximation and Resource Management in a Data Stream Management System," *Proc. CIDR*, Asilomar, CA, 2003.

[9]   Gehrke, J., Korn F., and Srivastava D., "On Computing Correlated Aggregates over Continual Data Streams," *Proc. ACM SIGMOD Conference on Management of Data*, Santa Barbara, CA, May 2001.

[10] Intanagonwiwat, C., Govindan R., and Estrin D., "Directed Diffusion: A Scalable and Robust Communication Paradigm for Sensor Networks," *Proc. International Conference on Mobile Computing and Networking*, Boston, MA, August 2000.

[11] Yao, Y., and Gehrke J. E., "The Cougar Approach to In-Network Query Processing in Sensor Networks," *ACM Sigmod Record*, Vol. 31, No. 3, September 2002.

[12] Shatdal, A., and Naughton J., "Adaptive Parallel Aggregation Algorithms," *Proc. ACM SIGMOD*, San Jose, CA, 1995.

[13] Madden, S., et al.. "The Design of an Acquisitional Query Processor for Sensor Networks," *Proc. ACM SIGMOD*, San Diego, CA, 2003.

[14] Ganesan, D., et al., "Complex Behavior at Scale: An Experimental Study of Low-Power Wireless Sensor Networks," Technical Report UCLA/CSD-TR 020013, UCLA Computer Science, 2002, at http://lecs.cs.ucla.edu/Publications/papers/Deepak-Empirical.pdf.

[15] Woo, A., Tong T., and Culler D., "Taming the Underlying Challenges for Reliable Multihop Routing in Sensor Networks," *Proc. SenSys '03*, Los Angeles, CA, November 2003.

[16] Johnson, D. B., and Maltz D. A., "Dynamic Source Routing in Ad Hoc Wireless Networks," *Mobile Computing*, Vol. 353, No. 1, 1996.

[17] Perkins, C. E., "Ad Hoc on Demand Distance Vector (AODV) Routing," Internet draft, October 1999.

[18] Park, V., and S. Corson. "Temporally Ordered Routing Algorithm (TORA) Version 1 Functional Specification," Internet draft, 1999.

[19] Chang, J. H., and Tassiulas L., "Energy Conserving Routing in Wireless Ad Hoc Networks," *Proc. 2000 IEEE Computer and Communications Societies Conference on Computer Communications (INFOCOM-00)*, 2000, Tel Aviv, Israel, pp. 22–31.

[20] Pottie, G., and Kaiser W., "Wireless Integrated Network Sensors," *Communications of the ACM*, Vol. 43, No. 5, May 2000, pp. 51–58.

[21] Ye, W., Heidemann J., and Estrin D., "An Energy-Efficient MAC Protocol for Wireless Sensor Networks," *Proc. IEEE Infocom*, New York, NY, 2002, pp. 1567–1576.

[22] Madden, S., *The Design and Evaluation of a Query Processing Architecture for Sensor Networks*. Ph.D. thesis, University of California, Berkeley.

[23] Crossbow, Inc., STARGATE technical report, 2004, at http://www.xbow.com/Products/Product_pdf_files/Wireless_pdf/6020-0049-01_A_STARGATE.pdf.

[24] Baldocchi, D., Wilson K., and Gu L., "How the Environment, Canopy Structure and Canopy Physiological Functioning Influence Carbon, Water and Energy Fluxes of a Temperate Broad-Leaved Deciduous Forestan Assessment with the Biophysical Model CANOAK," *Tree Physiology*, Vol. 22, 2002, pp. 1065–1077.

[25] Ratnasamy, S., et al., "GHT: A Geographic Hash Table for Data-centric Storage in Sensornets," *Proc. 1st ACM International Workshop on Wireless Sensor Networks and Applications (WSNA)*, Atlanta, GA, September 2002.

[26] Karp, B., and Kung H. T., "GPSR: Greedy Perimeter Stateless Routing for Wireless Networks," *Proc. Sixth Annual ACM/IEEE International Conference on Mobile Computing and Networking (MobiCom 2000)*, Boston, MA, August 2000.

[27] Borcea, C., et al., "Spatial Programming Using Smart Messages: Design and Implementation," *Proc. 24th International Conference on Distributed Computing Systems (ICDCS 2004)*, Tokyo, Japan, March 2004.

[28] Greenstein, B., et al., "DIFS: A Distributed Index for Features in Sensor Networks," *Proc. 1st IEEE International Workshop on Sensor Network Protocols and Applications*, Anchorage, AK, May 2003.

[29] Heinzelman, W., Kulik J., and Balakrishnan H., "Adaptive Protocols for Information Dissemination in Wireless Sensor Networks," *Proc. 5th ACM/IEEE Mobicom Conference*, Seattle, WA, August 1999.

[30] Ganesan, D., et al., "An Evaluation of Multi-resolution Search and Storage in Resource-Constrained Sensor Networks," *Proc. 1st ACM Conference on Embedded Networked Sensor Systems (SenSys)*, Los Angeles, CA, November 2003.

[31] Levis, P., and Culler D., "Maté: A Tiny Virtual Machine for Sensor Networks," *Proc. 10th International Conference on Architectural Support for Programming Languages and Operating Systems (ASPLOS X)*, San Jose, CA, 2002.

[32] Nath, S., et al., "IrisNet: An Architecture for Enabling Sensor-Enriched Internet Service," Technical Report IRP-TR-03-04, Intel Research Pittsburgh, June 2003.

[33] Zdonik, S., et al., "The Aurora and Medusa Projects," *Bulletin of the Technical Committe on Data Engineering IEEE Computer Society*, March 2003, Vol. 26, No.1, pp. 3–10.

[34] Ledlie, J., et al., "Open Problems in Data Collection Networks," *Proc. 11th ACM SIGOPS European Workshop*, Leuven, Belgium, September 2004.

[35] Kang, L., et al., "Using Virtual Markets to Program Global Behavior in Sensor Networks," *Proc. 11th ACM SIGOPS European Workshop*, Leuven, Belgium, September 2004.

[36] Deshpande, A., et al., "Model-driven Data Acquisition in Sensor Networks," *Proc. VLDB*, Toronto, Canada, 2004.

CHAPTER 15

Sensor-Network Security, Privacy, and Fault Tolerance

Jing Deng, Richard Han, and Shivakant Mishra

15.1 Introduction

Security is a key concern in many WSN applications. For military settings, dispersal of WSNs into an adversary's territory enables the detection and tracking of enemy soldiers and vehicles. For home or office environments, indoor sensor networks offer the ability to monitor the health of the elderly and to detect intruders via a wireless home security system. In each of these scenarios, lives and livelihoods may depend on the timeliness and correctness of the sensor data obtained from dispersed sensor nodes. Security methods are helpful in protecting the delivery and correctness of such sensor data.

The computing resources of microsensor nodes are highly limited, which influences the design of WSN security techniques. For example, both sensor motes [1] and nymphs [2] provide a standard hardware reference for today's microsensor nodes. Both utilize the ATMEL ATmega128L microcontroller, which features an 8-MHz processor, 4-KB SDRAM memory, and 128-KB internal flash memory. The limited memory and processor speed severely constrain the type of security algorithms that can execute on microsensor nodes. A further constraint is the radio, which typically operates at a transmission rate of 19.2 kbps. Another important limitation is the energy lifetime of these nodes, which is only a few days given continuous operation on two fresh AA batteries. As a result, WSN security features need to be energy efficient.

WSN security techniques are also heavily influenced by the framework and deployment environment of a sensor network. The typical structure of a WSN is illustrated in Figure 15.1. Sensor nodes collect sensor data from the environment. Sensor nodes also form a multihop wireless network that routes sensor data to one or more base stations. This routing topology forms a tree rooted at the base station. The base station collects data from sensor nodes and acts as a gateway to a wired backbone, forwarding the data to the end user of the WSN. In general, the computing resources of each base station are much greater than those of the sensor nodes. A recent trend has been to introduce intermediate nodes called aggregators, thereby increasing the heterogeneity of WSNs. These aggregators may summarize or aggregate data from downstream children in the routing tree [3] (e.g., by computing averages) in order to reduce bandwidth and save energy, and they may perform other tasks such as in-network processing [4] and acoustic-signal processing [5]. Aggregators may also have resources and capabilities that are intermediate between base stations and leaf sensor nodes. The large number of sensor nodes, moderate number of aggregators, and small number of base stations collectively form an asymmetric and hierarchical wireless multihop network. WSN security techniques are often tailored to exploit this asymmetry.

Early work on sensor-network security addressed the threats to WSNs and the requirements for WSN security [6, 7]. Later work analyzed various kinds of denial-of-service (DoS) attacks against WSNs and proposed countermeasures for these attacks [8]. Other threats were summarized in [9]. From 2001 to the time of this writing, research has grown exponentially in a variety of areas in sensor-network security. In the rest of this chapter, we will first describe threat models considered in WSN security in Section 15.2. Section 15.3 summarizes the general design goals for WSN security countermeasures. Sections 15.4 and 15.5 introduce lower-level fundamental primitives that underly WSN security. Section 15.4 discusses cryptographic primitives, while Section 15.5 discusses pairwise-key-establishment techniques for sensor networks. Section 15.6 introduces higher-level security concerns, namely secure and fault-tolerant WSN routing algorithms. Section 15.7 describes techniques for secure aggregation. Section 15.8 focuses on physical protection of sensor nodes. Section 15.9 discusses issues relating to privacy and anonymity in sensor networks.

15.2 Security Threats in WSNs

When WSNs are deployed in battlefields or used for monitoring in homeland security, WSNs can become the target of adversaries, as shown in Figure 15.1. These adversaries can pose many security threats [8, 9]. For our discussion, we assume the

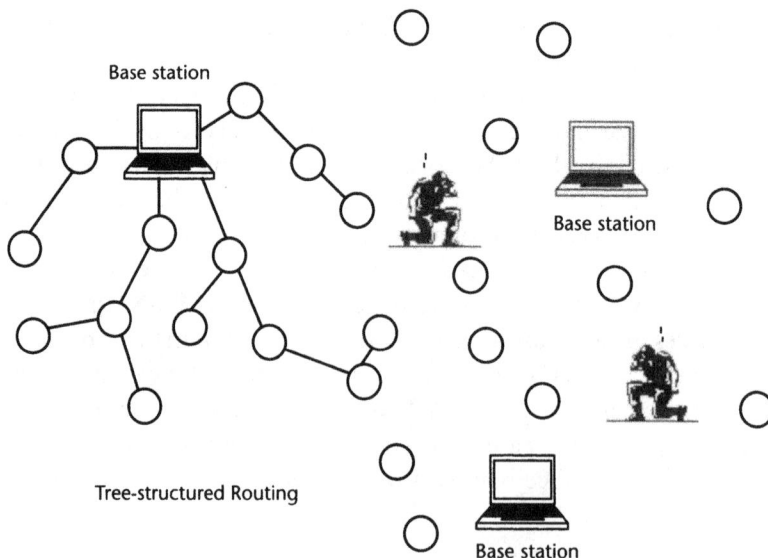

Figure 15.1 Sensor network with routing tree rooted in base station. Adversaries can appear anywhere in the network.

existence of an end user who typically receives sensor data through one or more base stations. This end user is typically remote from the actual in situ deployment of the sensor network. For example, each base station may be a satellite uplink to a remote end user who is monitoring the field of sensors through the base stations.

An adversary can deceive this end user by feeding base stations incorrect data from sensor nodes in a variety of ways. The contents of packets may be modified. Also, entirely new, false packets may be created and injected into the network by spoofing sensor nodes or aggregators. False, physical sensor events may also be generated by physically distorting the sensing environment (e.g., generating false heat sources or dummy acoustic noise). An adversary can also attack an aggregator node and generate forged aggregation results.

An adversary can also prevent the reception of sensor data by the end user in any of a variety of ways. First, an adversary can block the communication between a base station and sensor nodes. This can be accomplished by analog jamming of signals or by digital jamming in the form of DoS attacks that flood the network, base stations, or both. Targeted DoS attacks on strategic nodes in the WSN (e.g., partition points) can also effectively block communication of large parts of the WSN with the base station. Preventing communication between base stations and sensor nodes can also be accomplished by setting up incorrect routing information so that traffic goes to the wrong destination or loops. One form of incorrect routing is to spoof the base station and deceive nodes into rerouting all packets to the spoofed base station instead of the real base station. In the extreme, this effectively allows the adversary to control the entire sensor network and deny the end user access to any sensor data. Second, an adversary can seek to destroy the base station itself and, thereby, prevent reception of sensor data by the end user. Destruction of the base station can be accomplished by monitoring the volume and direction of packet traffic toward the base station so that the base station's location is eventually revealed. Destruction of the base station can also be accomplished by listening to RF signals to localize and triangulate the location of the base station.

A third threat is eavesdropping, which is made easier by wireless hop-to-hop communication. In this case, the adversary acts as an observer, seeking to understand what the end user knows about the adversary. In addition, eavesdropping can be used to track and deduce the location of the base station(s) for destruction.

Many of the above threat models are based on the idea that the adversary has compromised one or more sensor nodes. In order to mount analog or digital attacks, such as jamming, DoS flooding, packet modification, base station spoofing, or even eavesdropping, the adversary must know the spectrum of communication, as well as the digital networking protocols governing the WSN. These can be obtained most directly by capturing and compromising sensor nodes, which are deployed in situ in the battlefield and are therefore at physical risk. An adversary faces difficulty in deducing the WSN's spectrum of radio communication via simple frequency scanning because spread-spectrum techniques typically hide this signal in the noise quite effectively, especially direct-sequence CDMA. Even if an adversary knows the spectrum, analog jamming only has local effectiveness.

In contrast, the compromise of a sensor node is desirable because it enables an adversary to launch a large array of digital attacks that have the power to disable the entire WSN or deceive the end user. The adversary can extract secret information, such as keys. Once an adversary has this information, it may reprogram the compromised node into a malicious node and launch many types of far-ranging attacks on a sensor network, which will be described later.

Sensor-network research often assumes that there is an adversary capable of compromising a sensor node. This assumption is based on the expected mismatch between the protections afforded resource-poor sensor nodes and the tools at the disposal of the resource-rich adversary to break into a sensor node. For example, differential power attacks on resource-poor smart cards have been able to extract the card's secret key [10]. Recent work has shown the ease and speed with which today's sensor nodes can be compromised using simple off-the-shelf technology [11], which will be described in Section 15.8.

The above threats are magnified if an adversary is assumed to have several agents at its disposal who can each work in parallel to compromise different sensor nodes or disrupt the WSN in a widespread geographic manner. The agents may collude together in collaborative attacks on a WSN, but their numbers are typically assumed to be small relative to the size of the WSN. These are the most sophisticated forms of attack and include wormhole attacks, which are described later.

15.3   Design Goals and Challenges of WSN Security

Given these threat models, the uniqueness of the research challenge faced by designers of security techniques for WSNs is that three tough issues are confronted *simultaneously*: wireless communication, severe resource constraints, and in situ deployment. The design and implementation of secure mechanisms in WSNs must simultaneously address all three difficult research challenges.

First, wireless communication among the sensor nodes increases the vulnerability of the network to eavesdropping, unauthorized access, spoofing, replay, and DoS attacks. Second, the sensor nodes themselves are highly resource constrained in terms of limited memory, CPU, communication bandwidth, and especially battery life. These resource constraints limit the degree of encryption, decryption, and authentication that can be implemented on individual sensor nodes. Most of the security research to this point has made the standard assumption that today's representative microsensor nodes will be unable to support traditional security mechanisms, such as compute-intensive public-key cryptography. For example, early attempts to implement Rivest-Shamir-Adleman (RSA) cryptography on motes did not meet with success. Because of this, many traditional, secure wireless ad hoc routing schemes that employ public-key approaches were not deemed suitable for sensor networks [6, 12–15]. However, recent advances have demonstrated that, for example, elliptic curve cryptography is feasible on Mica2 motes, although with caveats, as described later. Third, WSNs face the added physical security risk of being deployed in the field so that individual sensor nodes can be obtained and subjected to attacks from a potentially well-equipped intruder in order to compromise a single resource-poor node.

Following a successful attack, a compromised sensor node could then be used to instigate many of the aforementioned malicious activities. In addition, physical risk means nodes can be destroyed, damaged accidentally, or depleted of energy over time. As a result, fault tolerance is a key concern in WSNs.

Other security regimes confront these challenges separately or not at all. For example, securing Internet servers usually assumes wired communication, abundant resources compared to sensor nodes, and physically secured locations. Smart cards typically face severe resource constraints and may be at physical risk but do not typically initiate wireless communication.

Security techniques designed for WSNs should therefore seek to attain the following general goals:

- Lightweight memory footprint;
- Computational efficiency;
- Energy efficiency;
- Bandwidth efficiency;
- Intrusion tolerance due to compromised nodes;
- Fault tolerance;
- Cognizance of wireless communication risks.

15.4 Cryptographic Primitives

15.4.1 SPINS: Security Protocols for Sensor Networks

A. Perrig et al. first proposed a set of cryptographic primitives for highly resource-constrained sensor nodes in 2001 [16]. SPINS was designed for the first generation of Rene motes, which had exceedingly scarce resources, namely 8 KB of flash memory for code space, 512 bytes of RAM, and 10 kbps of radio bandwidth. Only about 4,500 bytes of flash were available for security and applications. In comparison, the second generation of Mica2 motes has an order of magnitude more flash and RAM and twice the radio bandwidth. SPINS designed and implemented two security primitives for the first generation of sensor nodes, SNEP and μTESLA.

15.4.1.1 SNEP

Sensor network encryption protocol (SNEP) is a secure mechanism that efficiently protects end-to-end data confidentiality, integrity, authenticity, and freshness for unicast messages between two nodes, possibly separated by multiple hops. Every packet is encrypted with a stream cipher based on a symmetric key operating in block cipher counter mode. Every packet also contains a message authentication code

(MA to protect its integrity, with the counter incorporated in the message authentication code calculation to prevent replay attacks. To save bandwidth, the counter is not sent in the packet. The receiver synchronizes the counter with the sender by receiving a packet. If some packets are lost, sender and receiver need to resynchronize the counter of the packets.

15.4.1.2 μTESLA

Perrig et al. have also proposed a secure broadcast mechanism, Microtimed Efficient Stream Loss-Tolerant Authentication (μTESLA) [16]. Secure broadcast poses a challenge in WSNs because the earliest assumptions were that resource constraints dictated symmetric-key approaches. The simplest approach to distributing a global symmetric key to allow each node to decrypt or authenticate broadcast packets is vulnerable to compromise of even a single sensor node, which will reveal the global key to an adversary. Such compromise would allow arbitrary spoofing or alteration of broadcast packets, as well as DoS flooding attacks.

In μTESLA, when the base station broadcasts a message, it generates a message authentication code for that message with a symmetric key. That key is a number on a *one-way hash chain* (OHC), as shown in Figure 15.2. OHCs are especially useful for WSN security because they have a compact memory footprint yet enable rapid and secure verification. Every sensor node is preconfigured with the initial number of the OHC, called the *commitment*, so that when the base station later releases the key, each sensor node can verify whether the key is valid. To securely broadcast a message, say M_3, the base station first sends the message and its message authentication code generated by a key. At a later time, after every node has received the message, the base station releases the key according to a delayed key-disclosure schedule (e.g., the release of K_2 is delayed until release two time slots in the future). When a sensor node receives the key, it first verifies if the key is on the OHC by executing the one-way function. If verified, then the stored message's message authentication code is tested to verify whether it was generated by the base station. Before the key is released, an adversary will be unable to generate a spoofed packet that goes undetected because the adversary will not have access to the key. As the key is released, an adversary will be unable to generate a spoofed packet that passes muster because μTESLA only allows the released key to operate on packets that were broadcast and stored several time epochs in the past.

One problem in μTESLA is that when many broadcast messages are lost, sensor nodes have to spend a long time repeatedly executing the one-way hash function to verify the OHC number. To address this problem, D. Liu and P. Ning propose an efficient distribution of key-chain commitments for μTESLA [17]. Multilayer OHCs are proposed for OHC generation and maintenance in WSNs. The idea is to use an upper-layer OHC to manage the lower-layer OHC (i.e., the upper-layer OHC is used to bootstrap the initial number of the lower-layer OHC). That potentially reduces the computing time to verify the OHC number. Also, the whole OHC system is more robust to packet loss.

μTESLA's delayed key-release schedule is dependent upon the requirement that loose time synchronization be maintained for timing of epochs throughout the sensor network. Another issue in μTESLA is delayed message authentication. When a sensor node receives the broadcast message, it cannot verify the message until the

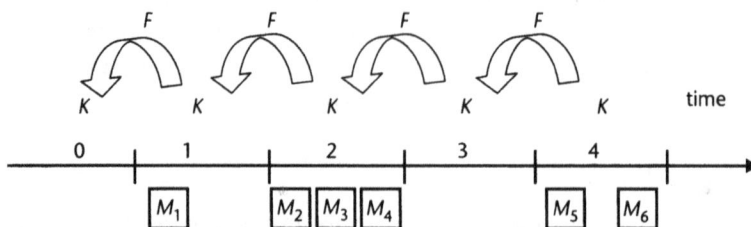

Figure 15.2 OHC, used in μTESLA to release keys and in INSENS as one-way sequence numbers.

base station releases the key of the message authentication code several time slots later. In the intervening time, an adversary can launch DoS attacks, flooding the network with unverifiable packets.

15.4.2   TinySec

The first fully implemented sensor-network link-layer security library is TinySec, built by Karlof, C., N. Sastry, and D. Wagner [18]. The TinySec project began in 2002, with TinyOS 1.0. As a link-layer security architecture, TinySec protects the authenticity, integrity, and confidentiality of messages between neighbor nodes. TinySec adheres to the principles of lightweight implementation in terms of low computing cost, low communication overhead, lightweight memory usage, and minimal energy consumption. TinySec's mechanisms are designed to be easy to learn and use. Two security operations are provided: authenticated encryption (TinySec-AE) and authentication only (TinySec-Auth). TinySec uses a four byte message authentication code to protect the integrity and authenticity of a message; it also uses Cipher Block Chaining (CBC) mode for encryption, with an initialization vector (IV) based on a two-byte counter and the header of the TinyOS packet. CBC mode is chosen because it is resilient to repeated IVs. Two lightweight cryptographic algorithms are implemented in TinySec: RC5 and Skipjack. TinySec consumes very little memory, adds one to five extra bytes compared with a standard TinyOS packet, and increases energy consumption and latency by less than 10%. TinySec is fully integrated with TinyOS official releases and has been used for other security research on sensor networks.

15.4.3   Public-Key Schemes in a Sensor Node

Early sensor-network security research believed that public-key schemes were too resource intensive for sensor nodes. Public-key schemes face memory limitations, slower CPUs that limit the speed of execution, limited communication bandwidth, and energy constraints. Both RSA and elliptic curve cryptographic (ECC) intensively use modular multiplication operations, which consumes large amounts of RAM, potentially causing stack overflow [19–21]. For RSA, the key size is also very large (at least 1,024 bits). Transmission of the key becomes an issue given the short packet sizes of sensor nodes. Recently, however, several research groups have made progress toward implementing public-key algorithms on microsensor nodes. R. Watro et al. implemented the RSA public-key cryptographic algorithm on Berkeley motes [20]. D. J. Malan, M. Welsh, and M. D. Smith implemented ECC public-key algorithm on Berkeley motes [21]. The ECC is used in the Diffie-Hellman protocol to generate a shared key for a pair of sensor nodes. The time for key generation is about 34 seconds. N. Gura et al. implemented both RSA and elliptic-curve algorithms on two 8-bit CPUs, the Chipcon CC1010 and Atmel ATmega128, which is used for Mica2 motes. In their implementation, the modular multiplication operations are carefully optimized so that the encryption time for ECC is about several seconds [22]. ECC is emerging as a candidate for public-key schemes in sensor networks since it employs a small key size (160 bits is secure enough for sensor node). In the future, we expect that public-key schemes will play an increasingly important role for sensor-network security.

15.5 Pairwise-Key Management

Pairwise-key management is an important security primitive for WSNs since pairwise keys can be used to secure hop-by-hop communication between neighbors and to bootstrap higher-level security mechanisms. For a small sensor network in which every node can communicate with the base station, it is not difficult to establish keys between each sensor node and the base station. However, for a large-size sensor network, pairwise-key management is not easy because of the severe resource constraints of sensor nodes. A single global key provides brittle security and is susceptible to the compromise of just one node. If there are n nodes in a WSN, then predistribution of $n - 1$ keys to each sensor node enables each node to establish pairwise keys securely with each of the other sensor nodes. However, as the size the network expands, this approach becomes more infeasible due to memory limitations and is also vulnerable to compromise.

To address these problems, a number of pairwise-key-predistribution schemes have been proposed that have modest memory requirements for each sensor node and are resilient to the compromise of one or more nodes. In random-key-distribution schemes [23–26], every node is randomly preconfigured with a number of keys in each sensor node from a large key pool. Then, after these nodes are deployed, each pair of neighbor nodes uses its preconfigured keys to set up a pairwise key. These schemes are scalable in terms of the size of sensor networks and are lightweight for resource-constrained sensor networks.

The first random-key preconfiguration scheme was proposed by L. Eschenauer and V. Gigor [23] and is shown in Figure 15.3. The entire system maintains a key pool, which contains a large number of keys. Each sensor node is predistributed a number of keys selected from the key pool. If a node u shares one or more keys with

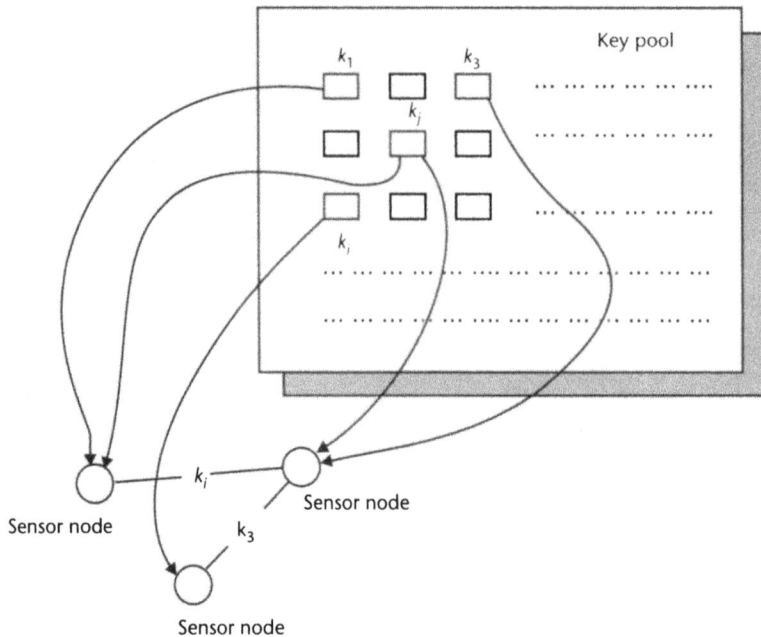

Figure 15.3 Random-key-predistribution schemes.

one of its neighbor nodes v, then the link between u and v can be securely established. However, if u is compromised, an adversary can only obtain the keys in u and not the other keys in the key pool. If the size of the key pool and the number of keys preconfigured in each sensor node are set properly, secure paths will be established with high probability such that the sensor network forms a connected graph. If a single node is compromised, then the adversary only has access to a small subset of the keys. However, an adversary can still use the keys in that compromised node to inject any number of malicious nodes into the network. Other legitimate nodes will not have a way to identify and restrain those malicious nodes.

H. Chan, A. Perrig, and D. Song [24] improved this scheme by introducing a q-composite random-key-predistribution scheme. Instead of using one shared key to set up the pairwise key, a pair of neighbor nodes uses their q common keys to set up their pairwise keys. This improves the resilience to node compromise.

W. Du et al. [25] and D. Liu and P. Ning [26] further improved the random-key-distribution scheme by integrating Blom's key predistribution mechanism [27] into random-key predistribution. They propose to use polynomials instead of random numbers as the key. The bivariate t-degree polynomial $f(x,y) = \sum_{i,j=0}^{t} a_{i,j} x^i y^j$, satisfying the property that $f(x, y) = f(y, x)$, is used to generate pairwise keys. If a node u is preconfigured with an $f(x, y)$-"generator" $f(i, y)$ and node v is preconfigured with $f(j, y)$, u and v can establish their pairwise key as $f(i, j) = f(j, i)$ by simply knowing each other's IDs. If node u is compromised, an adversary can only know $f(i, y)$, not $f(j, y)$. An adversary has to compromise at least $t + 1$ different $f(x, y)$-"generators" before knowing all keys generated by $f(x, y)$. These schemes have substantially improved network resilience against node capture, while enabling easy setup of pairwise keys between neighbor nodes with low communication overhead.

A different approach to pairwise keys was taken by S. Zhu, S. Setia, and S. Jajodia, who proposed a robust and lightweight method of setting up pairwise keys by using a temporary master key, called LEAP [28]. In this scheme, every node is preconfigured with an initial master key. Each sensor node uses a pseudorandom function and the master key to generate its own key and its pairwise keys with neighbor nodes. The master key is only kept in a sensor node for a short time and is then erased, so an adversary doesn't have enough time to compromise a node and capture the master key inside it. Similar to random-key schemes, LEAP also supports secure joining of new nodes. One advantage of LEAP is that an adversary cannot inject other malicious nodes, even if he can compromise existing nodes after the master keys are erased from these nodes. An issue with the temporary-master-key approach is that any node has to establish its pairwise keys within a short time while its master key exists in its memory. After the master key is erased, that node cannot establish any keys with other nodes.

15.6   Secure and Fault-Tolerant WSN Routing

By attacking data routing in a sensor network, an adversary can disrupt and damage a sensor network in a variety of ways. For example, a base station can fail to collect data from sensor nodes, and sensor nodes can waste their energy routing data in a loop. C. Karlof and D. Wagner [29] have analyzed numerous attacks on WSN routing. The major current routing protocols for WSNs, such as TinyOS beaconing, Directed Diffusion, geographical routing, and cluster-based routing, have been observed to lack security protections. Adversaries can launch various kinds of attacks against these routing protocols, including spoofed routing information, selective forwarding, sinkhole attacks, the Sybil attack [30], the wormhole attack [31], the HELLO flood attack, and an acknowledgement spoofing attack. In a sinkhole attack [29], a malicious node advertises that it is close to the base station, even though it may not be close, thereby attracting other node's data traffic. In a Sybil attack, a malicious node creates multiple false identities. In a wormhole attack, an adversary tunnels messages received in one part of the network and replays them in another part of the network. By launching a wormhole attack, an adversary can set up incorrect routing information among sensor nodes. Also, many protocols use a broadcast HELLO message to identify a node to its neighbors. If an adversary can send such a HELLO message with enhanced signal strength, it can capture many nodes as its neighbors, even though those nodes' data cannot reach the adversary. This HELLO attack is similar to the rushing attack in ad hoc network routing [32].

To defend against these attacks, Karlof and Wagner propose several strategies, including the use of a trusted base station as a verifier, a global-key scheme, and bidirectional verification. Employing the base station to detect Sybil attacks and HELLO attacks is suitable for a small-sized network but not for large sensor networks due to the heavy traffic overhead toward the base station. A global-key scheme can be used to defend against an outside attacker. In bidirectional verification, a node will not accept another node as its neighbor until it can send and receive data from that node. This technique can defend against the HELLO attack unless an adversary has powerful data-transmission equipment and very sensitive data-receiving equipment. However, the sinkhole and wormhole attacks still pose significant challenges.

15.6.1 INSENS Intrusion-Tolerant Routing

Many sensor-network routing protocols employ a tree-based routing structure rooted at a base station because of its simplicity and suitability for data collection. The intrusion-tolerant routing system for sensor networks (INSENS) proposed in [19, 33] provides a secure mechanism for building such a tree-like routing structure rooted in a base station. INSENS was designed with the principle that network routing should tolerate intrusions (i.e., INSENS provides robust communication between a base station and the sensor nodes in the presence of node failures, communication failures, and a small number of node compromises). This protocol comprises a route discovery phase and a data forwarding phase. The route-discovery phase ascertains the topology of the sensor network and builds appropriate forwarding tables at various nodes. First, the base station floods (limited flooding) a *request message* to all the reachable sensor nodes in the network. Second, each sensor node sends its neighborhood topology information back to the base station using a *feedback message*. Third, the base station authenticates the neighborhood information, constructs a topological picture of the network, computes the forwarding tables for each sensor node, and sends the tables to the respective nodes using a *routing update message*. Afterwards, the data-forwarding phase commences with forwarding of data from each sensor node to the base station and vice versa.

To address resource constraints, symmetric-key cryptography is employed, and the resource-rich base station is chosen as the central point for computation and dissemination of the routing tables. INSENS assumes that each sensor node is preconfigured with a pairwise key shared only with the base station. INSENS adheres to the following additional design principles.

15.6.1.1   One-Way Sequence Numbers to Limit DoS Attacks

First, to prevent DoS-style flooding attacks, individual nodes are not allowed to broadcast to the entire network. Only the base station is allowed to broadcast. The base station embeds in each broadcast message an OHC-derived one-way sequence (OWS) number so that individual nodes cannot arbitrarily spoof the base station and thereby flood the network. Sensor nodes are restricted to only unicasting a packet, and then only to the base station. This technique was first proposed by Y. Hu, A. Perrig, and D. Johnson [34] to authenticate the broadcast message from an ad hoc networking node.

Every node possesses a globally known cryptographic one-way hash function F and initial sequence number S_0, both distributed a priori. The base station will broadcast its first packet with sequence number S_1. Every node will be able to verify that S_1 could only have originated from the base station by executing F. An adversary will be unable to guess the next sequence number, in this case S_1, due to the one-way nature of F, despite having the current sequence number and F. The adversary will therefore be unable to generate valid new packets arbitrarily since they will be dropped because they have the wrong sequence number. This limits the ability of an adversary to launch DoS flooding attacks by posing as the base station.

It is still possible to mount a rushing attack or a wormhole attack while a new OWS number is propagating. In both attacks, an adversary will repeat an OWS number with modified packet contents at a distant location in the network. The rushing attack can be mitigated by the bidirectional verification [29, 35], which prevents an adversary from expanding its transmission power to outrush the propagation of the original correct broadcast message. The wormhole attack is more difficult to mount because it requires collusion between two adversaries. In both attacks, the disruption is limited in its scope because the OWS message is flooding outward and is presumed to reach most of the network before the modified malicious packet. The OWS approach thus provides an effective anti-DoS countermeasure.

15.6.1.2   Secure Topology Discovery via a Nested Message Authentication Code

To prevent advertisement of false routing data, control-routing information must be authenticated. In INSENS, only the base station is responsible for sending control messages and routing-table updates. One such control message is the *request* packet broadcast from the base station to solicit neighborhood topology information from each sensor node. When a node replies with its neighborhood information, these topology claims must be verified. A *nested message authentication code* is used for verification. Each broadcast request message contains a message authentication code in addition to an OWS. As the request message propagates outward from the base station, each node recomputes a new nested message authentication code based on its own key, as well as on the packet's contents, including the previous message authentication code. This uniquely identifies the nested or chained message authentication code as having been generated by a particular node along a particular path. Each node rebroadcasts the *request* packet, replacing the previous message authentication code with its own message authentication code and substituting its own ID as the immediate sender. Each node collects the message authentication codes heard from each of its neighbors and sends this list of message authentication codes with node IDs to the base station. An adversary will be unable to distort the topology by inventing an arbitrary number of neighbors because the adversary will be unable to generate authentic message authentication codes (i.e., will not know the path or keys). A consequence of this approach is that the base station receives correct knowledge of the topology, although it may only represent a partial picture due to malicious packet dropping. After verifying that the message authentication codes are consistent with the node IDs and paths, then the base station may compute routing tables and unicast them in breadth-first manner to each sensor node using the node's shared pairwise secret key.

15.6.1.3 Multipath Redundancy

To address the notion of compromised nodes, the base station in INSENS may compute redundant paths for the routing of data. The goal is to have disjointed paths so that even if an intruder takes down a single node or path, secondary paths will exist to forward the packet to the correct destination. Two paths are shown to be quite effective in reducing the number of blocked nodes for both random and grid topologies, given a single compromised node. INSENS is agnostic to the particular algorithm used to determine the multiple paths, although a centralized heuristic has been suggested [33]. Multipath redundancy enables INSENS to achieve intrusion tolerance.

15.6.2 Protecting the Base Station

Since all data is routed toward the base station, a central point of failure in a WSN is the base station. If an adversary can destroy the base station or otherwise disable it via DoS attacks, then the entire sensor network can be disabled. Routing mechanisms to protect the location and disguise the identity of the base station have been proposed [35]. As a form of intrusion tolerance, routing to more than one base station is introduced in case one base station is disabled. Hop-by-hop reencryption of each packet's header and data fields is proposed to change the appearance of a packet so it cannot be used to trace the direction toward or away from the base station. Uniform rate control is introduced so that the traffic volume nearer the base station is indistinguishable from traffic farther from the base station. Time decorrelation between packet arrivals and departures further increases the difficulty of tracing packets. The paper [35] uses the parent node to control traffic from its children nodes and sets a random delay for each packet.

Another way to defend against traffic analysis in WSNs is to introduce random data traffic into a WSN. J. Deng, R. Han, and S. Mishra [36] have proposed random routing paths, phantom traffic, and fractal data-propagation patterns to randomize the data traffic. Under these mechanisms, it is hard for an adversary to track the location of a base station by analyzing data packets nearby.

15.6.3   Fault Tolerance

Fault tolerance is an issue that overlaps with yet extends beyond the security domain. Benign failures include a sensor node that malfunctions, runs out of energy, or suffers accidental damage. Malicious failures include destruction of a sensor node by an adversary. In both cases, algorithms and protocols should be robust to failures. In outdoor deployments of WSNs, sensor nodes may randomly fail due to their hardware problems. Or, an area of sensor nodes can fail because of environmental disasters, such as flood or fire. Deng, Han, and Mishra have proposed a simple mechanism for sensor nodes to bypass failed nodes and forward data to a base station [37].

A procedure for tracing which sensor nodes have failed is proposed in [38]. The base station first discovers the topology of the network by having neighbor information piggybacked on normal sensor reports. Next, the base station groups all "silent" nodes into disjointed sets. Silent nodes belong to the topology but have not been heard from for awhile and may be dead or temporarily disconnected. These silent groups are traced by broadcasting routing-update probes to each group, collecting topology information, and then peeling away silent nodes whose status has been resolved from the outside in. This process iterates until the status of all silent nodes is resolved, that is, until they are either alive or dead.

15.6.4   Securing Hierarchical Sensor-Network Communication

Deng, Han, and Mishra have proposed a set of mechanisms for building secure communication in hierarchical sensor networks [39]. In such networks, local nodes belong to a group and communicate over multihop routes to an intermediate group-leader node, also called an aggregator or cluster head. The aggregators communicate over multihop connections to the base stations. In this approach, first the base station securely delegates authorization in the form of OWS numbers to an aggregator so that the aggregator can form and control its own sensor group. The OWS number is used by group members to verify aggregator control messages during setup. These control messages set up pairwise keys between the aggregator and each of its group member nodes. In addition, the aggregator sets up ripple keys for efficient command dissemination within a group (e.g., localized broadcast from the aggregator to the group member nodes).

15.6.5   Defending against DoS Attacks during Data Delivery

F. Ye et al. have proposed a statistical en route filtering (SEF) mechanism to filter out injected false data in sensor networks [40]. This mechanism prevents a compromised aggregator from sending forged aggregation results to the base station. In SEF, each sensing report is validated by multiple message authentication codes. If an intermediate node finds that it has a key that is used to generate one of the message authentication codes, that intermediate node will use that key to verify that message authentication code. If the message authentication code test fails, then that packet will be dropped. In this way, the bogus report will be dropped with high probability by at least one of intermediate nodes along the path to the base station and will not reach the base station. S. Zhu et al. have proposed using interleaved keys to filter out bogus packets [41]. In this scheme, the nodes in a sensor group and nodes on the path from the aggregator to the base station share keys in an interleaved manner. These interleaved nodes are used to generate message authentication codes for aggregation results and to filter out forged packets generated by outside adversaries or compromised sensor nodes.

15.7 Data Aggregation

An important emerging research topic in WSNs focuses on securing the role of aggregators. There are two aspects to this problem. First, the WSN should ensure that the aggregation result is correct or approximately correct. Although forged aggregation data generated by an outside adversary can be detected via authentication, identifying the forged aggregation result sent by a compromised aggregator node poses a more challenging problem. Second, the WSN should ensure that the aggregator's role in the routing hierarchy is secured (see Section 15.6.4). How can we make sure that the aggregator gives us a correct aggregation result? B. Przydatek, D. Song, and A. Perrig have proposed a framework for secure information aggregation in large sensor networks [42]. Their idea is to construct efficient random-sampling mechanisms and interactive proofs to probabilistically prevent a compromised aggregator from generating forged aggregation results. Their mechanisms can protect the aggregation results of median, MIN, MAX, and COUNT data-aggregation operations. However, the communication overhead between sensor nodes and the base station is very high. Other work has pointed out that certain aggregation functions (e.g., max) are more easily cheated by a group of malicious nodes providing false data than other functions (e.g., median) [43]. This work analyzes some common aggregation functions, such as average, min/max, median, and count, for their resilience to misleading data, using robust statistics as the mathematical tools [44].

Another way to counteract cheating from an aggregator is to employ multiple aggregators. Du et al. have proposed a witness node mechanism to protect the aggregation result [45]. Their idea is to use multiple aggregators to aggregate sensed data. While only one aggregator sends the aggregation result back to the base station, all aggregators will generate message authentication codes to protect the integrity and authenticity of the aggregation result. Unless an adversary compromises the majority of aggregator nodes, the correctness of the aggregation result is protected. The idea of using witnesses to protect aggregation results is also used in the en route filtering schemes proposed by Ye et al. [40] and Zhu et al. [41].

15.8 Physical Protection

Sensor nodes may be equipped with physical hardware to enhance protection against various attacks and with special software to detect physical tampering. Tamper-resistant hardware is an obvious though costly option and, hence, may not be appropriate in a variety of sensor-network deployments. Due to the volume of literature on tamper resistance, we do not cover these techniques here. In addition, various approaches have been proposed that employ ultrasound transmitters and detectors to identify the location of sensors [46], directional antennas to prevent wormhole attacks [47], and memory-checking algorithms to detect malicious code in sensor nodes [48].

Recent work has shown that today's standard sensor nodes, the Mica2 motes, can be compromised in under one minute [11]. In fact, the TinySec key can be readily revealed. A node is placed in a programming board, and standard readily available tools are employed to dump its code stored in Flash memory, its data stored in RAM, or both. This can be accomplished either through the serial interface or the Joint Test Action Group (JTAG) debugging interface. The time to compromise the node is dominated by the transfer time it takes to dump the entire Flash or RAM over a relatively slow serial line: about 45 seconds. In some cases, if the key is stored in a particular area of memory, then the attacker can target and dump only this area of memory, minimizing the transfer time and allowing node compromise in tens of seconds. Given the dumped memory, researchers were able to run a disassembler on the binary image, view the assembly code, and ultimately extract the TinySec key. While these results are not surprising given that the Mica2 lacks tamper-resistant hardware protection, they provide a cautionary note about the speed of a well-trained attacker.

If an adversary compromises a sensor node, then the code inside the physical node may be modified. A. Seshadri et al. have proposed a mechanism named SWATT to verify if the memory of a sensor node has been changed or not [48]. In SWATT, a verifier quickly and randomly tests the integrity of a sensor node's memory; so, if an adversary has really changed the code, then the verifier will be able to detect the modification with high probability.

A secure localization scheme, namely the ECHO protocol, has been proposed by N. Sastry, U. Shankar, and D. Wagner [46] and is shown in Figure 15.4. In ECHO, a sensor node broadcasts a message to claim its location. When the verifier receives the broadcast message, it sends a *nonce* to the sensor node. When the sensor node receives the *nonce*, it immediately echoes back that number to the verifier by using an ultrasound transmitter. The verifier can validate the distance between the

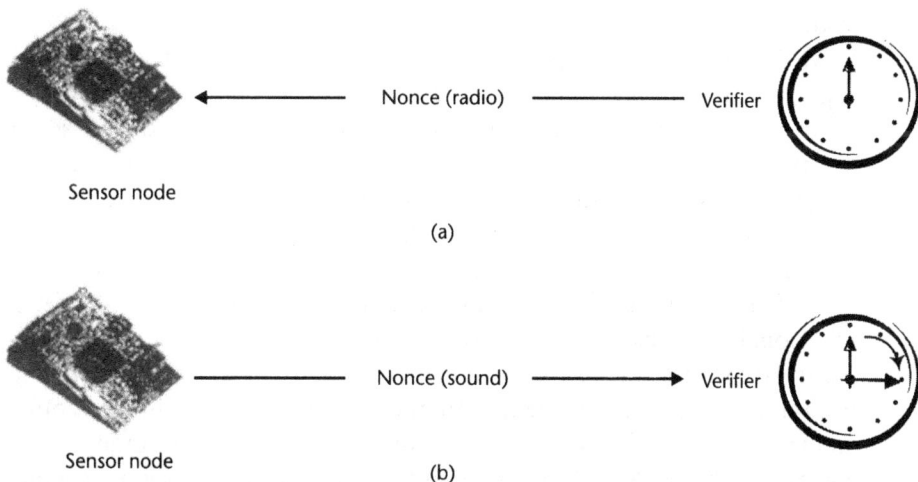

Sensor node

Nonce (radio) ———— Verifier

(a)

Sensor node

Nonce (sound) ———→ Verifier

(b)

Figure 15.4   ECHO verification scheme to validate localization claims. (a) Verifier sends a nonce to sensor node through radio. (b) Sensor node reply the nonce to verifier through ultrasound.

sensor node and itself by measuring the round-trip time between sending and receiving the *nonce*. An adversary cannot cheat and claim a shorter distance by starting the ultrasound response early because it will not have the nonce.

Directional antennas have been proposed as a way to defend against wormhole attacks by L. Hu and D. Evans [47]. If a sensor node has a directional antenna, it can determine the direction of a packet sender when it receives a packet. In addition, every sensor node exchanges its directional information with its neighbor nodes and verifies the consistency of its neighbor nodes' directional information. If node A and B are in different regions of the network and an adversary uses a wormhole to attempt to deceive A and B into believing that they are neighbor nodes, then the directional information collected by A and B will be inconsistent. A and B will therefore be able to detect and counteract a wormhole attack.

15.9 Privacy and Anonymity

While sensor networks may bring great benefits, they are also subject to privacy concerns, especially with respect to the involuntary collection and release of information about individuals. For example, by analyzing data about the temperature in a room collected by a sensor node, an adversary may be able to deduce personal habits such as when the person sleeps, wakes, or even uses his or her laptop. In some cases, the presence or absence of packets between wireless sensor nodes may be used by a criminal to infer the presence or absence of people in a home. A sensor-network-based location-tracking system helps supermarkets to understand the effectiveness of the store's marketing policies. However, sensor-network monitoring may also release privacy information about a customer or employee [49]. While encrypting sensor data may defend against eavesdropping attacks from an outside intruder, it is insufficient to prevent an inside user from obtaining private information. Furthermore, outside attackers may obtain private information from other sources, such as the data traffic of sensor nodes. To protect privacy, we need to develop secure mechanisms that ensure that either the sensor data collector or outside observer cannot obtain information he or she is not authorized to from the sensor network.

M. Gruteser et al. proposed a privacy-aware location-sensor-network scheme [50], which applies the k-anonymity concept [51] to enhance the privacy of a location-tracking system. The idea is to embed anonymity into the sensor network's infrastructure so that location queries to the sensor network would receive suitably anonymized location responses. In this scheme, the region covered by a sensor network is classified into different hierarchically organized subareas. For example, a building is classified into different levels, and each level is classified into different rooms. Each level has a sensor which is that level's coordination leader (CL), and each room has a sensor serving that room's CL. When sensor nodes report tracking data to an interested party, CL's at each stage in the hierarchy suitably anonymize the data (e.g., report the aggregate number of the individuals in the room instead of specific individuals' identities). If an individual desires to be anonymous within k people, then the WSN will respond that the individual can be found, for example, on a certain floor that currently has k or more people. In addition to the k-anonymity mechanism, we think many mechanisms, especially statistical mechanisms, could be applied to protect the privacy of sensed data.

C. Ozturk, Y. Zhang, and W. Trappe have proposed antitraffic analysis mechanisms to prevent an outside attacker from tracking the location of a data source since that information will release the location of sensed objects [52]. The randomized-data-routing and phantom-traffic-generation mechanisms are used to disguise the real data traffic so that it is difficult for an adversary to track the source of data by analyzing network traffic. Similar mechanisms are also used to prevent an adversary from finding the location of a base station by analyzing network traffic [22]. One key problem for these antitraffic analysis mechanisms is the energy cost incurred by anonymization.

15.10 Conclusion and Future Work

Achieving security in WSNs is a challenging endeavor. This chapter has surveyed threat models against WSNs, which are especially vulnerable to compromised nodes. Low-level security primitives, such as pairwise keys, have been outlined, followed by a description of higher-level approaches to building secure routing and secure data aggregation. The chapter has presented approaches to protecting the physical security of sensor nodes and concluded with an investigation of privacy and anonymity in WSNs.

Emerging areas of security research in sensor networks include the privacy and anonymity work described above, as well as secure localization, secure time synchronization, intrusion detection, and resource-constrained public-key cryptography.

References

[1] Hill, J., et al., "System Architecture Directions for Network Sensors," *Proc. 9th International Conference on Architectural Support for Programming Languages and Operating Systems (ASPLOS '00)*, Cambridge, MA, November 2000.

[2] Abrach, H., et al., "Mantis: System Support for Multimodal Networks of In Situ Sensors," *Proc. WSNA '03*, San Diego, CA, September 2003.

[3] Zhao, Y., et al., "Computing Aggregates for Monitoring Wireless Sensor Networks," *Proc. 1st IEEE International Workshop on Sensor Network Protocols and Applications (SNPA '03)*, Anchorage, AK, May 2003.

[4] Bonfils, B. J., and Bonnet P., "Adaptive and Decentralized Operator Placement for In-Network Query Processing," *Proc. 2nd International Workshop on Information Processing in Sensor Networks (IPSN '03)*, Vol. 2634 of Lecture Notes in Computer Science, Palo Alto, CA, April 2003.

[5] Kumar, R., Tsiatsis V., and Srivastava M., "Computation Hierarchy for In-Network Processing," *Proc. 2nd ACM International Workshop on Wireless Sensor Networks and Applications (WSNA '03)*, San Diego, CA, September 2003.

[6] Carman D. W., Kruss P. S., Matt B. J. "Constraints and Approaches for Distribution Sensor Network Security," NAILabs. Technical Report 00-100, Sept. 2000.

[7] http://www.ece.cmu.edu/~adrian/projects.html.

[8] Wood, A., and Stankovic J., "Denial of Service in Sensor Networks," *IEEE Computer*, Vol. 35, No. 10, 2002, pp. 54–62.

[9] Perrig, A., J. Stankovic, and Wagner D., "Security in Wireless Sensor Networks," *Communications of the ACM*, Vol. 47, No. 6, 2004, pp. 53–57.

[10] Chari, S., et al., "Power Analysis: Attacks and Countermeasures," *Monographs in Computer Science, Programming Methodology*, Springer-Verlag, New York, NY, 2003, pp. 415–439.

[11] Hartung, C., Balasalle J., and Han R., "Node Compromise in Sensor Networks: The Need for Secure Systems," Technical Report CU-CS-988-04, Department of Computer Science, University of Colorado, Boulder, 2004.

[12] Kong, J., et al., "Providing Robust and Ubiquitous Security Support for Mobile Ad Hoc Networks," *Proc. IEEE International Conference on Network Protocols (ICNP '01)*, 2001.

[13] Papadimitratos, P., and Haas Z., "Secure Routing for Mobile Ad Hoc Networks," *Proc. SCS Communication Networks and Distributed Systems Modeling and Simulation Conference (CNDS '02)*, San Antonio, TX, 2002.

[14] Zhou, L., and Haas Z., "Securing Ad Hoc Networks," *IEEE Network*, Vol. 13, No. 6, November/December 1999.

[15] Kong, J., et al., "Adaptive Security for Multi-layer Ad Hoc Networks," *Wireless Communications and Mobile Computing*, Special Issue, Vol. 2, No. 5, August 2002.

[16] Perrig, A., et al., "SPINS: Security Protocols for Sensor Networks," *Proc. 7th Annual International Conference on Mobile Computing and Networking (MOBICOM)*, Rome, Italy, July 2001, pp. 189–199.

[17] Liu, D., and Ning P., Efficient Distribution of Key Chain Commitments for Broadcast Authentication in Distributed Sensor Networks, *Proc. 10th Annual Network and Distributed System Security Symposium*, San Diego, CA, February 2003.

[18] Karlof, C., Sastry N., and Wagner D., "Tinysec: A Link Layer Security Architecture for Wireless Sensor Networks," *Proc. 2nd ACM Conference on Embedded Networked Sensor Systems (SenSys '04)*, Baltimore, MD, November 2004.

[19] Deng, J., Han R., and Mishra S., "The Performance Evaluation of Intrusion-Tolerant Routing in Wireless Sensor Networks," *Proc. IEEE 2nd International Workshop on Information Processing in Sensor Networks (IPSN '03)*, Palo Alto, CA, April 2003.

[20] Watro, R., et al., "Tinypk: Securing Sensor Networks with Public Key Technology," *Proc. 2004 ACM Workshop on Security of Ad Hoc and Sensor Networks (SASN '04)*, Washington, D.C., October 2004.

[21] Malan, D. J., Welsh M., and Smith M. D., "A Public-Key Infrastructure for Key Distribution in Tinyos Based on Elliptic Curve Cryptography," *Proc. 1st IEEE Communications Society Conference on Sensor and Ad Hoc Communications and Networks (SECON '04)*, Santa Clara, CA, October 2004.

[22] Gura, N., et al., "Comparing Elliptic Curve Cryptography and RSA on 8-bit CPUs," *Proc. 2004 Workshop on Cryptographic Hardware and Embedded Systems*, Cambridge MA, August, 2004.

[23] Eschenauer, L., and Gigor V., "A Key-Management Scheme for Distributed Sensor Networks," *Proc. Conference on Computer and Communications Security (CCS '02)*, Washington D.C., November 2002.

[24] Chan, H., Perrig A., and Song D., "Random Key Predistribution Schemes for Sensor Networks," *Proc. IEEE Symposium on Security and Privacy*, Berkeley, CA, May 2003.

[25] Du, W., et al., A Pairwise Key Pre-distribution Scheme for Wireless Sensor Networks. *Proc. 10th ACM Conference on Computer and Communications Security (CCS '03)*, Washington, D.C., October 2003.

[26] Liu, D., and Ning P., "Establishing Pairwise Keys in Distributed Sensor Networks," *Proc. CCS '03*, Washington, D.C., October 2003.

[27] Blom, R., "An Optimal Class of Symmetric Key Generation Systems," *Advances in Cryptography: Proc. EUROCRYPT 84*, Vol. 209 of Lecture Nodes in Computer Sciences, Interlaken, Switzerland, 1985.

[28]   Zhu, S., Setia S., and Jajodia S., "Leap: Efficient Security Mechanisms for Large-Scale Distributed Sensor Networks," *Proc. 10th ACM Conference on Computer and Communications Security*, Washington, D.C., October 2003.

[29]   Karlof, C., and Wagner D., "Secure Routing in Wireless Sensor Networks: Attacks and Countermeasures," *Proc. 1st IEEE International Workshop on Sensor Network Protocols and Applications*, Anchorage, AK, May 2003.

[30]   Douceur, J., "The Sybil Attack," *Proc. 1st International Workshop on Peer-to-Peer Systems*, Vol. 2429 of Lecture Notes in Computer Science, Cambridge, MA, March 2002.

[31]   Hu, Y. C., Perrig A., and Johnson D. B., Packet Leashes: A Defense against Wormhole Attacks in Wireless Networks. *Proc. IEEE Infocom*, San Francisco, CA, April 2003.

[32]   Hu, Y., Perrig A., and Johnson D. B., "Rushing Attacks and Defense in Wireless Ad Hoc Network Routing Protocols," *Proc. 2nd ACM Workshop on Wireless Security (WiSe '03)*, San Diego, CA, September 2003.

[33]   Deng, J., Han R., and Mishra S., "Insens: Intrusion-Tolerant Routing in Wireless Sensor Networks," Technical Report Technical Report CU CS-939-02, Department of Computer Science, University of Colorado, November 2002.

[34]   Hu, Y., Perrig A., and Johnson D., "Ariadne: A Secure On-Demand Routing Protocol for Ad Hoc Networks," *Proc. 8th Annual International Conference on Mobile Computing and Networking (MobiCom '02)*, Atlanta, GA, 2002.

[35]   Deng, J., R. Han, and S. Mishra. "Intrusion Tolerance and Anti-traffic Analysis Strategies in Wireless Sensor Networks," *Proc. IEEE 2004 International Conference on Dependable Systems and Networks (DSN '04)*, Florence, Italy, June 2004.

[36]   Deng, J., Han R., and Mishra S., "Countermeasures against Traffic Analysis Attacks in Wireless Sensor Networks," Technical Report CU-CS-987-04, Computer Science Department, University of Colorado, Boulder, December 2004.

[37]   Deng, J., Han R., and Mishra S., "A Robust and Light-Weight Routing Mechanism for Wireless Sensor Networks," *Proc. 1st Workshop on Dependability Issues in Wireless Ad Hoc Networks and Sensor Networks (DIWANS 2004)*, Florence, Italy, June 2004.

[38]   Staddon, J.,. Balfanz D, and Durfee G., "Efficient Tracing of Failed Nodes in Sensor Networks," *Proc. 1st Workshop on Sensor Networks and Applications (WSNA '02)*, Atlanta, GA, 2002.

[39]   Deng, J., Han R., and Mishra S., "Security Support for In-Network Processing in Wireless Sensor Networks," *Proc. 1st ACM Workshop on Security of Ad Hoc and Sensor Networks*, Fairfax, VA, October 2003.

[40]   Ye, F., et al., "Statistical En Route Detection and Filtering of Injected False Data in Sensor Networks," *Proc. IEEE INFOCOM*, Hong Kong, China, 2004.

[41]   Zhu, S., et al., "An Interleaved Hop-by-Hop Authentication Scheme for Filtering of Injected False Data in Sensor Networks," *Proc. 2004 IEEE Symposium on Security and Privacy*, Oakland, CA, May 2004.

[42]   Przydatek, B., D. Song, and A. Perrig, "Sia: Secure Information Aggregation in Sensor Networks," *Proc. ACM SenSys '03*, Los Angeles, CA, November 2003.

[43]   Wagner, D., "Resilient Aggregation in Sensor Networks," *Proc. 2004 ACM Workshop on Security of Ad Hoc and Sensor Networks (SASN '04)*, Washington, D.C., October 2004.

[44]   Hampel, F. R., et al., *Robust Statistics: The Approach Based on Influence Functions,* New York: John Wiley & Sons, 1986.

[45]   Du, W., et al., "A Witness-based Approach for Data Fusion Assurance in Wireless Sensor Networks," *Proc. IEEE GLOBECOM*, San Francisco, CA, December 2003.

[46]   Sastry, N., Shankar U., and Wagner D., "Secure Verification of Location Claims," *Proc. 2003 ACM Workshop on Wireless Security*, San Diego, CA, September 2003.

[47]   Hu, L., and Evans D., "Using Directional Antennas to Prevent Wormhole Attacks," *Proc. 11th Annual Network and Distributed System Security Symposium (NDSS)*, San Diego, CA, February 2004.

[48]	Seshadri, A., et al., "Swatt: Software-based Attestation for Embedded Devices," *Proc. IEEE Symposium on Security and Privacy*, Oakland, CA, May 2004.

[49]	"Infrared Person Tracking," http://www.research.ibm.com/ecvg/misc/footprint.html.

[50]	Gruteser, M., et al., "Privacy-Aware Location Sensor Networks," *Proc. USENIX 9th Workshop on Hot Topics in Operating Systems (HOTOS IX)*, Lihue, HI, 2003, pp. 163–167.

[51]	Samarati, P., and Sweeney L., "Protecting Privacy When Disclosing Information: *k*-anonymity and Its Enforcement through Generalization and Suppersion," Technical Report SRI-CSL-98-04, Computer Science Laboratory, SRI International, 1998.

[52]	Ozturk, C., Zhang Y., and Trappe W., "Source-Location Privacy in Energy-Constrained Sensor Network Routing," *Proc. 2004 ACM Workshop on Security of Ad Hoc and Sensor Networks*, Washington, D.C., October 2004.

CHAPTER 16

Habitat Monitoring with ZebraNet: Design and Experiences

P. Zhang, C. M. Sadler, T. Liu, I. Fischhoff, M. Martonosi, S. A. Lyon, and D. I. Rubenstein

Environmental sensor networks are increasingly becoming practical and feasible ways of gathering data across broad landscapes and on a rich collection of animal and plant species. Through a mixture of fixed and mobile sensors, data can be collected across huge areas with relatively low infrastructure costs, using ad hoc networking techniques. The key to these approaches is in developing energy-efficient sensing and computing hardware and communication protocols for interpreting, fusing, and aggregating sensed data and for delivering it to researcher base stations.

The ongoing Princeton ZebraNet project is studying wildlife-tracking systems based on ideas from mobile ad hoc networks. We have designed energy-efficient hardware tracking nodes that integrate computing, wireless communication, and nonvolatile storage, along with GPS and other sensors, and we are building communication-protocol software to aggregate the archived position data and forward it peer-to-peer toward the researcher base station. Our first test deployment was in January 2004 on zebras at Sweetwaters Game Reserve near Nanyuki, Kenya, to study both fine-grained and long-term movement and migration decisions. This chapter describes the biology research goals of ZebraNet, discusses many of the ZebraNet hardware- and software-design decisions, and gives some initial information from our first deployment.

16.1 Introduction

Environmental sensor networks have been a topic of significant research attention in recent years because they offer the potential to get fine-grained experimental observations about natural phenomena that have been thus far very difficult to study [1–4]. Habitat monitoring, with its focus on dynamic interactions within and between a variety of scales, is an ideal application of sensor networks because answering fundamental biocomplexity-research questions on animal interactions on landscapes that are changing in response to normal as well as anthropogenic processes requires large amounts of diverse data, collected and correlated across large temporal and spatial scales.

Sensor networks consist of many (tens, hundreds, or more) nodes, each of which combines sensing capabilities with a small computer and storage. Each sensing node also communicates, usually wirelessly, with other nodes or a researcher base station to share and archive data. Sensing capabilities are quite diverse and might consist of anything from simple temperature or orientation sensors, to cameras, microphones, or GPS systems [5–8]. Sensor networks enable researchers to do continuous, long-term, autonomous sensing of many different aspects of an environmental system.

While sensor networks offer the potential to make great strides forward in biocomplexity research, their application has been limited thus far by significant technical challenges in implementing them. For the most part, environmental sensor-network deployments have involved modest collections of sensors at fixed positions in a relatively small ecological area under study. These include Intel/University of California, Berkeley's deployment on Great Duck Island, Maine, and UCLA's deployment at the James Reserve.

Sensors that are fixed in their position limit the type and amount of data that can be collected by a sensor network. With this in mind, in 2002, we began the ZebraNet project: research on mobile sensor networks for tracking wildlife migrations over large and sparsely populated areas [3]. The ZebraNet research project is an interdisciplinary project combining computer systems and biology thrusts. On the computer systems side, we are researching application and protocol issues for building sensor networks that are effective in aggregating data collected over very large (hundreds or thousands of square kilometers) and sparsely populated areas. We are also researching strategies for building power-efficient sensor hardware. In January 2004, ZebraNet hardware prototypes were deployed at the Sweetwaters Game Reserve near the Mpala Research Centre in Kenya [9].

These prototypes are the outcome of the computer systems side of the research and, in turn, drive the biology side of the research, where the nodes enable biologists study both fine-grained and long-term decision making in certain species (notably zebras).

16.2 Habitat Monitoring and ZebraNet's Biology Goals

16.2.1 Wildlife Tracking and Habitat Research: Background

Research on animal behavior depends critically on having good technology for wildlife tracking. For understanding both the mechanistic and functional basis of many types of animal behavior—especially those associated with foraging, moving, mating, and sociality—individually identifiable animals must be located and followed repeatedly and on a consistent basis. Often detection and subsequent tracking is visual, but for nocturnal species or those moving over large distances (tens of kilometers per day or hundreds of kilometers per season) or even for those whose lifestyle is simply secretive, detection and monitoring must be done remotely, typically by radio tracking.

Ordinarily, radio tracking involves collaring animals with very high frequency (VHF) radio transmitters and using triangulation from a series of radio fixes to find them [10, 11]. If aerial reconnaissance is employed, then large distances can be covered and many animals can be located in a reasonable amount of time. But if detection requires searching on the ground, then fewer animals can be monitored, and these are only monitored during daylight hours. In addition, visual/VHF approaches typically preclude learning animal data such as body temperature or heart rate. Despite these limitations, the VHF system is the one most often employed because collars and the antenna-receiver system only cost a few hundred dollars apiece and the continuing cost of transport required for each spatial fix is low.

In a nutshell, the goal is to track both position and other data on what animals do within varying social and environmental contexts. We want to track animals nearly continuously (not simply during flyovers), and we wish to perform this tracking without assuming any fixed towers or antennas are needed. Simple satellite-based tracking collars overcome some of the deficiencies of VHF approaches: position data can be gathered both day and night, and for a fee, this data can be acquired from companies that manage the system. But the usage costs of such collars are quite high. Even more problematic, the bit rates for communicating to and from the satellite are poor, and so, the collars provide only infrequent position sampling and do not track important environmental information such as ambient temperature, body temperature, heart rate, or activity pattern. More advanced GPS-based collars have been introduced that address this shortcoming by storing short logs and by recording head movements (to determine if the animal is eating or not) and ambient temperature in addition to animal location. Fundamental limitations of these approaches include insufficient storage for long events and a reliance on fixed communication towers for data uploads. Our current ZebraNet system is designed to overcome these problems through a peer-to-peer (or collar-to-collar) approach that percolates data toward a base station via pairwise transfers.

16.2.2   Habitat Monitoring and Mobile Sensors

Predators and prey are inextricably linked. Preliminary observations show that zebra density increases with vegetation height and decreases with the abundance of lions and hyenas. But lion abundance also increases with grass height. So, even if zebras appear to avoid lions, nutritional needs and the lure of large amounts of vegetation bring zebras and lions together. Until now, it has been difficult to monitor how predators and prey respond to each other and how the game of hide and seek plays out. Although visual observation is possible, close follows for long periods of time disrupt natural behavior. Using traditional VHF radio fixes for short periods cannot provide sufficient spatial and temporal resolution to determine who is driving the movements of these coupled species. After dark, collecting VHF data is unrealistic.

Tracking devices such as the ZebraNet collars can sample the locations and activities of animals at fine-grained intervals (e.g., 8-minute sampling). By gathering such fine-grained data, we can determine if and when each species' activities initiate changes in the other species' behavior. Truly fine-grained analysis of coupled movements and activities for large and wide-ranging predators and prey has never been attempted before. At the regional, or landscape, level, we can also begin to monitor how zebra movements and change in density are shaped by the bottom-up features of vegetation abundance and quality and the top-down forces of predator abundance and perceived risk. Such research can help guide effective wildlife-management plans and identify at which trophic level interventions are likely to be most effective.

With the expansion of human settlements and the intensification of land use, as well as the conversion of pastoral lands to horticulture, seasonal migrations of wildlife are being impacted [12]. Curtailment of traditional migration routes, as well as the creation of rich feeding alternatives for herbivores and predators alike, increases the potential for human-wildlife conflict [13–15]. As the project proceeds, it may also be possible to fit collars on livestock and (along with fixed sensors positioned near human settlements) help identify the ways in which humans and wildlife interact.

16.3   ZebraNet System Overview

The ZebraNet system has been designed specifically to support the research of biologists working at the Mpala Research Centre in Kenya [9] by applying the latest sensor-network technology to the field of animal tracking. ZebraNet consists of sensor nodes built into collars on zebras, which take positional readings using a GPS unit and propagate them from zebra to zebra until infrequent communications percolate data to base stations [3]. Given that the zebras are fairly mobile and spread over a large distance, we expect our system to form an extremely sparse network, with roughly one collar per 10 km^2. Our system goals are as follows:

- *Detailed, accurate position logs.* Take frequent positional readings that are fine-grained enough to give the biologists an accurate view into the daily migration pattern of a set of zebras.
- *High data recovery.* Propagate data to a mobile base station through pairwise communications, with latency being less important than eventual success.
- *Autonomous operation.* Survive long term in the wild with no human contact except for occasional drive-by radio uploads of data.

A ZebraNet hardware node includes GPS, a simple microcontroller CPU, a wireless transceiver, and nonvolatile storage to hold logged data. ZebraNet does not rely on constant communication access to a base station or other nodes. Instead, it uses periodic node discovery and node-to-node communication to propagate data toward the base station in a store-and-forward manner.

The ZebraNet hardware is composed of energy-efficient components ideal for use in mobile sensor networks. A photograph of the most recent ZebraNet hardware (top and bottom) is presented in Figure 16.1. A block diagram of the main system components is given in Figure 16.2. Finally, an overview of the main hardware and software system layers in ZebraNet is depicted conceptually in Figure 16.3. The major functional components on the board are the microcontroller, GPS, external Flash, radio, and battery with solar chargers.

To control the hardware, we selected the Texas Instruments Ultra-Low-Power MSP430F149 16-bit microcontroller. This chip has 2 KB of RAM, 60 KB of internal Flash memory, and two serial interfaces [16]. It runs off an uninterruptible power supply as we expect it to run continuously. The microcontroller operates in a dual-clock configuration. It uses an 8-MHz clock when accessing sensing, storage,

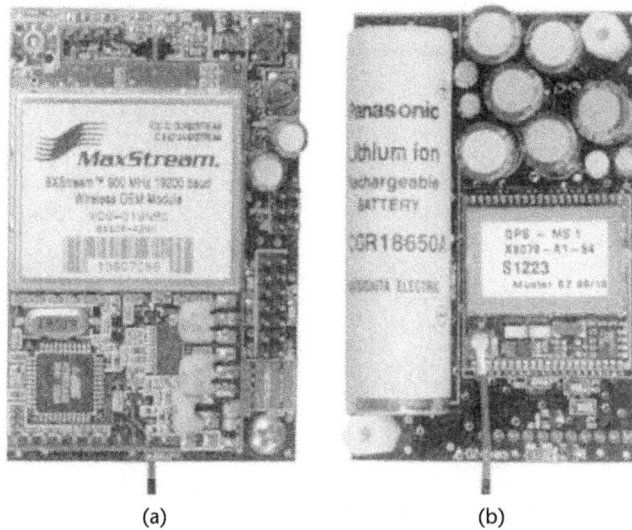

(a)   (b)

Figure 16.1   Photo of a ZebraNet node (2″ × 3″ × 1.25″): (a) the side of the card with the radio, and (b) the GPS chip and the battery.

or communication peripherals and a 32-KHz clock at all other times. The 32-KHz clock consumes half as much power as the 8-MHz clock and can be used instead of putting the processor to sleep.

The system acquires position readings every 8 minutes. Biologists' observations of zebra behavior show that 8 minutes is a fine enough interval to achieve useful tracking. This sampling also still allows the GPS to be powered down most of the time. When the GPS is turned on, we leave it on for 1 minute or until it acquires an acceptable lock, whichever comes first. Figure 16.4 shows the power consumption when GPS turns on, acquires a lock, and promptly turns off. The second GPS lock is delayed 4 minutes due to the radio communication time.

Radio communications occur every 2 hours in a time-slotted manner. The GPS provides the real-time clock used for time synchronization between the nodes, and the system keeps accurate time between GPS readings. Every 2 hours, the radio turns on, attempts to find other nodes in range, and if successful, transmits position data in their time slot. After 5 minutes, the communication session ends, and the radio powers down. Any unsent data will wait for the next communications interval. Figure 16.4 shows the power consumption when the radio turns on, sends a packet, and waits for a response from its neighbors until radio communication ends.

The ZebraNet system is controlled by the Impala middleware layer [17, 18]. Impala allows for a combination of scheduled and spontaneous events, and in this implementation, GPS-sensing and radio-communication times are prescheduled. Several protocols are possible, but in our field deployment, the information is propagated through the network using a flooding protocol, which ideally allows every zebra to know where every other zebra has been. This allows the mobile base station to receive data from all of the collared zebras by encountering just one.

16.3.1   Evolution of Our Hardware

Based on the system design parameters, the main design constraint is energy. ZebraNet nodes are built using low-cost, energy-efficient, off-the-shelf

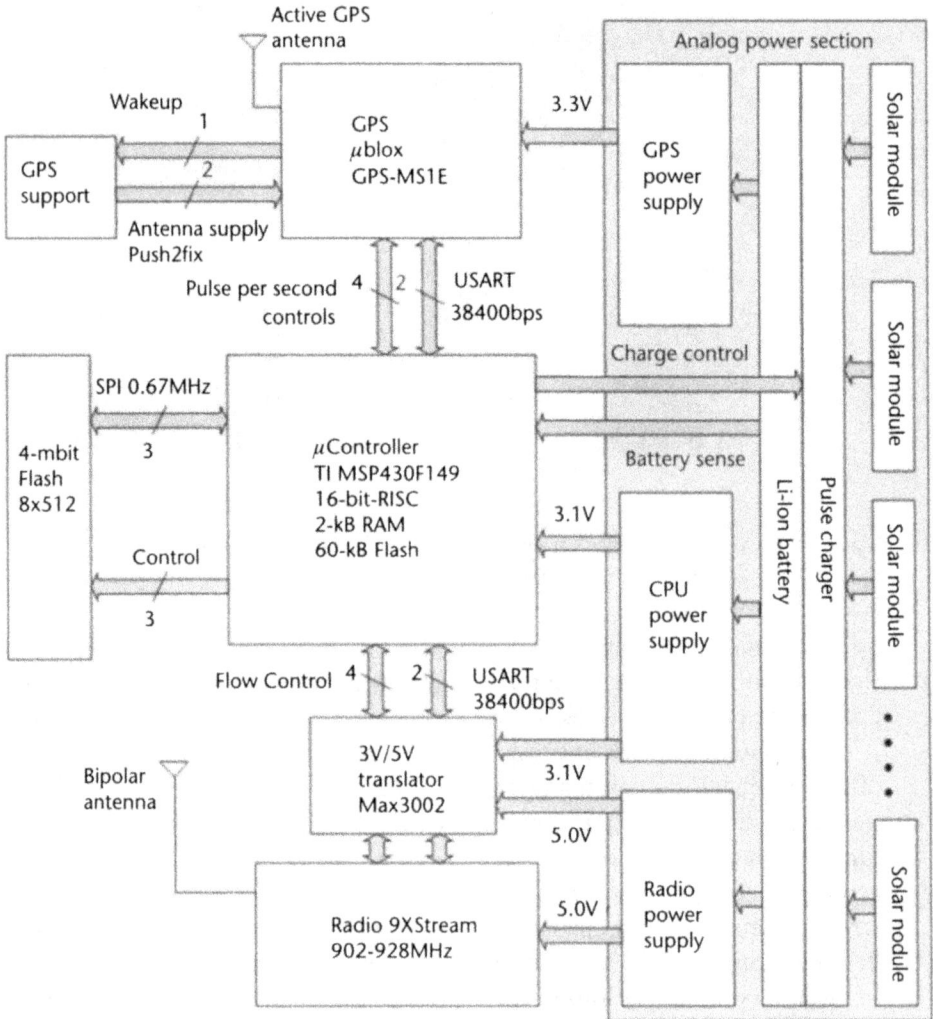

Figure 16.2 Block diagram of components, interface connections, and power supplies for the ZebraNet hardware version 3.

components. From conception to realization, the ZebraNet node has gone through several design iterations. Table 16.1 shows the major design changes between the major iterations of the design. Version 0.1 was a proof-of-concept design [3]. In this version, a few design schemes were tested, such as dual radios, where we included two radios, each with different baud rates and ranges. Cycle stealing was employed to use unused cycles of the GPS processor. However, the power consumption of these schemes proves to be too great.

The major step from Version 0.1 to Version 1 was the move to include a low-power microcontroller that controls all the peripherals and an external 2-Mbit Flash memory to store data. (Version 0.1 scavenged both Flash and CPU cycles from the GPS chip.) Having dedicated processing and storage greatly simplified software development and eased design changes by allowing peripheral changes with only firmware-level updates. Version 1 did not have the solar-charging circuitry but did

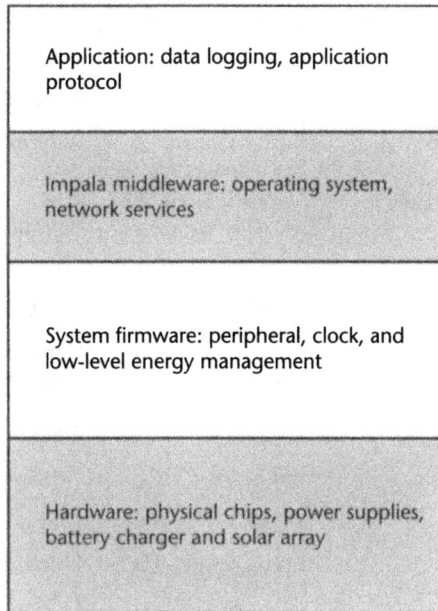

Figure 16.3   ZebraNet system structure overview.

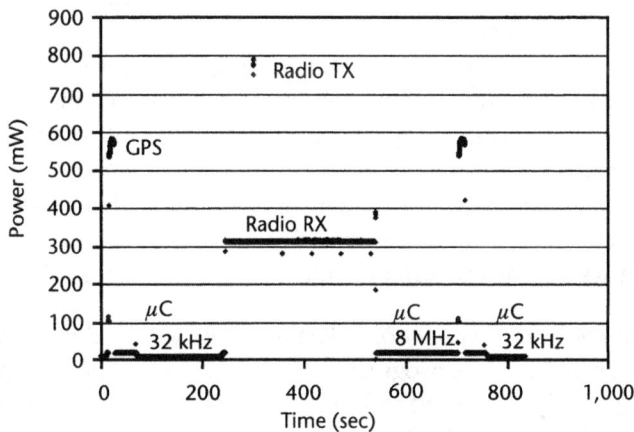

Figure 16.4   Power consumption of our system during a periodic data-sampling and communication time. The GPS and radio transmit (Tx) points are the highest power. Radio receive (Rx) is moderate in its power consumption. The lowest-power portions are the microcontroller (C) alone, running at either 32 KHz or 8 MHz.

include various exploratory designs of high-efficiency power supplies. Their impact on the system performance and efficiency is presented in Section 16.5.

Although similar to Version 1, Version 2 was designed to require half as much board area. The design mostly explored issues with the layout and cross-talk interference of onboard components. The power supplies were improved to lower noise.

Version 3 is a complete system design. The microcontroller-centric design and the power supplies were further improved. We also further increased the Flash capacity to 4 Mbit. A photograph of this most recent version of the sensing

Table 16.1   Design Summary for Different Versions of ZebraNet Hardware Nodes

	Version 0.1 [19]	Version 1	Version 2	Version 3
	August 2002	Feb. 2003	July 2003	Nov. 2003
Power supply	Off-board	Buck-boost and boost converters	Same as version 1	Two buck-boost converters
Noise reduction	Bypass capacitor	Standard low-ESR capacitors	Os-con capacitors	Liquid crystal (LC) post filters and common mode choke
Radio	Two-radio system	2.4 GHz	900 MHz	900 MHz
GPS	Cycle-stealing from onboard CPU	Off-board antenna power	Ultra-low-noise linear regulator	Same as version 2
Data flash	5 Mbit	2 Mbit	4 Mbit	4 Mbit
Battery charger	None	Off-board	Off-board	Pulse-charging
System weight	1,151g	145g	136g	138g
Battery	N/A	Li-Ion, 2A-hr, 45g	Li-Ion, 2A-hr, 45g	Li-Ion, 2A-hr, 45g

hardware is shown in Figure 16.1. It is powered using a lithium-ion battery that is recharged with a solar array. This version was built into collars and deployed in central Kenya in January 2004. We deployed seven nodes on plains zebras at the Sweetwaters Game Reserve near Nanyuki, Kenya. Based on the results from the first deployment of the system, we are currently making improvements with a broad-scale, long-term deployment in mind. More information about our deployment experiences are provided in Section 16.5.

One fundamental aspect of sensor systems is their ability to overcome the numerous constraints imposed by the hardware and the applications scenarios. In particular, we list the most important design constraints below.

- *Data and program memory*. The data memory in the microcontroller is only 2 KB. This affects the program behavior in many aspects, especially in data buffering. As they are used to keep system states and to handle large flows of network data, data buffers often consume large amount of memory and, therefore, must be carefully allocated. Additionally, the program memory is only 60 KB. This requires that software programs be concise.

- *Energy*. The energy budget is tight as we use a solar array to recharge the battery and to provide the energy essential to achieve the sensing and communication tasks. As is estimated, we are able to charge the battery fully in 50 hours of daylight. This number can vary in either direction, however, depending on the orientation of the solar cells in relation to the sun. Therefore, efforts must be made to save energy maximally, and resorts must be provided to preserve the system when the energy level is severely low.

- *Device access*. Device accesses must be carefully scheduled to avoid conflicts that are likely to happen due to hardware limitations. For example, due to voltage-regulation challenges, the GPS and the radio should not be turned on at the same time for interference-avoidance purposes. Additionally, the GPS and the Flash share the same serial connection to the microcontroller and, therefore, cannot be accessed simultaneously.

- *Radio packet size.* The physical packet size of our radio hardware is only 64 bytes, an order of magnitude smaller than the Ethernet packet size, for example. This means the multilevel packet header in the traditional TCP/IP model will entail a significant communication overhead. Therefore, we need a special network protocol that requires a low overhead to accomplish the essential network-communication services.

- *Flash data storage.* For the Flash memory, new data cannot be written to an address before the data currently at that address is erased, and the smallest erasable unit is a 264-byte page. This means writing data to one location will affect data at other locations. Therefore, a global Flash organization is required to achieve efficient data storage.

- *GPS sensing time.* The time for the GPS unit to acquire an accurate position lock is typically 10 to 20 seconds. This considerable delay in data acquisition implies that an asynchronous access and control model is preferred to a synchronous model for operating this sensing device.

16.4 The Impala Middleware System

As depicted in Figure 16.3, the ZebraNet hardware is managed by several custom-designed software layers. In particular, the Impala system comprises the main operating system and network services for ZebraNet. It works with a thin, underlying firmware layer to provide a range of services to sensor-network applications. In this section, we discuss Impala's design in more detail.

The Impala operating system and middleware service model is driven by several issues applicable to ZebraNet and to general sensor-network applications as a whole. The first issue is that the long-term sensing and communication tasks of sensor-network applications require dependable scheduling of regular operations. Sensor networks are designed to run for indefinite periods without human intervention. Many sensing and communication operations occur on a predictable timetable. ZebraNet, for example, executes GPS position sensing and wireless radio communication periodically. In addition to these operations, the system must perform many other routine system computations and maintenance. Therefore, Impala must provide clean mechanisms to schedule recurring operations.

A second issue is that sensor-network applications require efficient handling of irregular events. Fundamentally, sensor nodes are event-driven systems. Events such as sensor-data capture and network-data reception occur frequently and are the primary triggers of system computations. An event may result from a single or a sequence of hardware interrupts. Depending on the API, promiscuous hardware interrupts can be made transparent to applications and delivered only as a few types of abstract events. However, an appropriate event abstraction should balance simplifying application programming with maintaining the granularity of application-level processing. Additionally, events may be handled by different components of the system and, therefore, require efficient event filtering and dispatching.

Third, sensor-network applications require specialized network support. As data gathering is the primary goal of sensor networks, sensor nodes often use aggressive flooding strategies to maximize the chance of finding a path to the desired destination. The resultant multicasts and broadcasts are common communication patterns. Transmissions must be reliable in scenarios where data integrity is critical. Data can be unreliable, however, in cases where packets can be lost without compromising our goals. For example, peer-discovery messages are considered unreliable as the nodes are mobile and may not be in range of the other nodes. Additionally, due to the severe resource constraints and limited hardware capabilities of sensor nodes, efforts must be made to minimize the overhead in communication, buffering, and processing.

The fourth issue is that the complexity of sensor-network systems requires dynamic software adaptation. The scale of sensor-network systems can be on the order of thousands of nodes; therefore, coordinating the communication and computation across the system is complex. Depending on node topology, network connectivity, and node mobility, over its lifetime the system may encounter a number of different scenarios, for each of which a different communication protocol may be appropriate. As such, it is nearly impossible for a single protocol to be appropriate all the time. Some amount of adaptivity is crucial for applications to handle an interesting range of possible parameter values properly.

Finally, the long-term deployment and inaccessibility of sensor-network systems require automatic remote software updates. It is inevitable that software updates will be required during the lifetime of a sensor network. Because sensors are typically deployed in large numbers in inaccessible places, updates must be deployed wirelessly. Therefore, Impala needs to support automatic remote software updates so that new software can be plugged in at any time. ZebraNet offers very clear motivation for remote software updates since we clearly do not want to have to tranquilize and recapture a collection of collared animals each time we need to update the software.

Compared to other sensor operating system work, such as TinyOS [20], Impala's differences can be summarized in terms of providing additional core services beyond the minimal layer provided by TinyOS. These services include base networking support, as well as hooks for software adaptation and update. By embedding these services into the operating system itself, these services can be provided in a lean and system-oriented manner that lessens the need for application programmers either to custom-write these services themselves or to further layer prewritten services on top of the operating system.

We have built two implementations of Impala. The first implementation was a prototype running on HP/Compaq iPAQ pocket PC handhelds [17]. Subsequently, we implemented Impala again on the real ZebraNet hardware nodes [18]. This latter implementation and description focuses on its operating system functionalities in hardware/software interfacing, system operation scheduling, event handling, and network communication support.

16.4.1 Impala System Layers

Figure 16.5 shows the static view of Impala with three system layers: the uppermost application layer, the Impala layer, and the firmware layer. Services and events are

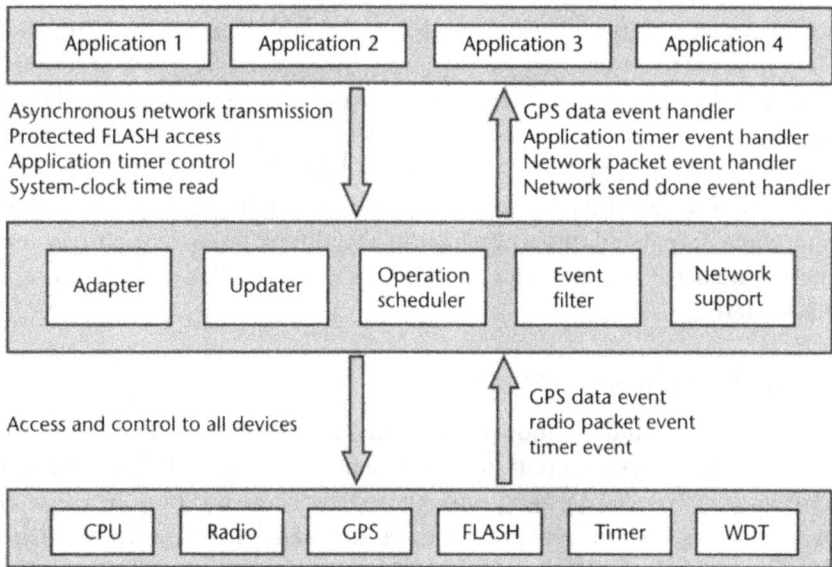

Figure 16.5   Impala system architecture: layers and interfaces.

the major interfaces between layers. Through the service interface, the firmware layer exports numerous hardware access and control functions to the Impala layer. The Impala layer, however, protects these firmware functions from direct use by the application layer and only exports the ones needed by applications in a reduced or protected form. It also exports its own network interface to the application layer.

16.4.2   Regular Operation Scheduling

The dynamic view of Impala comprises *regular* computing and maintenance operations required by the long-term sensing and communication tasks and *irregular* events incurred by the inherent event-driven attribute of sensor-network applications. For regular operations, Impala acts as an operation scheduler that schedules and coordinates system operations based on application goals, hardware constraints, and energy budget. For irregular events, it acts as an event filter that captures and dispatches events to different system components and initiates chains of processing.

Impala uses timers to trigger various operations. GPS-aided time calibration allows networkwide operation synchronization. Since ZebraNet sensor nodes have ongoing access to global GPS time, sensor nodes can be easily synchronized. This is especially important for network communication in which all nodes need to turn on and off their radios simultaneously and transmit in assigned time slots to avoid collisions. In addition, long-duration events, such as GPS, also influence operation scheduling. Where there may be large variation, a split-transaction approach is used.

The stringent energy budget requires energy conservation whenever possible. The battery capacity of the ZebraNet nodes can support the full level of system activities for one to three days. The solar array can extend this time indefinitely, but solar cell area is limited, so we need to conserve energy whenever possible. Impala achieves this with two approaches. First, ZebraNet has an 8-minute GPS-data-sampling interval to ensure that we capture significant movements of zebras while minimizing redundant data records. In between these intervals, Impala works to put system components into low-power modes to save energy. Second, although the energy supply of our system is designed to fulfill the energy consumption under typical conditions, we still need to preserve the system in the case of energy deprivation. Therefore, Impala adapts its operation scheduling to current battery levels. It skips energy-intensive phases such as communication or GPS sensing if the energy level is inadequate.

16.4.3   Event-Handling Model

Impala's event-handling model is designed to attack three fundamental issues. First, sensor-network systems require an efficient event-based API. Events originate from hardware interrupts. Dealing with these interrupts not only involves considerable programming efforts but also requires detailed hardware knowledge. Therefore, Impala has an event abstraction that encapsulates miscellaneous hardware interrupts into abstract events to simplify application programming, while maintaining the granularity of application-level processing.

Impala implements four types of abstract events that are essential for ZebraNet applications. An event is generated and enqueued by an event signaler, dequeued and dispatched by Impala's event filter, and processed by an application event handler. Figure 16.6 shows the abstract events and Impala's event-handling components. A network-packet event represents the arrival of a network packet. Impala's network interface generates this type of events after it receives a packet from the radio firmware and examines the validity of the data. A network send-done event represents the completion or failure of a network-message transmission. The network interface generates this type of events after it has completed the transmission or a failure has occurred. An application-timer event represents the time to execute a prescheduled application operation. The timer firmware generates this type of event after the application timer expires. A GPS-data event represents the capture of a GPS position fix. The GPS firmware generates this type of event after it analyzes the information output from the GPS unit and identifies a position fix.

Because concurrency is an important attribute of sensor-network systems, Impala has a hierarchical event-handling model that processes simple hardware interrupts in short or atomic routines and handles complex software events in long or nonatomic routines. This not only achieves concurrency among multiple flows of processing but also allows low-level processing to interleave with and, if needed, override high-level processing.

Finally, event prioritization is desirable in sensor-network systems. Some events are urgent and require immediate processing, such as the network-packet events. Some events are time constrained but are not sensitive to small delays, such as the application-timer events. Other events are highly latency tolerant, such as the GPS-data events. Therefore, event prioritization allows events with different time constraints to be processed in the desired order. As shown in Figure 16.6, Impala's event filter maintains an event queue for each type of event and associates each queue with a priority for event processing.

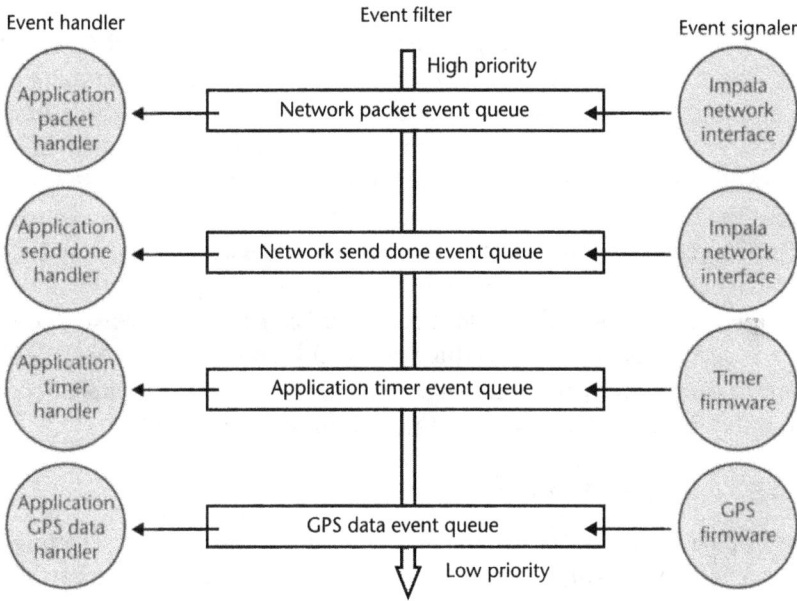

Figure 16.6 Impala event-handling model.

16.4.4 Communication Characteristics and Impala Networking

The network interface, as a middleware service, is crucial in mobile wireless sensor systems. As in many other mobile WSNs, ZebraNet uses peer-to-peer communication. Unlike many others, the sparse connectivity caused us to choose pairwise store-and-forward routing rather than common path-based approaches. To support the application layer, which studies various store-and-forward routing strategies, Impala's network interface focuses on the networking model within one hop.

The special communication pattern of sensor-network applications like ZebraNet changes the message model. Sensor nodes often use an aggressive flooding strategy to maximize the chance of finding a path to the base station. This leads to the common use of multicast and broadcast protocols. Furthermore, transmissions must be reliable in some cases but, for energy reasons, unreliable transmissions are preferred in other less critical uses. Impala's network services must support these different uses.

Impala uses session-based transport control. A session is a message designated by the application to have network-transaction semantics. Sessions can vary from 1K to 32K bytes, can be unicast, multicast, or broadcast, can transmit data from Flash or from application RAM buffer, and can use reliable or unreliable transmission. We chose connectionless session transmission because connection-oriented approaches seemed a poor match for the unpredictable motion of sensor nodes in our system. Connectionless sessions also reduce computation and communication overhead.

16.4.5 Time-slot-based Media Access Control

Since ZebraNet sensor nodes have ongoing access to globally synchronized GPS time, sensor-node activities can be easily synchronized. Impala takes advantage of this time synchronization and uses simple, round-robin, time-slot-based MAC. We chose this partly for simplicity (code size and energy) but also because the round-robin nature of the approach means that the MAC layer always knows which nodes should be acknowledging reception in each time slot. This allows simple yet efficient time-out and retransmission mechanisms, as described in the next subsection.

In the time-slot-based MAC, each node is statically assigned a unique time slot for transmission in an iteration. A sensor node uses its time slot both to transmit data packets and to acknowledge previously received packets. Figure 16.7 shows an example of time-slotted transmissions between multiple sensor nodes. ZebraNet only expects tens of nodes, so this nonscalable solution is acceptable and more efficient. In a larger network, one might choose to use a hybrid time or contention algorithm in which a small number of nodes share a time slot.

16.4.6 Impala Evaluations

The Impala middleware system, including the firmware layer, the Impala layer, and a baseline application, has been implemented on the ZebraNet hardware nodes. To evaluate Impala's overhead and performance, we have conducted some preliminary measurements and analysis as described below.

Static memory footprint. Since our system faces severe memory constraints, it is important for us to minimize code size and RAM usage. Figure 16.8 shows the program-memory and data-memory footprints of different system layers. For program memory, the network interface requires 5,712 instruction bytes and is the largest component in the Impala layer. The Flash, GPS, and timer modules are the major components in the firmware layer. The application layer is lean because we only implemented one basic application.

For data memory, the network interface in the Impala layer claims 51 bytes of data memory. In the firmware layer, the GPS module requires a 125-byte buffer to receive information from the GPS unit, and the radio module requires a 64-byte buffer to receive packets from the radio.

As depicted in Figure 16.8, our code currently consumes less than one-third of the total program memory, and we statically allocate less than one-sixth of our RAM, which leaves ample memory for dynamic allocation and for future expansions to our system.

Network interface performance. We evaluate Impala's network interface in its processing overhead for packet reception and transmission, its communication

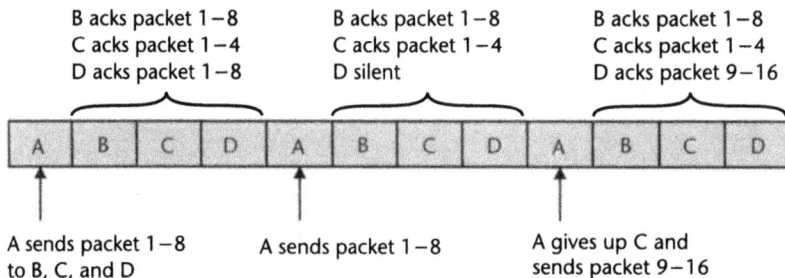

Figure 16.7 Data sends and related acknowledgments in the time-slot model.

Program memory footprint

Data memory footprint

(a)   (b)

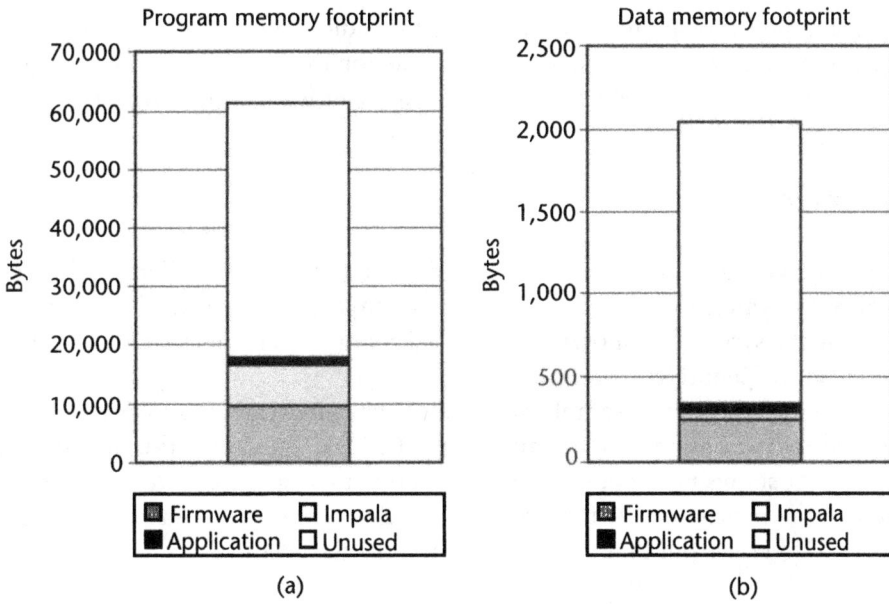

Figure 16.8   (a) Program-memory footprint, and (b) data-memory footprint.

overhead caused by packet headers, and its communication latency in reliable multicast.

For packet reception, Impala propagates an incoming network packet from the radio hardware, to the radio firmware, to the network interface through interrupts and callouts. Then, the packet is enqueued by the network interface until the event filter dequeues and delivers it to the application. Table 16.2 shows the processing time to receive a network packet in each system layer.

On the transmission side, the network interface provides the applications with an asynchronous operation for network transmission. As we described in Section 16.6.4, sessions can be dropped into the network interface by the application at any time. The time to drop an unreliable broadcast session, such as a peer-discovery message, is 496 cycles. The time to drop a reliable multicast session, such as a GPS data session, is 901 cycles.

When the networking time comes, the network interface will update session states, compute the information to send, copy data between Flash and RAM, and

Table 16.2   Packet Reception Processing Time by System Layers

System Layer	Time	Processing Breakdown	Time
Application Software	111819 cycles	Packet write to FLASH	110172 cycles
		Non-FLASH related computation	1647 cycles
Network and Event Filter Middleware	1058 cycles	Packet processing by the network interface	970 cycles
		Packet delivery and remove by the event filter	88 cycles
Radio Firmware	3470 cycles	Processing packet synchronization bytes	56 cycles per byte
		Processing intermediate packet bytes	49 cycles per byte
		Processing the last packet byte	369 cycles per byte
Radio Hardware	255585 cycles	N/A	N/A

invoke the radio firmware to transfer data to the radio send buffer. All these operations are in parallel with the actual data transmission by the radio hardware. Figure 16.9 shows the time spent on each system component for transmitting a packet.

16.5 Deployment

During January 2004, seven complete nodes were successfully deployed at the 100-km^2 Sweetwaters Game Reserve in central Kenya. During the deployment, data gathered by the system has shown fine-grained zebra movements that had not been available before ZebraNet.

Figure 16.10 shows an example of collected zebra-location data. The datapoints are plotted over a habitat classification on a Landsat image for the Sweetwaters area. The plot shows how fine-grained the spatiotemporal data is. In addition, the location of the points reveals that, at least for this zebra, her behavior at night is somewhat different from that exhibited during the day. Not only is she moving to slightly more brushy habitats (filled squares) after dark, but she is also moving most rapidly as darkness sets in. During the day, most of her GPS fixes are located in areas of open grassland or light acacia bush (white and light grey regions of Figure 16.10). But at night, a significant number of her sightings occur in the bushy habitats (darker areas in Figure 16.10). Since these areas are also characterized by the steepest slopes, these findings suggest that zebras may be trading off reduced acuity for increased agility relative to predators, who are forced to cut and weave on uneven and sloping surfaces. Since fixes were collected every 8 minutes, meaningful zebra velocities could be computed for the first time with remotely recorded data. At least for this zebra, walking during the day, apart from walking to water late in the

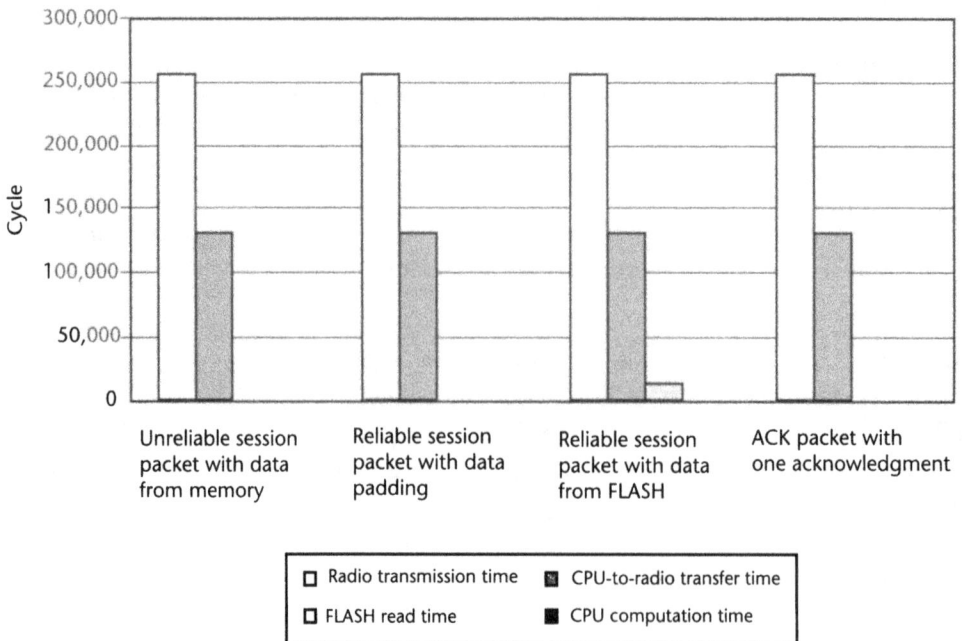

Figure 16.9 Packet-transmission processing time by system components.

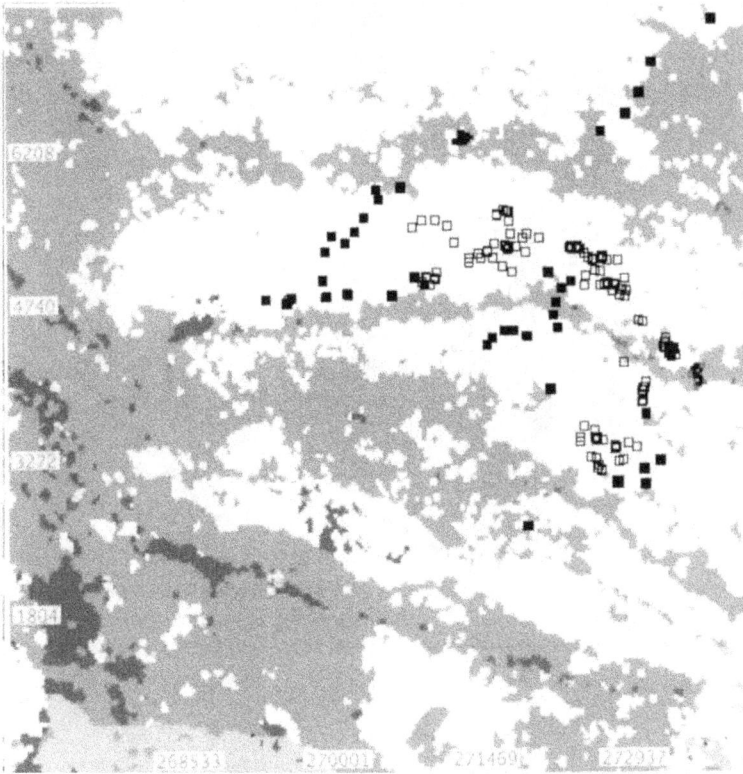

Figure 16.10 An example of GPS location data collected over a one-week period. The pictured area is roughly 10 km × 10 km. (*Source:* Nasser Olwero of Mpala Research Centre produced the habitat classification map on which we have plotted the position data.)

morning, was done at a leisurely pace. At dawn and dusk (crepuscular periods), however, her speed increased to very high rates as she changed habitat types.

Figure 16.11 shows a photograph of a collared zebra. For reliability and ruggedness, the radio antenna is embedded between the layers of butyl belting that form the collar. A dipole antenna is formed from a strip of copper tape and braid. Since the radio uses a range of frequencies, the dipole's length is tuned for the center frequency of the radio. To reduce absorption into the animal's neck and thereby improve the radio range, we also include a ground plane of woven conductive cloth separated roughly 1 cm from the antenna by a foam dielectric.

Like the radio antenna, the GPS antenna is also sandwiched between the butyl collar layers. Our design experiments show that it operates quite successfully through the butyl layer. We intended for a v-shaped wedge positioned at the animal's throat to seat the collar upright at all times, but unfortunately, during the deployment, we noticed that the collars would rotate, sometimes misdirecting the GPS antenna.

The photograph also illustrates the proper orientation of the solar modules. Each collar has an array of 14 solar modules, connected in parallel. Each module, in turn, is the series, voltage-controlled connection of three solar cells, as described in Section 16.3. Under optimal conditions, each collar's 14 modules can together generate roughly 100mA at 5V under full sun. However, since the solar cells wrap

Figure 16.11   Plains zebra at Sweetwaters Game Reserve with ZebraNet collar. The collar is made of a white butyl belting material. The darker spots visible on the upper half of the collar are the solar modules, which lie between the two layers of butyl belting, with openings exposed to allow sunlight in.

somewhat around the zebra's neck, we do not reach full charging efficiency and design for roughly half that amount.

We have learned many design lessons from the deployment. In addition to physical design issues related to collar orientation, some communications issues also have also arisen. For example, energy constraints caused us to design in a 2-hour interval between communication times. We hope in the future to implement more event-driven communication intervals to interrogate nodes and to allow for opportunistic data transfers between collars when they are within range. Decreasing the interval, while maintaining the power profile, however, will require the radio to have a significantly lower power profile.

To accomplish this, we plan to replace the radio with separate receiver and transmitter chips. This will allow more flexibility and control over receive and transmit power consumption than what is available in transceiver modules. In receive mode, the current radio drains approximately 50mA at 5V. In our next version, we plan to achieve a drain of less than 10mA at 3V with chips similar to the Xemics DP1201A [21]. This can be accomplished while maintaining the high receiver sensitivity of −107 dBm due to recent advances in low-power single chip receivers and by lowering the baud rate. We also plan to use a lower-frequency band (around 150 MHz) to reduce path loss of the radio and extend the radio range even further. Since zebras spend most of the time grazing with the collar close to the ground, lower frequencies are less affected by ground absorption and by obstructions such as small hills. A lower-power radio will also allow us to use a linear regulator and reduce energy consumption while increasing performance.

GPS accuracy also comes into question as some outliers were found. In addition to the filtering scheme previously discussed, we plan to switch to a GPS with a power profile that will allow it to be kept on for longer periods of time. In our upcoming version, we plan to use to the Xemics XE1600 chipset, which consumes an order of magnitude less energy than our current chip.

In the near future, we also plan on switching microcontrollers within the TI MSP430 family. The MSP430F1611, which will be available soon, has 10 KB of RAM, which will be extremely helpful when we replace our flooding protocol with a more selective routing protocol. According to its data sheet, the new microcontroller consumes 330 μA at a voltage of 2.2V and a clock rate of 1 MHz [22], which is 50 μA more than the MSP430F149 under the same conditions [23]. Although we cannot be sure how much more energy will be consumed once the chip is integrated into our system, we feel that the power savings from reducing transmissions with selective routing protocols will almost certainly conserve more energy than the new microcontroller will expend.

16.6   Related Work

We have developed a system that combines aspects from the sensor-networking community and from the mobile ad hoc–networking community. In this section, we look at how our system compares to new technologies in both of these distinct fields.

Sensing hardware.   A number of energy-efficient sensor nodes have been developed in the past few years [24–27]. The two devices that most closely parallel our nodes are Berkeley's Mica2 mote and UCLA's Medusa-MK2.

The Mica2 mote has a 4-MHz, 8-bit processor and uses the same off-chip Flash memory chip and serial interfaces as our nodes. Medusa-MK2 features a dual-microcontroller scheme, which uses the same processor as the Mica2 in situations that require minimal computational power, and a 40-MHz Advanced Risc Machine (ARM) processor to operate its onboard GPS unit and other attachable high-energy-consuming sensors.

However, once deployed, both of the aforementioned nodes are intended to remain close together to form a densely populated network. This allows them to use extremely low-power radios with a very limited range. ZebraNet nodes, on the other hand, are intended to be extremely mobile and distributed over a large area. Having a sparsely populated mobile network demands a more powerful radio with a much larger range.

In addition, due to the high power consumption of our radio and GPS, we cannot hope to run the system continuously for months at a time on one set of batteries. Nor can we reduce the duty cycle of data collection and still achieve our objectives. To compensate, we use a rechargeable battery with solar cells distributed around the collar.

Sensing operating systems and middleware.   Various operating systems and middleware layers have emerged to control sensor nodes [20, 28–31]. Two such systems that closely relate to Impala are TinyOS and Maté.

TinyOS, the popular operating system designed to run on the motes, has many low-level characteristics in common with Impala. For example, both Impala and

TinyOS place an emphasis on event handling through hardware interrupts and the utilization of on-the-fly processing to conserve memory.

One big difference between our system and TinyOS is a result of the differences in the nodes on which they will be used. The mote is designed to accommodate a variety of interchangeable sensors. This is reflected in TinyOS's emphasis on concurrency-intensive operations that allow the system to handle multiple flows of data from independent sensors simultaneously. Impala uses a combination of polling and interrupt handling to allow for a similar interleaving of scheduled and unscheduled events. In the ZebraNet system, however, Impala takes advantage of the fact that we have a fixed number of sensors that work in a predetermined fashion by using hardware timers to schedule all major events. This allows us to save a great deal of power through the use of the dual-clock scheme and the timely use of energy-hungry components.

Maté is a virtual machine that lies on top of the operating system and is designed to provide a layer of security and a basis for automatically updating nodes via virally propagated programs. ZebraNet does not need the added security Maté provides, but the ability to perform viral software updates was implemented in the original version of Impala [17] designed for palmtop computers and will be implemented on the ZebraNet nodes in the near future.

Protocols and routing schemes.   Our peer-to-peer routing scheme has roots in a number of proposed routing methods designed to make communication in mobile ad hoc networks more efficient. DSR [19] and Ad Hoc on Demand Distance Vector (AODV) [32] send out route-discovery messages that perform similar roles to our peer-discovery messages. The difference is that DSR and AODV attempt to discover a complete route to a destination, whereas our algorithm only attempts to discover a node's immediate neighbors. This modification is essential to our system because under normal circumstances, there will not be a complete route to the base station; rather, data is expected to propagate slowly from node to node until the base station comes into range.

Directed Diffusion [33] uses a data-centric scheme in which messages are passed through the network in a series of independent neighbor-to-neighbor communications very similar to our peer-to-peer transmissions. However, this scheme would not work well in our system because our network has a very low connectivity and our topology is changing too fast for important messages to arrive in a timely fashion.

Sensing application studies.   In addition to ZebraNet, there are other concurrently running efforts to use sensors to monitor wildlife or in other mobile applications. The VAFalcons project places solar-powered satellite transmitters on Falcons and uses satellite telemetry to determine the animals' position [34]. Similarly, the Pacific Ocean Salmon Tracking Project places acoustic tags on juvenile salmon [35]. The sound emitted from the tags is recorded by receivers strategically placed along known migration routes and can be used to reconstruct the exact movements of the fish closely. Telenor's Electronic Shepherd project bears considerable similarity to ZebraNet in its goals but has lower-range radios. All of these projects, moreover, rely on a fixed infrastructure to gather data; none harnesses the sort of peer-to-peer node interactions that ZebraNet uses to improve data collection and connectivity.

A stationary sensor network composed of motes running TinyOS has been deployed on an uninhabited island off the coast of Maine to monitor the nesting habitats of certain birds, along with environmental conditions such as temperature and humidity [7, 36]. The group conducted a successful multiple-month experiment in which data was collected and transmitted through the network using a CSMA MAC layer to protect against collisions. This deployment, along with a similar deployment at the James Reserve in Idyllwild, California [37], is providing the sensor-network community with a great deal of insight into the numerous issues relevant to a real-world deployment.

16.7 Conclusion

This chapter has described our experiences building and deploying the ZebraNet system for habitat monitoring in Africa. ZebraNet utilizes mobile sensor-networking technologies to monitor zebra migrations on energy-constrained hardware. We have developed a highly mobile, energy-efficient sensing system that determines fine-grained positional data and propagates it through the network. To reduce the energy consumption on the board, we have implemented several system-level energy-management techniques. Even with these methods, the high-power of the GPS and the long system-lifetime target requires that our hardware utilize a solar array for recharge.

One of the most sobering lessons learned in wildlife tracking today is the degree to which the technology limitations of widely available observational tools (tracking collars and other sensors) limit the ability of biologists to understand our environment. The goal of the ZebraNet project is to answer key questions about interspecies interactions and resource impacts in the context of African wildlife. Written at the halfway-point of a multiyear project, this chapter has described the design decisions, system status, and deployment experiences of our project thus far.

References

[1]   Carribean Conservation Corporation, 2002, at http://www.cccturtle.org/satwelc.htm.

[2]   Cerpa, A., et al., "Habitat Monitoring: Application Driver for Wireless Communication Technology," *Proc. ACM SIGCOMM Workshop on Data Communications*, San Jose, Costa Rica, April 2001, pp. 20–41.

[3]   Juang, P., et al., "Energy-Efficient Computing for Wildlife Tracking: Design Trade-offs and Early Experience with ZebraNet," *Proc. 10th International Conference on Architectural Support for Programming Languages and Operating Systems (ASPLOS-X)*, San Jose, CA, October 2002.

[4]   NASA Satellite Tracking of Threatened Species, 2002, at http://www.nasa.gov/vision/earth/lookingatearth/elephants_space.html

[5]   Estrin, D., et al., "Connecting the Physical World with Pervasive Networks," *IEEE Pervasive Computing*, Vol. 1, No. 1, 2002, pp. 59–69.

[6]   Kahn, J., Katz R., and Pister K., "Next Century Challenges: Mobile Networking for Smart Dust," *Proc. Fifth Annual International Conference on Mobile Computing and Networking (MOBICOM '99)*, Seattle, WA, August 1999.

[7]   Mainwaring, A., et al., "Wireless Sensor Networks for Habitat Monitoring," *Proc. ACM International Workshop on Wireless Sensor Networks and Applications (WSNA '02)*, Atlanta, GA, September 2002.

[8]   Pottie, G., and Kaiser W., "Wireless Integrated Network Sensors," *Communications of the ACM*, Vol. 43, No. 5, 2000, pp. 51–58.

[9]   Mpala Wildlife Foundation, Mpala Research Centre, at http://www.mpala.org/ researchctr.

[10] Lotek Corporation, 2002, http://www.lotek.com.

[11] Amlaner, C. J., and Macdonald D., (eds), *A Handbook on Biotelemetry and Radio Tracking*, Oxford: Pergamon Press, 1980.

[12] Newmark, W., "The Role and Design of Wildlife Corridors with Examples from Tanzania," *Ambio*, December 1993, Vol. 22, No. 8, pp. 500–504.

[13] Borner, M. "The Increasing Isolation of Tarangire National Park," *Oryx*, 1985, Vol. 1.

[14] Prins, H., "Nature Conservation as an Integral Part of Optimal Land Use in East Africa: The Case of the Masai Ecosystem of Northern Tanzania," *Biological Conservation*, Vol. 40, 1987.

[15] Woodroffe, R., and Ginsberg J., "Edge Effects and the Extinction of Populations inside Protected Areas," *Science*, Vol. 280, 1998.

[16] Texas Instruments, MSP430x1xx Family Ultra-Low-Power Microcontroller User's Guide, 2002, at http://www.ti.com.

[17] Liu, T., and Martonosi M., "Impala: A Middleware System for Managing Autonomic, Parallel Sensor Systems," *ACM SIGPLAN Symposium on Principles and Practice of Parallel Programming (PPoPP '03)*, San Diego, CA, June 2003.

[18] Liu, T., et al., "Implementing Software on Resource-Constrained Mobile Sensors: Experiences with Impala and ZebraNet," *Proc. 2nd International Conference on Mobile Systems, Applications, and Services*, Boston, MA, June 2004.

[19] Johnson, D., and Maltz D., "Dynamic Source Routing in Ad Hoc Wireless Networks," Editors: Imielinski, Korth, H., *in Mobile Computing*, Kluwer Academic Publishers, Hingham, MA, 1996, pp. 153–181.

[20] Hill, J., et al., "System Architecture Directions for Networked Sensors," *Proc. 9th International Conference on Architectural Support for Programming Languages and Operating Systems*, Cambridge, MA, April 2000.

[21] Xemics, DP1201A, 433.92MHz Drop-In Module Product Brief, March 2004, http://www. xemics.com.

[22] Texas Instruments, MSP430x16x Mixed Signal Microcontroller, 2002, at http://www. ti.com.

[23] Texas Instruments, MSP430x14x Mixed Signal Microcontroller, 2002, at http://www. ti.com.

[24] Hill, J., and Culler D., "Mica: A Wireless Platform for Deeply Embedded Networks," *IEEE Micro*, Vol. 22, 2002, pp. 12–24.

[25] Min, R., et al., "Energy-centric Enabling Technologies for Wireless Sensor Networks," *IEEE Wireless Communications*, Vol. 9, No. 4, 2002, pp. 28–39.

[26] Rockwell Science Center, "Wireless Sensor Networks," http://wins.rsc.rockwell.com.

[27] Savvides, A., and Srivastava M., "A Distributed Computation Platform for Wireless Embedded Sensing," *Proc. International Conference on Computer Design (ICCD)*, Friedburg, Germany, 2002.

[28] Gay, D., et al., "The nesC Language: A Holistic Approach to Networked Embedded Systems," *Proc. Programming Language Design and Implementation (PLDI)*, San Diego, CA, 2003.

[29] Levis, P., and Culler D., "Maté: A Tiny Virtual Machine for Sensor Networks," *Proc. 10th InternationalConference on Architectural Support for Programming Languages and Operating Systems (ASPLOS-X)*, San Jose, CA, October 2002.

[30] Madden, S., et al., "The Design of an Acquisitional Query Processor for Sensor Networks," *Proc. ACM SIGMOD*, San Jose, CA, 2003.

[31] Sun Microsystems, Java 2 Platform, Micro Edition, November 2002, http://java. sun.com/j2me.

[32] Perkins, C. E., and Royer E. M., "Ad Hoc On-Demand Distance Vector Routing," *Proc. 2nd IEEE Workshop on Mobile Computing Systems and Applications (WMCSA)*, New Orleans, LA, February 1999.

[33] Intanagonwiwat, C., Govindan R., and Estrin D., "Directed Diffusion: A Scalable and Robust Communication Paradigm for Sensor Networks," *Proc. Sixth Annual International Conference on Mobile Computing and Networking (MOBICOM '00)*, Boston, MA, August 2000.

[34] The Center for Conservation Biology, VAFALCONS, 2002, http://fsweb.wm.edu/ccb/vafalcons/falconhome.cfm.

[35] Census of Marine Life, POST: Pacific Ocean Salmon Tracking Project, 2003, http://www. postcoml.org.

[36] Szewczyk, R., et al., "Lessons from a Sensor Network Expedition," *Proc. 1st European Workshop on Wireless Sensor Networks*, Berlin, Germany, January 2004.

[37] Center for Embedded Networked Sensing, Research Infrastructure: James Reserve Local Area Power System and Network Enhancements, at http://research.cens.ucla.edu/portal/page?_pageid=59,43783&_dad=portal&_schema=PORTAL.

CHAPTER 17

Sensor Webs in the Wild

Kevin A. Delin

17.1 Introduction

In October 2001, a new era in wireless sensor systems began when the NASA sensor web, deployed at the Huntington Botanical Gardens in southern California, went on-line. For the first time, it was possible for a person with nothing more than a computer, an Internet connection, and a standard browser to watch streaming, real-time data generated by an ad hoc wireless-networked system permanently embedded in an outdoor environment. Wireless-networked systems, sensor webs in particular, are only just beginning to change the ways in which we can sense, monitor, and control large spatial areas.

The sensor web's capabilities are useful in a diverse set of outdoor applications, ranging from precision agriculture to perimeter security to effluent tracking. Wireless networks of sensors are often marketed as replacements for running wire to sensing points. Naturally, this holds true for sensor webs as well, where the individual pods communicate among themselves wirelessly. However, it is more significant that the sensor web, with its unique, global information-sharing protocol, forms a sophisticated sensing tapestry that can be draped over an environment. This Sensor Web approach allows for complex behaviors and operations, such as on-the-fly identification of anomalous or unexpected events, mapping vector fields from measured scalar values and interpreting them locally, and single-pod detection of critical events, which then triggers changes in the global behavior of the Sensor Web. This chapter describes the Sensor Web, a technology enabled by the confluence of the massive computer and telecommunications markets. In addition, it discusses the design and deployment of such systems in the wild, away from the temperature- and humidity-controlled rooms characteristic of offices and factories.

17.2 The Sensor Web: A Different Type of Wireless Network

In 1997, the Sensor Web was conceived at the NASA Jet Propulsion Laboratory (JPL) to take advantage of the availability of increasingly inexpensive yet sophisticated mass-consumer-market chips for the computer and telecommunication industries and to use them to create platforms that share information among themselves and act in concert as a single system. This system would be embedded in an environment to monitor and even control it. The purpose of a sensor-web system is to extract knowledge from the data it collects and use this data to react intelligently and adapt to its surroundings. It links a remote end user's cognizance with the observed environment.

In its most general form, a sensor web is a macroinstrument comprising a number of sensor platforms [1]. As shown in Figure 17.1, these platforms, or pods, can be orbital or terrestrial, fixed or mobile. Coordinated communication and interaction among the pods provides a spatiotemporal understanding of the environment. Specific portal pods provide end-user access points for command and information flow into and out of the sensor web. At NASA/JPL, the focus has been on in situ sensor webs, with the resulting system output viewed over the Internet.

Wireless networks are not a new approach to environmental monitoring, and it is common to find systems where remote sensors in the field communicate with central points for data processing in a star-network formation. *The Sensor Web, however, is a temporally synchronous, spatially amorphous network,* creating an embedded, distributed monitoring presence which provides a dynamic infrastructure for sensors. By eschewing a central point on the network, information flows everywhere throughout the instrument (see Figure 17.2).

So far, this sounds like a typical ad hoc, self-configuring network. Often, the ideas of hopping information around such a network are framed in terms of the power advantage gained by doing so. While this advantage certainly exists, the sensor-web concept goes a step further: The individual pods making up a sensor web

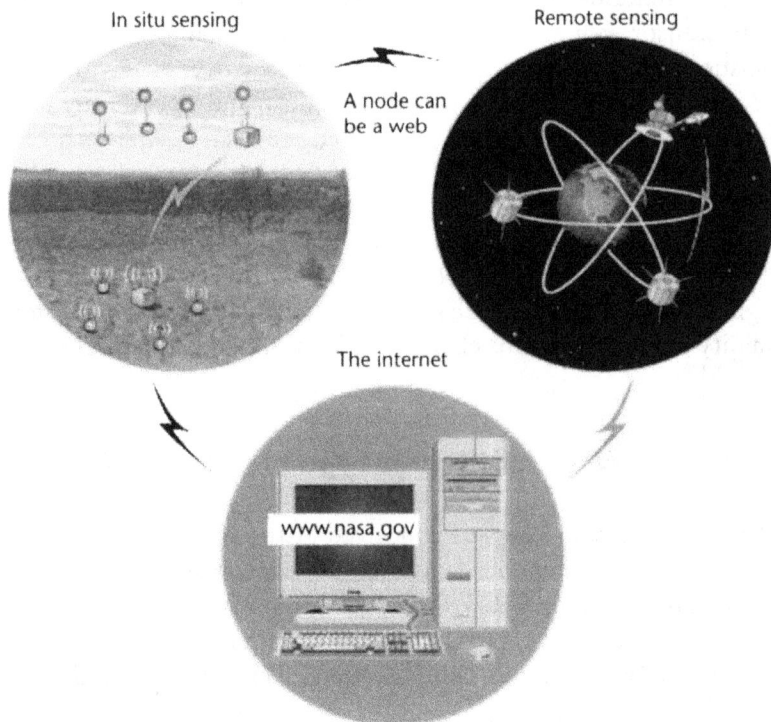

Figure 17.1 Generalized concept of the sensor web, including both orbital and terrestrial platforms. Note the recursive nature of the sensor web, as individual nodes on a particular sensor web may be sensor webs themselves.

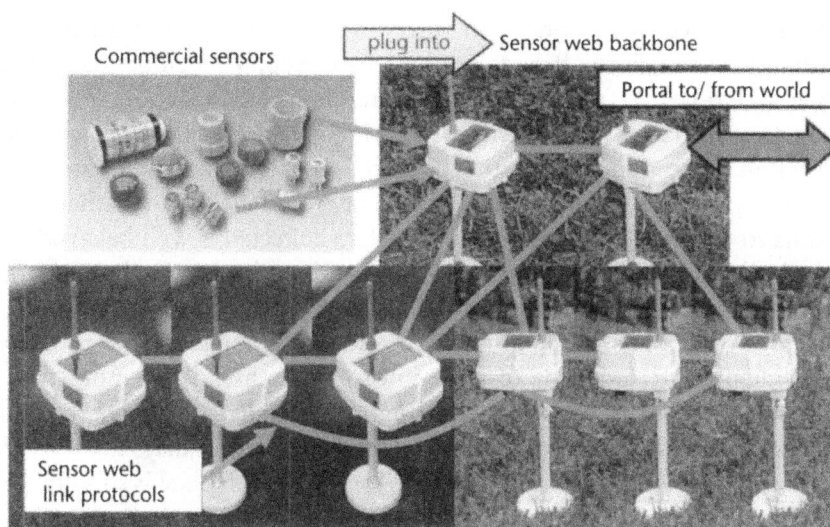

Figure 17.2   The sensor web forms an informational backbone that creates a dynamic infrastructure for the sensors in the sensor-web pods. The total sensor-web instrument includes both the individual pods and the space between them.

are not just elements that can communicate with one another; they are elements that must communicate with one another. Whereas wireless networks are typically discussed as confederations of individual elements (like computers connected to the Internet), the sensor web is a single, autonomous, distributed instrument. The pods of a sensor web are akin to the cells of a multicellular organism; the primary purpose for information flow over a sensor web is not to get data to a gateway or an end user but rather to the rest of the sensor web itself.

By design, the sensor web spreads collected data and processed information throughout its entire network. As a result, there is no design criterion for routing as in other wireless systems. Routing, by definition, is the focused movement of information from one point to another. In contrast, data collected by a sensor web is spread everywhere, rendering meaningless the concept of routing. Instead, the communication protocol on a sensor web is relatively simple and is structured for both omni- and bidirectional information flows. Omnidirectional communication implies no directed information flow, while bidirectional communication lets individual pods (and end users) command other pods, as well as receive information from them. Consequently, information on the sensor web can result from four types of data: (1) raw data sensed at a specific pod, (2) postprocessed sensed data from a pod or group of pods, (3) commands entered into the distributed instrument by an external end user, and (4) commands entered into the distributed instrument by a pod itself. The sensor-web processes this internal information, draws knowledge from it, and reacts to that knowledge.

Since there is no specific routing of information, all pods share everything with each other. After each measurement is taken, both raw and processed information from each pod is moved throughout the sensor web to all other pods before the next measurement is taken. Because the sensor web is a single, distributed instrument, its internal operations are synchronous from pod to pod (again in contrast with more common wireless networks). In this way, a total snapshot associated with that instant is available to all pods on the sensor web. This global data sharing allows each pod to be aware of situations beyond its specific location. Pods may therefore combine data across the sensor web to identify a moving front and determine its speed and direction—a task that a single-point measurement cannot accomplish. Pods may also use neighbors to examine the stochastic nature of their local measurements to determine whether or not the data collected is well behaved. Such macroscopically coordinated data processing would not be as straightforward if each pod were semiautonomous on the network, as in typical wireless sensor systems. There is a degree of stiffness to the information flow over the sensor web compared to the individually directed node-to-node information threads on more typical wireless systems. Sensor-web pods may be thought of as individual, synchronized pixels in a much larger instrument that can take snapshots at regular intervals of the entire environment in which it is embedded. Moreover, each pixel is simultaneously aware of the overall picture as well as its local readings.

Since there is no distinction between instructions originating from the end users and those from other pods within the system, the sensor web is both a field-programmable and self-adapting instrument. Bidirectional communication also allows portal pods to be linked (by Internet or satellite) so that the end user of a particular sensor web may be another web. As a result, a local sensor web can leap beyond the bounds of its own spatial confinement—an important consideration for very large-scale environmental studies.

17.3   Sensor Web Pods

A sensor-web pod consists of five basic modules:

1. The radio. This module links each pod to its local neighborhood. The NASA/JPL sensor-web pods use radios operating in the 900-MHz license-free ISM band with an upper range of about 200m or more. Communication occurs in bursts at 56 kbaud. (Implicitly, we assume that none of the in situ sensor webs discussed here are deployed underwater, where acoustic modems severely limit bandwidth and communication range relative to these ISM radios.)

2. The microcontroller. This module contains the system's protocols, communicates with the attached sensors, and carries out data analysis as needed.

3. The power system. The NASA/JPL system uses a battery pack with solar panels to keep the batteries charged. The combination of solar panels and micropower electronic design have kept sensor-web pods operating in the field for years without requiring maintenance.

4. The pod packaging. This key module is often overlooked, especially for sensor-web applications in the wild. The package must be light, durable, inexpensive, and sealed against such elements as rain, snow, salty sprays, dust storms, and local fauna. In addition, it must provide for easy and rapid mounting.

5. The sensor suite. This module is completely determined by the specific application. Ideally, the sensor suite will be the prime determining factor for the size, cost, and power requirements of a sensor-web pod, making the sensor-web infrastructure attractive for any application. What is considered an inexpensive or small sensor-web pod in one application may not be viewed as such in another.

In the hyperbolic world of high-technology, engineering metrics are often based more on sound bites than sound principles. We have been conditioned by decades of experience with Moore's law (and the technology revolution associated with it) to think that smaller is always better. There are certainly practical reasons for limiting the size of a sensor-web pod. In an outdoor environment, smaller and lighter pods are easier to deploy since more can fit into, say, a backpack. However, shrinking pods to infinitesimal sizes is undesirable for a typical, outdoor sensor-web system. Consider the impact of size with respect to three key sensor-web-pod design issues: power, antenna size, and transducers.

An important design consideration for typical, outdoor sensor webs is pod longevity. In many cases, deployment is only practical during certain seasons, so intraseason maintenance must be avoided. As a result, maximizing the available power by cleverly using batteries, energy harvesting, or both is critical. Batteries are often rated in terms of their energy density (watt-hours per unit volume) because cells can be added in series to increase total available voltage. The larger the volume of the sensor-web pod, the more volume is available for power from any particular battery technology.

There are only two ways to maintain a given available battery power level, while allowing the pod volume to shrink: improve the battery technology or reduce energy use within the pod. While there are numerous efforts to provide higher-energy-density power sources than are typically available (e.g., lithium-ion batteries), none are yet commercially available for consumer use. In addition, many experimental batteries have limited lifetimes, and any suitable battery technology must be essentially zero maintenance and environmentally robust (especially to changes in temperature, both seasonal and diurnal). As for improving energy efficiency, the laws of physics require a certain power output to broadcast a given distance. Therefore, although one can lower the energy per bit involved in computation, wireless communication puts a hard limit on how much energy will be required for the system to operate for a given pod-to-pod distance.

Now consider energy harvesting, which is typically accomplished via solar power charging secondary batteries. Here, too, the smaller the platform, the smaller the solar panels used to reenergize the system and the smaller amount of energy that can be harvested for a given panel. Clearly, beyond a certain size, the smaller one designs a sensor-web pod for a given set of operating parameters, the more one gives up in longevity with respect to available power.

Antennas are also directly related to platform size. Again, the laws of physics dictate the appropriate antenna geometry for a given operating frequency to ensure a proper impedance match into the radiated space. As a result, while the onboard processor and radio electronics may shrink, the antenna may not if a particular communication range is required. Without proper coupling, radiation efficiency is reduced, and radio power must be increased to maintain range. We therefore find that since most outdoor sensor webs require pod-to-pod ranges of at least tens of meters, indiscriminate shrinkage of the individual antennas clearly compromises the telecommunication subsystem.

Lastly, consider the sensors themselves. Many sensors used in outdoor field applications, although compact and inexpensive, are not Micro-Electro-Mechanical System (MEMS) devices and, therefore, cannot be integrated into the sensor-web pod at the chip level. Examples include gas sensors (which often require a certain volume of sample gas) and seismometers (which require a certain mass for appropriate mechanical resonant frequencies). As a result, for a wide variety of sensor-web applications in an outdoor environment, the sensors will be additional components added to the basic pod platform. Clearly, there is little to be gained by continually shrinking the platform if the sensors themselves remain the limiting size element. Moreover, shrinking the platform may actually complicate the design if it becomes difficult to integrate the sensors into the pod.

As shown in Figure 17.3, the NASA sensor-web pods have been developed in several sizes, ranging from that of a gumball to that of a couple of decks of playing cards. Significantly, the gumball-sized pod dates back to 1998 [1], demonstrating that even then, it was relatively easy to make small platforms so long as only simple measured parameters (e.g., temperature, humidity) and short pod-to-pod communication distances (i.e., on the order of meters) were required. Such small pods are ideal for building or factory monitoring but less practical for outdoor environments for the reasons discussed above. From this discussion, it is apparent that while smaller pods are desirable, shrinking pods beyond a certain point leads to diminishing returns.

17.4   Sensor Web Deployments

With the objective of performing genuine in situ monitoring, the NASA/JPL Sensor Web Project has aggressively fielded many instruments. Sensor webs have been deployed in a large variety of demanding, real-world locations for lengthy periods of many months or even years. Here, we summarize some representative deployments. A more complete list, along with real-time streaming data from current deployments, is available over the Internet [2].

Huntington Botanical Gardens, San Marino, CA.   Sensor webs have been at the Huntington Botanical Gardens in San Marino, California, since the deployment of Sensor Web 2.0 in June 2000. Sensor Web 3.0, the first permanent WSN system to provide continuous real-time streaming data to users over the Internet, was deployed in October 2001 in a nursery area, remaining there until August 2002. At that point, it was replaced by Sensor Web 3.1, which also expanded the spatial extent of the coverage into the public areas of the gardens. In all of these systems, canonical botanical parameters, including light levels, air temperature, and air humidity, are measured by the pods in 5-minute intervals.

Significantly, the deployment of Sensor Web 3.1 occurred in two stages. Pods 0 (the mother or portal pod) through 11 were deployed in August 2002. The purpose of this initial deployment was to string the pods out over a large area and determine

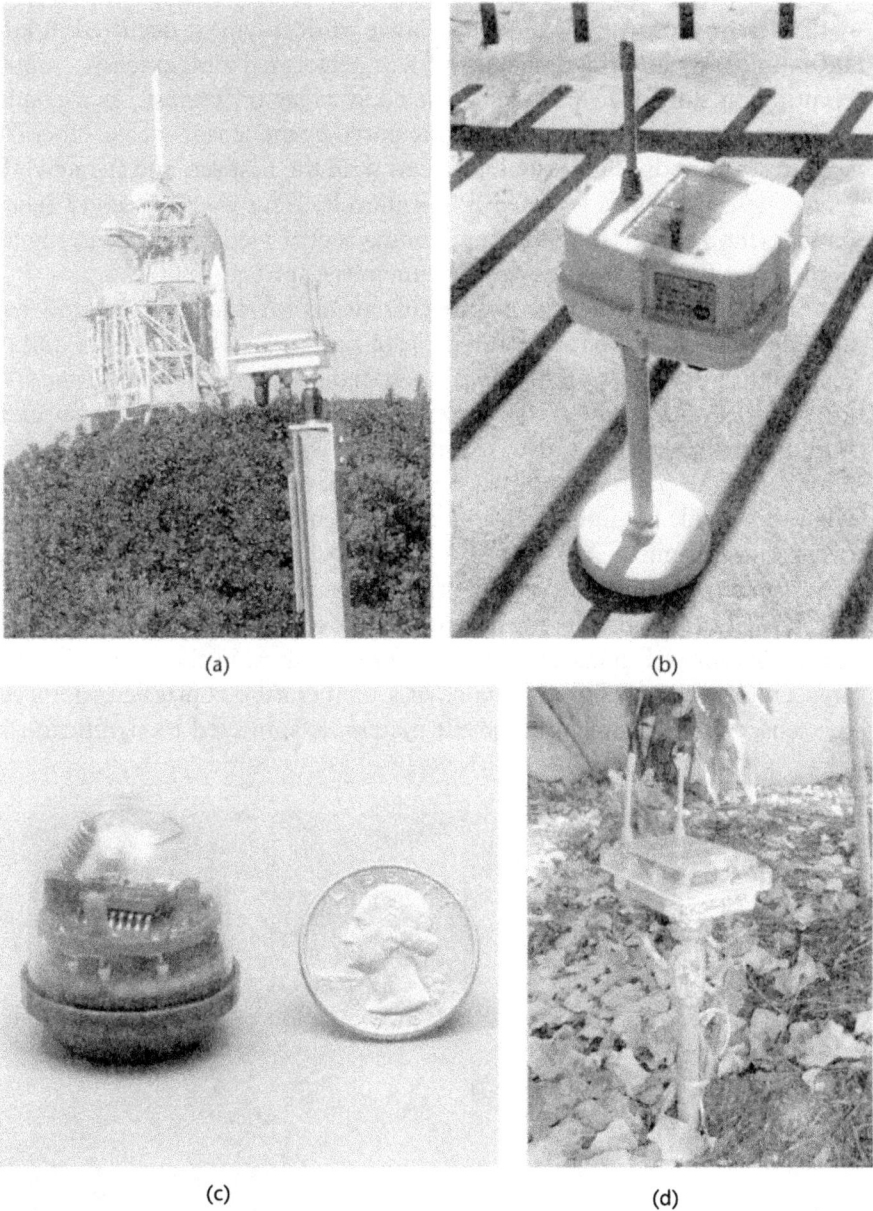

(a)

(b)

(c)

(d)

Figure 17.3 Various sensor-web pods: (a) Functioning Sensor Web 1.0 pod, circa 1998. Note the small size, which includes antenna, battery, and temperature and light sensors. (b) Sensor Web 3.1 pod deployed at the Huntington Botanical Gardens, circa 2002. It is about the size of two decks of playing cards. The pod is mud spattered from rain and watering and has a chewed antenna. Subterranean sensors (soil moisture and temperature) can be seen going into the ground. (c) Sensor Web 3.1 pod deployed less than a quarter mile from Space Shuttle Launch Pad 39A at NASA's Kennedy Space Center (note shuttle on pad). This sensor web measured environmental conditions in the lagoons surrounding the launch site. (d) Sensor Web 5.0 pod, circa 2004. This new generation of sensor-web pods is more compact and more power efficient than previous ones, a direct result of exploiting Moore's law in its design.

the robustness of the connectivity. It typically took four or five hops for the data to move to and from the extreme points on the network, several hundred meters apart. (Shorter-duration system tests have demonstrated that 12 or more hops are possible with present protocols.) The second stage of deployment occurred in late January 2003. In contrast to the first stage, it was performed with extensive input from the Huntington staff to help monitor their key areas of interest. As a result, the pod placement was more confined. As anticipated by the sensor-web protocol design, the second set of pods seamlessly integrated with the first set, and the new 19-pod system coalesced within a few measurement cycles. This was the second demonstration of augmenting an existing and functioning sensor-web deployment, the first having occurred at Lancaster Farms in Virginia in the spring of 2002.

Several sensor-web pods in the current deployment measure soil temperature and moisture with modular subterranean probes (see Figures 17.3 and 17.4). The Huntington Gardens staff has used the system to remotely monitor the state of their greenhouses and to ensure that watering (from both sprinkler and rainfall patterns) is uniform across various areas. Rains in February 2003 showed the surprising variations in these parameters, even on scales of a meter. This suggests using sensor webs to provide continuous resource-management information on both micro- (of order 1m) and meso- (of order 100m) spatial scales, which is unattainable from most remote (e.g., satellite or airborne) measurements.

The Huntington Gardens deployment covers a large number of microenvironments. Sensor web pods are located in greenhouses, in the open sun, and in shady areas, while the portal pod remains in a temperature-controlled room. As a result, the sensor web, a single-instrument system, is subjected to significant differential

Figure 17.4 Sensor Web 3.1 pods 12 (rear, upper center of picture) and 13 (foreground, lower left corner) deployed at the Huntington Gardens around cycads. One cycad trunk is partially seen on the right side of picture; the fronds of a second cycad are seen just to the right of sensor-web pod 12. Subterranean soil temperature and moisture are attached to both pods. Differing subterranean soil-moisture conditions between these two pods alerted gardeners that the two plants were not receiving identical watering. It was discovered that ground brush around one of the cycads was acting as a sponge, retarding water entry into the ground, and was subsequently removed.

heating, a stress not encountered in typical indoor deployments. Nevertheless, it is possible to compensate the electronics to handle this issue while maintaining system integrity and synchronicity. In addition, one pod was found with an antenna chewed, probably by a bird or coyote. This abuse did not interrupt pod functionality. Sensor Web 3.1 at the Huntington has run continuously for over 2 years, showing the robustness of both the system protocols and pod packaging.

Antarctic wilderness. In cooperation with researchers hunting for meteorites, a sensor web was deployed on the East Antarctic Ice Sheet [3]. International treaty prevents leaving unattended equipment in the remote fields, so this deployment was limited to 3 weeks starting in December 2002. Nevertheless, the sensor web had to prove itself in an extreme test environment with low temperatures, constant wind, and dry air. In addition, the deployment was made under very hard conditions, with researchers fully suited up for the bitter weather as shown in Figure 17.5.

The deployment was a test of the system in preparation for extended studies of biological activities in cryogenic environments, especially those on Mars. The more hostile the environmental conditions, the more widely distributed any biological blooms are expected to be in both time and space. The continual monitoring presence of the sensor web, coupled with its abilities to do event-triggered sensing, make it an ideal instrument for such studies. For example, if one pod detects that its local conditions are more favorable to a thaw, which might activate a bloom, the entire sensor web could begin making measurements more frequently to better understand the short-term event. Thus, most data is accumulated only when a significant event

Figure 17.5 Sensor Web 3.0 deployment on the East Antarctic Ice Sheet. Bamboo is used as the mounting stake because of its known resilience to the harsh Antarctic conditions. Wires connect the pod to the temperature sensors at the surface. Note the second pod located in the snowmobile in lower right corner.

occurs, further conserving power and memory resources on the sensor web. In addition, the sensor web provided data from across the region that minimized stochastic components resulting from specific pod placement.

The 14-pod sensor web was distributed about the home-base site over a distance of 2 km. Parameters such as air and soil temperature, humidity, and light flux were measured across the sensor web at 5-minute intervals. System operations could be monitored continuously at the home-base site. Typical temperatures over the deployment were lower than –10°C with extremes below –20°C. From this deployment, it was discovered that temperature differences between sensors exposed to the air and those buried in a moraine could differ consistently by 10°C or more. The sensor-web system not only performed well under the harsh, dry conditions but was also easy to set up. (As is often the case, no members of the NASA sensor-web team assisted in the deployment.) Under severe weather conditions, such issues are critical for mission success. Significantly, these pods were identical to the ones that were deployed at the Huntington Gardens; there were no special modifications made for the different environment.

Flood Basins, Tucson, AZ.   We have deployed a sensor web at the Central Avra Valley Storage and Recovery Project (CAVSARP) facility located west of Tucson, Arizona [4]. The facility is located in a desert environment in the semiarid southwestern United States, where the artificial recharge basins experience repeated flood cycles. The controlled flooding conditions at the CAVSARP facility are ideal for the investigation of various hydrologic processes. There are several technology-related reasons for this site choice as well. The CAVSARP facility allows us to continue our efforts to develop the sensor web as a tool for the study of spatiotemporal phenomena. For example, the sensor web can track the moving flood front, follow the infiltration of water into the ground, and provide information to map and characterize the lateral and vertical extent of the floodwaters. Moreover, the extreme temperature variations of the Arizona desert (both diurnal and seasonal) provide yet another test of the sensor web's robustness. Again, this sensor-web hardware is essentially identical to that previously deployed, with no special modifications made for the new environment.

A single basin, measuring approximately $700 \times 2,400$ ft^2, was strategically outfitted with 13 pods, their number and placement being determined by science requirements rather than technological limitations. The pods were mounted on stakes to elevate them above the flood waters, which can rise as high as 7 ft (see Figure 17.6). (While the pods themselves are watertight, pod-to-pod radio communication would not be possible if they were submerged.) Each pod, in addition to collecting air temperature, humidity, and light levels, also collects two soil-moisture readings (one at the surface and one 0.5m below) and a surface soil-temperature reading. This is accomplished by wires that run from the pod into the ground. Measurements are made at 5-minute intervals with the results being fed to the Internet in real time (via the portal pod 0).

This sensor web has been collecting data since its deployment in November 2003. The real-time data stream is available via the Internet [2], with a sample screen-capture shown in Figure 17.7. Unlike remote techniques, which can only observe the basins for relatively short durations on finite schedules, the sensor web's data stream provides continuous information for tracking surface-water motion and

Figure 17.6   NASA/JPL team members deploy a Sensor Web 3.2 pod in a flooding basin outside Tucson, Arizona. Extended stands allow the pod to stay operational above water during a flooding event.

ground infiltration. The spatial and temporal patterns of wetting and drying can thus be fully monitored with results incorporated into hydrological models and compared with space-based and airborne investigations [5]. This analysis is ongoing. As a result, this sensor web can both augment and ground-truth the remote data traditionally used in hydrologic studies.

The repeatable nature of the flooding/drying dynamics is apparent from the sensor-web data shown in Figure 17.7. The soil moisture measurements are made with Watermark sensors [6], in which electrodes embedded in a granular matrix have a lower resistance when the surrounding soil is wetter. As a result, the raw data reveals the motion of the flooding water as sharp drops in resistance. (The diurnal cycles seen in the raw data are sensor artifacts and can be corrected with soil-temperature measurements [7].) It only takes a few hours for the flood front to traverse the basin from the inlet in the northwest corner to the basin center, but it takes a much longer time (about 20 hours) to reach the basin's southwest and southeast corners. Note, too, that the water reaches the southern border relatively evenly (as

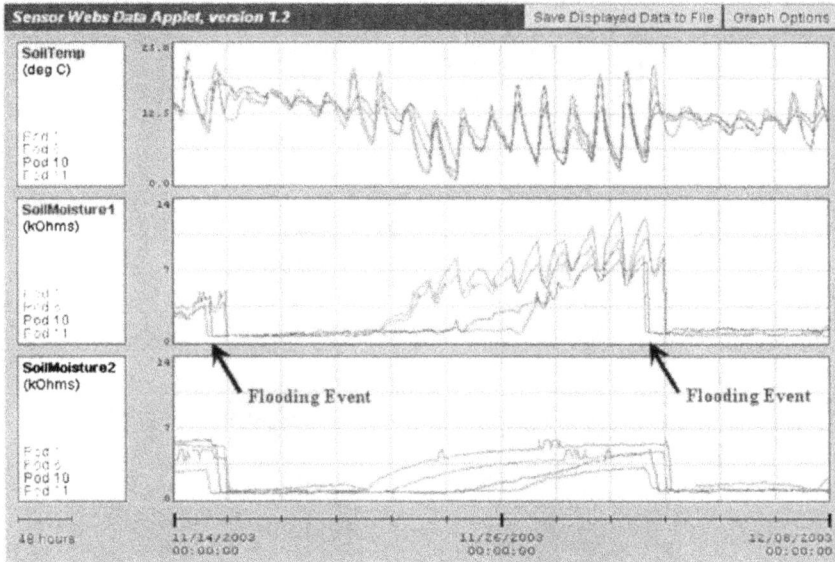

Figure 17.7   Screen-capture of Internet data from the CAVSARP facility: (a) surface temperature (°C), (b) surface moisture, and (c) soil moisture at 0.5m depth (relative units; lower values imply wetter soil). Diurnal cycles in moisture measurements are artifacts and can be corrected with local soil temperature. When the basin floods, sensor-web pod 6 (basin inlet, northwest corner) is the first to detect the water front, followed by pod 10 (basin center), pod 11 (southeast basin corner, diagonal from pod 6), and, lastly, pod 1 (southwest basin corner). The direction of inundation can be determined from this data. Note that the data also shows that the basin dries out in reverse order of flooding. The traces correlate with water discharge into basin, inundation, infiltration, drying, and the beginning of another cycle.

indicated by pods 1 and 11), which is expected from basin construction. Moreover, it is also clear from Figure 17.7 that the drying front traverses the reverse route, albeit at a much slower speed. Not surprisingly, the surface dries more thoroughly than the deeper portions of the ground.

17.5   Deployment Operations

From the experience of deploying sensor webs in a multitude of environments with varying conditions, it is apparent that the ease with which the system is deployed is just as critical for acceptance by end users as are its technological aspects. With the exception of applications in battlefield theaters, most outdoor sensor-web applications require the system to be deployed in a manner that does not harm the monitored environment. For example, end users have expressed concerns that if sensor-web pods are too small, local fauna may try to ingest them and choke. End users also want to avoid littering local environments with hundreds of pieces of microelectronic gear. It is therefore worthwhile to consider sensor-web technology from an operations point of view.

Most applications require tracking specific pod locations with high precision. It is therefore highly unlikely that any WSN will simply be sprinkled over a large area. In addition, coupling sensors into the environment will also prevent such a passive deployment. For example, neither subterranean nor seismic sensors can be deployed by a sprinkling technique, as both require laborious efforts for appropriate sensor mounting. There are also applications, particularly those involving agriculture, where pod placements must be compatible with existing operations, such as harvesting. For example, sensor-web pods must be mounted out of the way of threshers. Consequently, the mounting and placement of sensor-web pods will be an active operation and likely to be done by hand in most instances.

In addition, we have encountered few applications in the wild where an average pod-to-pod distance is less than 10m. It is therefore necessary for the sensor-web pods to have a wireless range limited only by the government-mandated ISM-band specifications. More limited wireless communication will not be suitable if the density required to maintain network connectivity exceeds the desired measurement density required to monitor the region. Therefore, a critical metric is the *sensor-web-pod density* rather than the more usually specified pod number. Ideally, the density of a sensor web should be just greater than that required by the application itself to ensure connectivity and internal communication redundancy within the instrument.

The methods used to mount the sensor-web pods depend not only on the application but also the particular field site. Pod placement very close to the ground can limit transmission distance. Nevertheless, the Huntington Garden pods are within 10 inches of the ground and have sufficient communication power to keep an adequate pod-to-pod distance. Often, for logistical reasons, the sensor-web pods tend to be mounted higher off the ground, with the attendant benefit of increasing the wireless range. Local terrain is rarely level, which also tends to increase transmission distances. We have typically used posts (for horizontal surfaces) and brackets (for vertical surfaces) to mount the pods. These types of mounts are both small enough and light enough to bring into the field yet provide proper robustness for fixing the sensor-web pods rigidly in place for long durations.

Sensor-web-pod placement is probably the most application-specific issue: each deployment is different, depending on local terrain, even if applications have identical purposes. It is always necessary to have a clear understanding of the terrain to ensure proper pod density for both measurement and web connectivity. General purpose algorithms for "optimal" pod placement, which, by definition, do not account for local geometry or the specifics of an experiment, tend to be of limited use. In fact, pod placement has had the least systematic study and will likely remain an open issue for each end-user community rather than the technologists developing the sensor-web systems.

17.6 The Future

The primary focus of sensor-web development thus far has been to demonstrate that the technology is stable, robust, and attractive to potential end users. For a user community to adopt it, however, the sensor web needs to be more than just well engineered; it must also be easily deployed and maintained and provide valuable output. The overall simplicity of the sensor-web system as an operational instrument is demonstrated by the fact that most sensor webs are deployed and operated in a variety of environments without requiring assistance from the NASA/JPL team.

Having demonstrated the sensor web's core capabilities with a myriad of deployments, we are now moving sensor-web development into a new phase, focusing as much on applications as technology. The continuous, virtual monitoring and reacting capabilities have wide ranging uses for resource management, pollutant tracking, and perimeter monitoring. This step requires us to take full advantage of the large-scale awareness already built into the sensor-web protocols. In this way, the output of, for example, a hydrologic sensor web would not consist of a collection of scalar measurements (soil moisture) but rather a single vector (water motion) with pod-to-pod data fusion occurring within the sensor web itself. This will truly give the sensor web the capacity to make sophisticated, autonomous decisions. As awareness of this unique distributed instrument and its capabilities spreads into other user communities, the sensor web is expected to become a dominant WSN architecture.

17.7 Acknowledgments

I wish to acknowledge the significant contributions of Scott Burleigh, Shannon Jackson, Dave Johnson, Ken Manatt, Myche McAuley, and Richard Woodrow in developing the sensor-web instrument. Fielding the sensor webs would not have been possible without the active participation of my field collaborators, including Julio Barrantes, Eric Bender, Greg Bonito, Renee Brown, Nancy Chabot, Scott Collins, James Dohm, Jim Folsom, Ralph Harvey, Felipe Ip, Doug Moore, Danielle Rudeen, Jonathan Sax, and Theresa Trunnelle.

References

[1]   Delin, K. A., "The Sensor Web: A Macro-instrument for Coordinated Sensing," *Sensors*, Vol. 2, 2002, pp. 275–280.

[2]   See http://sensorwebs.jpl.nasa.gov. This site also contains the Sensor Web papers referenced in this chapter.

[3]   Delin, K. A., et al., "Sensor Web in Antarctica: Developing an Intelligent, Autonomous Platform for Locating Biological Flourishes in Cryogenic Environments," *Proc. 34th Lunar and Planetary Science Conference*, Houston, TX, March 2003.

[4]   Delin, K. A., et al., "Sensor Web for Spatio-temporal Monitoring of a Hydrological Environment," *Proc. 35th Lunar and Planetary Science Conference*, League City, TX, March 2004.

[5]   Rucker, D. F., et al., "Central Avra Valley Storage and Recovery Project (CAVSARP) Site, Tucson, Arizona: Floodwater and Soil Moisture Investigations with Extraterrestrial Applications," *Proc. 35th Lunar and Planetary Science Conference*, League City, TX, March 2004.

[6]   Available from Irrometer Co., http://www.irrometer.com.

[7]   Shock, C. C., "Soil Water Potential Measurement by Granular Matrix Sensors." In B. A. Stewart and T. A. Howell, (eds.), *The Encyclopedia of Water Science*, Marcel Dekker, New York, NY, 2003, pp. 899–903.

CHAPTER 18

Defense Systems: Self-Healing Land Mines

William. M. Merrill, Lewis Girod, Brian Schiffer, Dustin McIntire, Guillaume Rava, Katayoun Sohrabi, Fredric Newberg, Jeremy Elson, and William Kaiser

18.1 Introduction

This chapter provides an overview of the WSN developed to support the Self-Healing-Minefield Defense System and the development history for that system. The self-healing minefield (SHM) is an autonomic system of antitank landmines, where each antitank mine monitors its neighbor's state, senses threats to itself and its ability to contribute to the system obstacle function, and responds autonomously to those threats by moving. To support this capability each SHM node includes a mobile ad hoc network, a distributed self-contained acoustic location system, and acoustic, acceleration, and network-status distributed sensing.

The SHM is an application-driven WSN. This chapter provides a system-level description of the SHM solution developed by our team from June 2000 to April 2003. The SHM program was created by the U.S. Government's Defense Agency Research Planning Association (DARPA) Applied Technology Office (ATO) to develop and demonstrate one of the first of a wide class of applied wireless-networked embedded systems [1–3]. This chapter describes the SHM system, its development, and its demonstrations from a system perspective, emphasizing our implementation and experiences with this sensor-network deployment.

The SHM system merges "the capabilities of networked communications and autonomous systems with the tactical needs of modern antitank landmines" [4]. An SHM field consists of up to thousands of nodes (mines), each including processing, sensing, and networking capabilities that sense their state and autonomously react to changes in the system state. The system provides an obstacle to disrupt the maneuver of enemy vehicles, which is robust against mounted and dismounted minefield breaching (without, unlike current U.S. systems, the use of antipersonnel landmines to protect antitank mines). This chapter summarizes the embedded hardware and software systems integrated into the SHM system to provide an autonomous, applied WSN and discusses our demonstration of a 96-node SHM system.

Each SHM node is both a development platform and a component of a demonstrated self-healing network and autonomous distributed system. SHM nodes include an embedded computing platform, a software infrastructure to enable development and testing of large collections of nodes, and a suite of software services. In the SHM program, Sensoria Corporation, a subcontractor to Science Applications International Corporation (SAIC), managed the program and developed high-level mine-control algorithms. Other team members included the Universal Propulsion Company, which developed the rocket engines used to enable mines to move, and Ensign Bickford, which developed the safe-and-firing (SAF) system controlling each node's rocket engines. Within this chapter, in order to focus on the networked, embedded-distributed-computing features of the SHM system, we will focus solely on the system developed by the authors at Sensoria Corporation.

This chapter provides an overview of the WSN component of the SHM system, organized as follows. Section 18.2 provides an overview of the motivational goals for developing the SHM components. Sections 18.3 and 18.4 provide an overview of the hardware and software systems developed for each SHM node to support distributed operation of this WSN. Section 18.5 summarizes the design considerations used in creation of this multilayer distributed system. Finally, Section 18.6 provides examples of measured SHM operations.

18.2   SHM System Need for a WSN and Program History

To provide perspective on the SHM WSN, this section describes how a WSN enabled the SHM defense system. The goal of the SHM program was to demonstrate a system of self-aware antitank mines that could detect and heal minefield breaches. To enable this, each mine (node) in the SHM system includes networking and sensing capability to monitor the state and disappearance of neighboring nodes and to coordinate internode perceptions and actions. Specific system goals, desired and demonstrated, are shown in Table 18.1. To enable the SHM system to operate robustly without human oversight required each node to communicate and locate (relative to itself) its neighbors. This was the need that the WSN provided in the SHM system.

The SHM system was developed over a 3-year period and culminated in a demonstration of 96 nodes. A representation of the development progress is shown in the number of interacting, autonomous nodes at each system demo, as displayed in Figure 18.1. As an increasing number of peer-to-peer nodes interacted, the system capability and complexity increased to a local scale designed for 50 nodes. The final system demonstrated met the stated goals of the program and, as discussed in more

Table 18.1   SHM System Requirements

Requirement	Goal	Demonstrated
Individual nodes could detect and respond to a breach within:	10s per node/s	10s with 0.02 false alarms
The number of nodes (location and status) each node was aware of:	50	96
Time to self-form and self-locate from initial deployment:	5–15 minutes	10+ minutes
Geolocation accuracy	< 1m without GPS	< 1m without GPS
Mobility	10m distance, 2m height	10m distance, 2m height
Form factor	12-cm diameter by 7.6 cm	12-cm diameter by 7.6cm

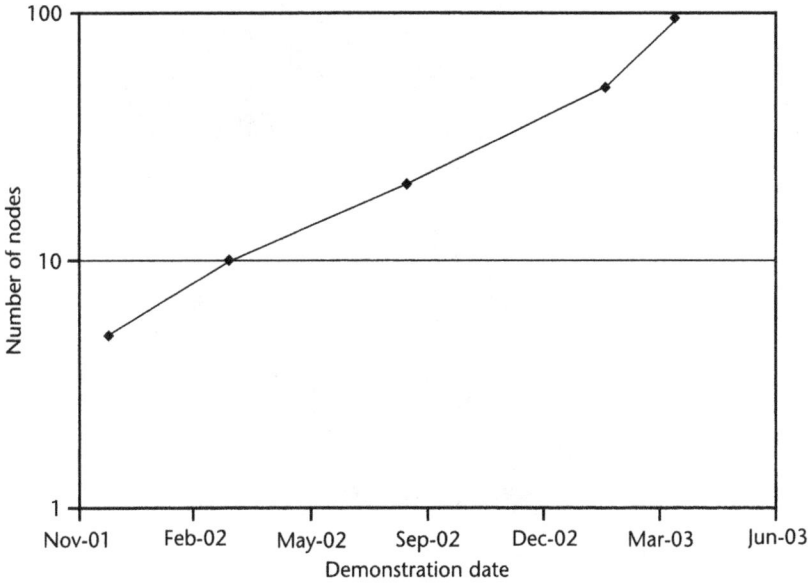

Figure 18.1   Schedule versus size of SHM demonstrations. All demos shown include networking and geolocation.

detail in the following sections, provided an effective self-healing vehicular obstacle. For further details on the SHM system, its motivation, and efforts to develop next generation land-mine systems refer to [5].

18.3   SHM Hardware

Each SHM node was designed as a networked, embedded computer with every node including the same potential capabilities. Thus, even as large fractions of the total number of nodes were destroyed, the system could adapt and respond. Every node contained significant processing, sensing, and networking capability within an energy-constrained package. A picture of an SHM node is shown in Figure 18.2. Each SHM node contains a CPU (based on Hitachi SH4 7751); a wide variety of interfaces to support subsystems; an integrated RF communication system; an integrated four-channel acoustic, three-axis magnetic, and three-axis acceleration sensing system; four acoustic speakers, eight rocket thrusters for mobility; a safing and arming system; two 2.4-GHz Frequency Hopping Spread Spectrum (FHSS) radio modems with four antennas; a switch board to optimize radio antenna usage based on node orientation; and a detachable daughter board enabling Ethernet connectivity to each node. Within the DARPA program, all aspects of the envisioned final system were incorporated except the warhead.

The SHM core electronics package was designed for a 12-cm-diameter, 4.4-cm-tall cylinder. Details on this core electronics package are shown in Table 18.2. When combined with the eight rocket thrusters and the integrated thruster SAF system, the total node size is 7.6 cm in height and 12 cm in diameter, as required to utilize the planned deployment mechanism (the VOLCANO™ mine-dispenser system).

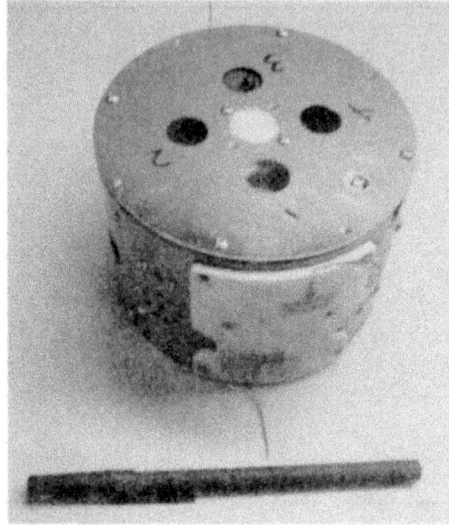

Figure 18.2   SHM node shown next to a pen.

Table 18.2   Summary of SHM Communication Unit Features

Component	Description	Specifications
Processor	Hitachi SH4 7751	300-MIPS CPU, 1.1-GFLOP, FPU
Processor memory	16-MB Flash, 64-MB RAM	Accessible from the OS
Operating system	Linux	2.4.16 kernel
Preprocessor	(1) Power control	Subsystem monitoring and power control, enabling processor to subsystem interface
	(2) Subsystem interface	
Integrated power conditioning	Supplies power to each subsystem	Ranges from 0.15W to 5.1W as modules are slept
Integrated acoustic subsystem	Analog-to-digital system with four input and output channels	Two dual AC97 CODECS each with (1) Stereo sigma delta 48-Ksps inputs, 46-dB gain (2) Stereo 18-bit DAC, up to 95-dB attenuation
Digital I/O	14 lines of GPIO and 4 LEDs	
Integrated antennas	Four 2.4-GHz Nitinol quarter-wave monopole antennas	2.4–2.49-GHz, switched between two on top and two on the bottom

The SHM hardware was designed to provide development flexibility by making available significant processing resources (a 32-bit microprocessor) and a convenient development environment (standard Linux operating system supporting C programming with a convenient Ethernet development/debugging channel). The development flexibility these choices provided proved critical during the development of the scalable SHM system algorithm and software development. Specifically, this flexibility enabled rapid changes in the implementation of the multilateration algorithms, convenient use of available open-source modules, rapid response to changing test scenarios through modifications of bash and Perl scripting, and significant changes in the acoustic-ranging implementation by leveraging standardized interfaces such as an AC97 codec and open-source driver.

18.4 SHM Software

Each SHM node, autonomously upon powering up, sets up a communication network with its neighbors, collaborates to determine its and its neighbors relative position, monitors each neighbor's status, and then responds to neighbor disappearances (minefield breaches). An overview of the SHM software components that provide autonomous operation is provided in Table 18.3. The SHM software architecture is built of multiple layers over the Linux kernel; further details on the individual software components are available in the references presented in Table 18.3. Prior to discussing the performance of the SHM software in the system, our SHM design philosophy, used during the development of each software process, is presented in Section 18.5.

18.5 SHM Design Methodology

To develop the WSN components of the SHM system, our design was based on general considerations incorporated into our design methodology, common to embedded autonomic defense systems. Based on our experience with these defense systems, the SHM system was developed to provide the following:

- *Autonomous unattended operation.* This includes autonomous or semiautonomous configuration at deployment, as well as autonomous, hierarchical decision making to identify events or threats, respond locally, fuse information, and in certain cases notify users.
- *Resilience to intermittent communications.* Wireless-communication loss is expected from node mobility, fading (including with static topologies due to

Table 18.3 Software Modules Operating on the SHM Communication Units

Software	Description
Distributed networking	This forms, maintains, and enables the mesh communications network. Components include network assembly, link monitoring, a multihop distributed database, point-to-point and point-to-multipoint route creation/monitoring, and efficient multihop flooding [6, 7].
Time synch	This leverages radio heartbeat packets to provide microsecond timing synchronization between nodes, shared over multiple communication hops in the system [8, 9].
Relative geolocation	This coordinates the autonomous relative position determination of all nodes. It is separated into an acoustic-ranging component to determine range and angles between local node pairs and a multilateration procedure to enable every node to build a large set of node coordinates [10].
Breach heal	This determines the response to a node's disappearance (detects minefield breaches).
GUI and debugging support	Software modules coordinate with multiple software GUIs and provide debugging information. Interfaces provide the state of each SAF and thrusters over the radio and of system operation over Ethernet and enable generic access via the radio or Ethernet interfaces.
Up/down	This collaboratively verifies and reports node disappearances.
HW drivers	These include drivers for the radios, the SAF to control rocket firing, the orientation subsystem, the preprocessor power control, and the acoustic-sampling section.

frequency changes), and changes within the system due to node failure and duty cycling. This results in both intermittent connectivity and dynamically varying latency in communication, even within a generally static system.

- *Energy constraints on communication and processing.* Processing is much more efficient than communication on a per information bit comparison [11].
- *Application in a wide variety of environments and missions.* The same system may be deployed in various environments, particularly after successful utilization in a first environment.

These competing requirements, in addition to those specific to the SHM, resulted in the development of a solution incorporating a large number of complementary components. Within SHM, the requirements for the flexibility needed for rapid development, system autonomy under various conditions, an infrastructure-free geolocation system, collaborative node-breach detection and response, graceful system degradation as up to half the nodes are destroyed, and a robust self-contained, self-healing network all influenced multiple components of the SHM system. Where possible, we constrained the interaction between component solutions to reduce design problems masked by complexity; however, the scope of the nonoverlapping requirements resulted in these algorithms combining to form a complex SHM system.

Large, distributed software systems face the challenge of software defects that are not covered by unit and system testing and may surface only after deployment. The software-development process, static and dynamic program analysis tools, code review and walkthroughs, white box testing, black box testing, and system integration testing cannot eliminate all potential program errors and, in this case, were often limited given the evolving nature of the SHM research program. As a result, the SHM system used a number of specific additional techniques to detect and recover from transient and permanent software faults causing memory access violation, memory exhaustion, and CPU exhaustion. These include the following:

- Following process termination, resources are automatically freed by the Linux operating system.
- The SHM process manager automatically restarts a terminated process. Application libraries detect closed file descriptors of terminated processes and subsequently reopen and reinitialize these processes.
- Per-process resource limits are activated (e.g., CPU time, heap memory, file descriptors).
- The SHM process manager is launched from the operating system root process (init), which in turn will restart the process manager every time it terminates.
- A watchdog timer in a secondary 16-bit microprocessor determines when the operating system deadlocks or detects a kernel panic situation. If this timer expires, power to the main processor is cycled.

In addition to the techniques described above, general constraints were integrated into the design of each of the software processes of Table 18.3 to add robustness to the system. These include the following:

- *A soft-state approach to communication [12], which allows for inconsistency in interactions, particularly for broadcast and multicast messages.* Each process can to operate with partial information (missed messages) with the message-redundancy energy-cost trade-off optimized by considering the total energy to ensure an application QoS, rather than per-message energy cost. Soft state-based communication survives transient failures of the sender and receiver by sending periodic updates and timing out cached information, adding to system robustness. Examples of the use of this technique include the use of wireless communication by each application on a best-effort basis, the use of available measurements of node separation (via time of flight of acoustic pings) to obtain relative node positions, and the use of only partial information on neighbors status and location when a node decides to move to heal a breach in the minefield.

- *Providing local explicit control so that information passed over the network may be used to enhance decisions.* To ensure graceful degradation, each local node makes its own decisions. Thus, each node can still respond to a locally perceived threat, even with limited information about the network. This local control is driven by the need for autonomous, graceful degradation. Examples of these local decisions include determining when to send out an acoustic ping to enable geolocation, when to hop, and when to break off communication with a node's current network neighbors and attempt to reform the network via communication with other nodes.

- *Multilayer self-sensing and self-healing.* To enable robust network and system operation, each component provides feedback on its operation. This feedback enables system verification and developer debugging and improvement, as well as enabling the system to optimize autonomously the collection of services available. Examples of this multilayer sensing and self-healing include using the availability and quality of links to enable network formation and routing, using the measured signal-to-noise ratio of each one-way, time-of-flight acoustic ranging in weighting node-location accuracy, and using the anticipated effectiveness of a planned firing of a rocket engine to move the node and fill in a breach in the SHM obstacle.

- *Modularization of services using open interfaces.* For example, each software process utilizes a language-independent device-file interface provided by the Framework for User Space Devices (FUSD) [13]. This enables flexibility and facilitates group development encompassing multiple disciplines. This also limits the impact of process failures.

- *Designing to facilitate testing at scales appropriate for the application.* Testing within SHM included algorithmic simulations to scales of thousands of nodes, testing of the developed node-software base in a workstation emulation environment (using the implemented SHM code) to the hundreds of node's scale, as well as testing up to the scale of 100 nodes on the SHM hardware in a representative outdoor environment.

Integration of these principles was a key component enabling the SHM system to respond autonomously to varying degrees of system failure and resulted in comprehensive *self*-healing and enhanced robustness in system operation, as shown in the demonstrated results discussed in Section 18.6.

18.6   SHM Operation

The SHM system was tested at Fort Leonard Wood (FLW), Missouri, over one month in March and April 2003. This section provides an overview of the demonstrated performance of the SHM system with examples of network, time synchronization, and geolocation performance. During each day of testing, from 50 to 95 individual SHM nodes were operated off internal batteries for multiple trials per day. Each trial demonstrated and monitored autonomous networking, relative position determination, collaborative decision making, and identification of system breaches. During these tests, nodes were set up outdoors, in a field 35m deep by 190m across (for all 96 nodes). The results of system operation presented below were obtained via noninvasive Ethernet connections to a fraction of the nodes. Nodes reported their status periodically when connected to Ethernet. The Ethernet connections to nodes allowed observation of autonomous operation of the nodes, as nodes were deployed, powered up, and then autonomously configured their network and determined their neighbor's locations. For example, as the nodes built a coordinate table location of all their neighbors, each node's local coordinate table monitored if the node was connected to Ethernet.

An example of the Ethernet monitor's display during an outdoor test of 74 nodes is shown in Figure 18.3. Within this figure, each rectangular box denotes an SHM node. The arrows indicate direct communication links reported, the boxes

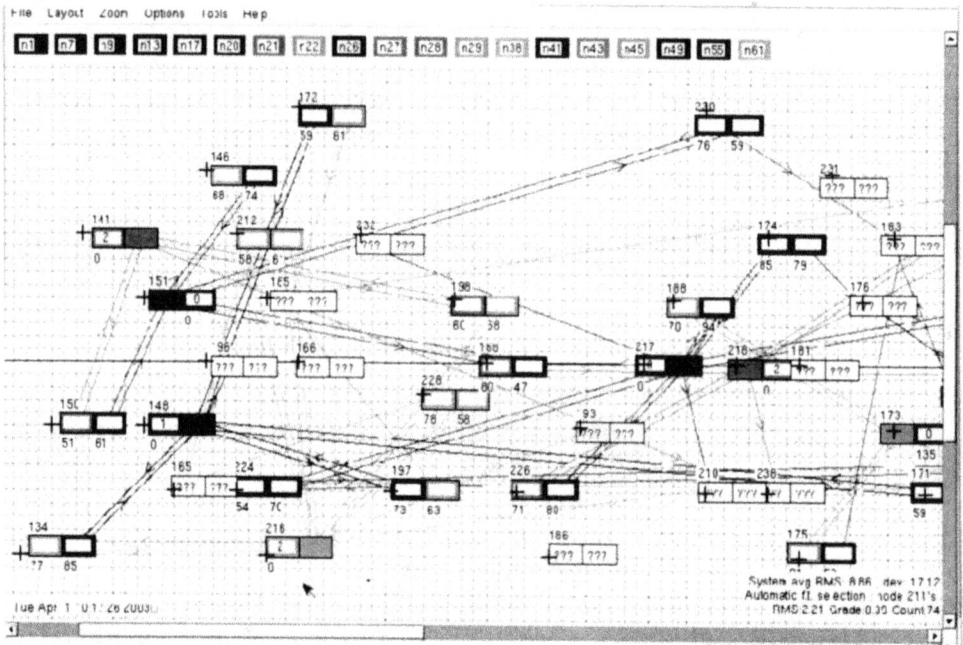

Figure 18.3   Example display using an Ethernet monitoring channel. Shows a subset of this 74-node test.

with questions marks denote nodes not connected to the wired Ethernet (inferred from information reported by those connected), and the measured ground-truth position of nodes is shown with crosses. The scale of the grid within Figure 18.3 is 1m, with approximately one-third of the 35m × 150m test field shown.

To illustrate the SHM system's operational capabilities, measured properties of the system are presented here. These include the time for initial stable networks to form (Figure 18.4), the time for nodes to determine the coordinates of their neighbors (Figure 18.5), the traffic sent from one node's radio as a function of time (Figure 18.6), the distribution of nodes' traffic rate over the wireless network (Figure 18.7), the accuracy of time synchronization between neighboring nodes (Figure 18.8), the distribution of errors in the range and angles measured

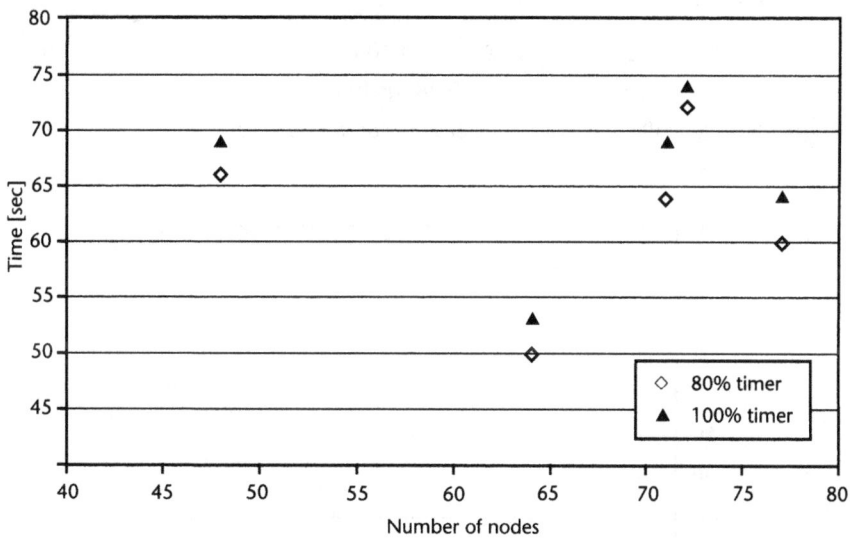

Figure 18.4   Time for 80% and 100% of the nodes to form a stable network on one radio as a function of the total number of nodes in various tests. Only one network formation at each number of nodes is recorded.

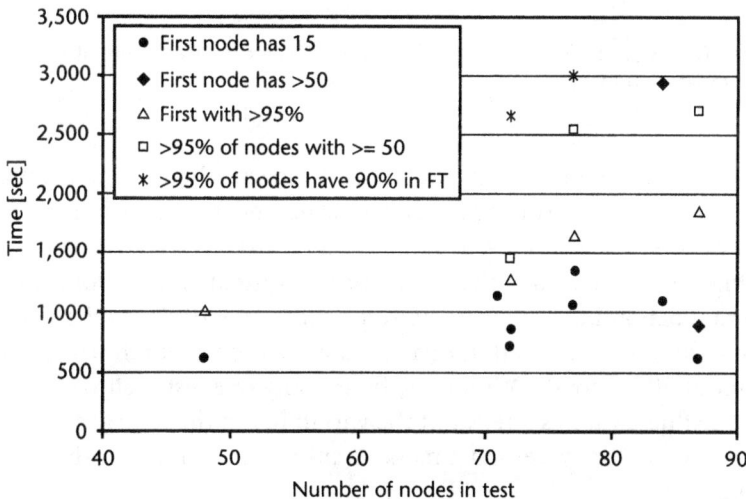

Figure 18.5   Time to determine a node's neighbor's locations during a few of the tests at FLW.

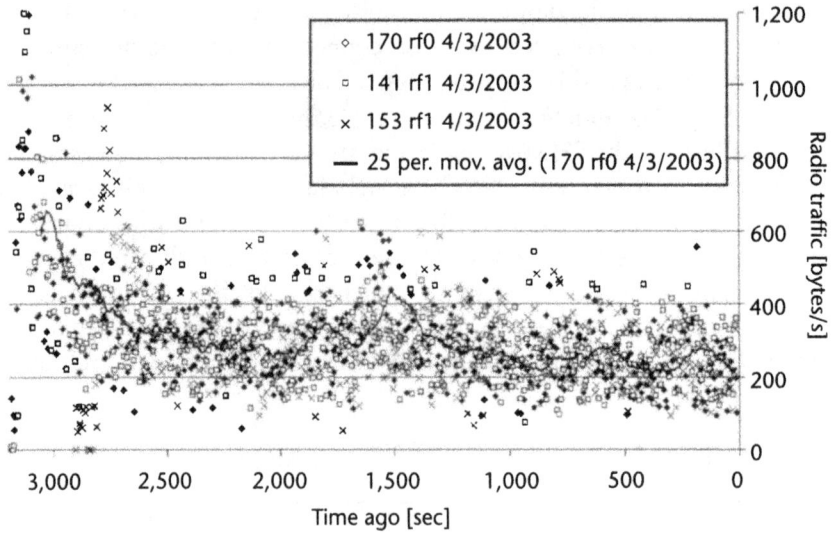

Figure 18.6 RF traffic sent over three base radios during a test. Each point is a 6-second average.

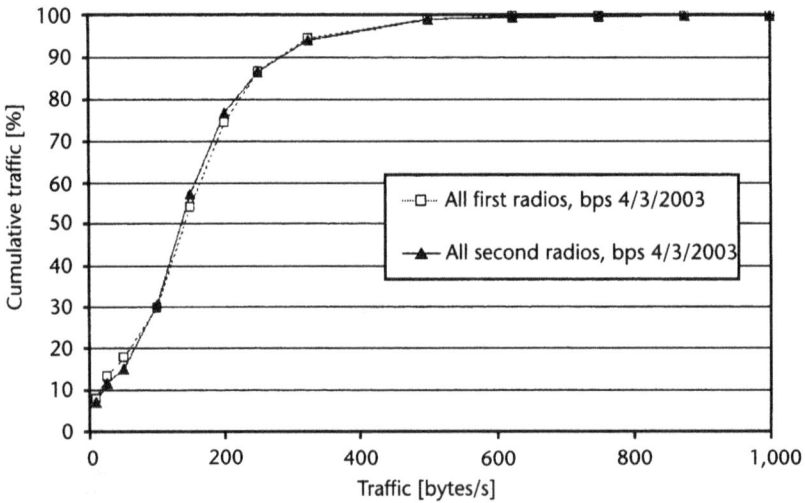

Figure 18.7 Cumulative distribution function (CDF) for the traffic sent at 6-second intervals over all the radios during a test.

between nodes (Figures 18.9 and 18.10, respectively), and an example of the variation in positions determined for neighboring nodes over various runs (Figure 18.11). Each of these figures is discussed below.

Figure 18.4 displays the times for the system to autonomously form the first communication infrastructure between 80% and 100% of the nodes in a test. Following this initial time, additional communication-link refinement occurs, generally for about 300 seconds. Within the figure only one test is shown at each point; thus, the time fluctuation is a result of the variability in the time as the network is formed based on randomly seeded timers, resulting in variation about the single times shown.

One-hop-synchronization RMS error

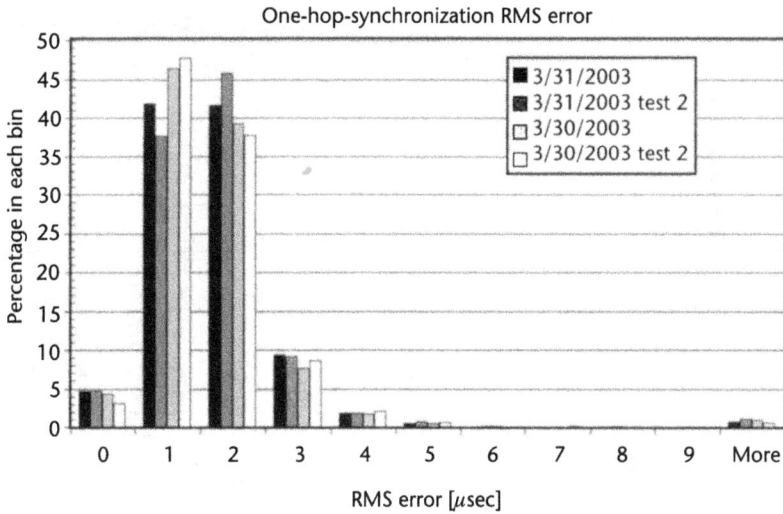

Figure 18.8   CPU-to-CPU root-mean-square synchronization error reported for each set of synchronized nodes.

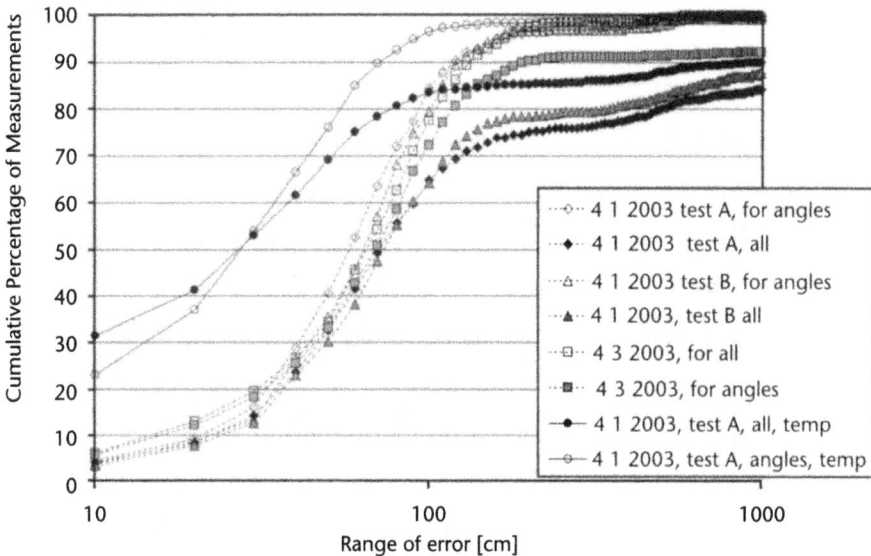

Figure 18.9   CDF for range errors. A distinction is made between ranges from which an angle was calculated (these require higher confidence) and all ranges collected. One temperature-corrected dataset is also shown.

Following initial network formation, the SHM nodes determine the relative location of their neighboring nodes. Shown in Figure 18.5 is the time during multiple tests for nodes to determine the location of set numbers of their neighbors. This figure is plotted as a function of the total number of nodes in the test and shows time from node start (including network-formation time). Generally, tests with 50 nodes assembled a complete neighbor-location list on most of the nodes within 15 to 20 minutes; however, tests with increased numbers of nodes increased this time, particularly in windy or high-noise conditions. In all tests run, neighbor coordinates were eventually acquired by all nodes as, even in the worst conditions, the nodes would

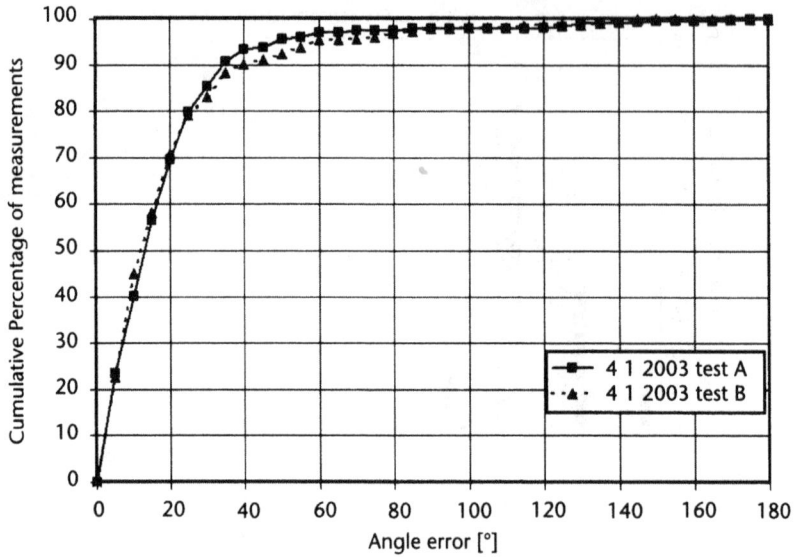

Figure 18.10   Cumulative distribution for the angle errors in data collected on April 1, 2003, at FLW.

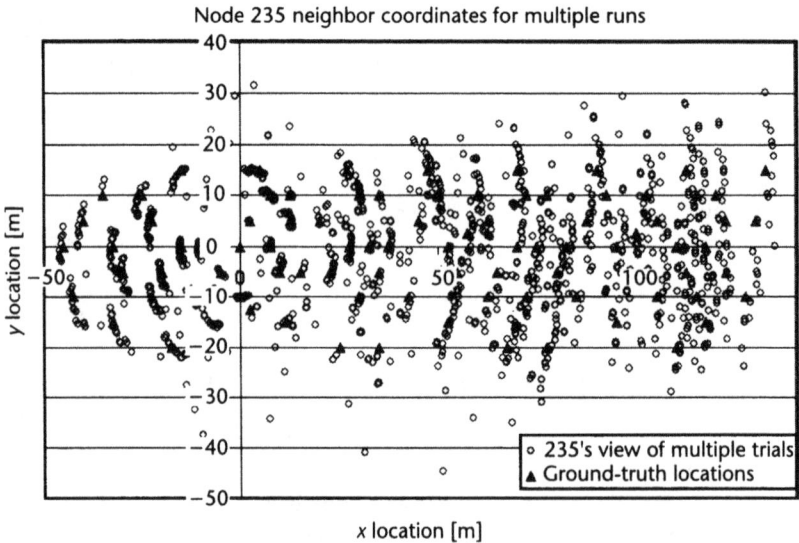

Figure 18.11   Display of multiple field tables collected by node 235 over a week at FLW.

continue to collect data until they were able to calculate the positions of their neighbors.

During testing, each node monitored the amount of traffic sent over the SHM radios and reported this over the Ethernet diagnostic link when connected. The radio modems used within the SHM system form local clusters with a base coordinating up to 10 remotes. Since each node participates in two independent clusters, and clusters are formed to overlap (as illustrated in Figure 18.3 and discussed further in [6, 7]), the SHM nodes are able to form a scalable interconnected network. An example of the traffic sent as a function of time over the radios acting as RF bases is shown in Figure 18.6. The observed traffic for radios operating as remotes is about half of that for bases. Figure 18.6 includes the headers used by the radio driver to identify incoming packets.

To place the RF traffic of Figure 18.6 in perspective, we calculate the capacity of bases and remotes for the SHM system. For bases, the maximum transmitted packet size is set to 75 bytes, while for remotes it is 46 bytes, with each radio's frequency hop lasting 14.227 ms (each base and remote can send one packet per frequency hop). Thus, the SHM radios have a sending capacity of 5.27 kbps for bases and 3.23 kbps for remotes. It is clear that the SHM traffic is well below the radio capacity limit. This is illustrated in the cumulative distribution function (CDF) for the traffic, averaged over 6-second intervals, shown in Figure 18.7. The CDFs in this figure are shown independently for the first and second radios on the nodes, with the similarity in CDF expected, as both radios on a node are treated similarly.

The SHM system leverages an RBS [8, 9] to support the acoustic time-of-flight range measurements used to determine each node's position. RBS uses receivers of broadcast pulses to time-stamp incoming packets and then compares timestamps. Each timestamp is assumed to vary with Gaussian error (after outliers are filtered out) and the Gaussian variance recorded as the root mean square (RMS) error for synchronization between nodes. RMS values are reported at each node to a precision of 1 μs. The distribution of RMS error between each node pair's CPUs for four FLW tests is shown in Figure 18.8. Error's reported above 9 μs were orders-of-magnitude larger as a result of the clock's restarting recently enough on a node to result in significant differences in the time-stamping of packets from one source.

Within the SHM system, local time synchronization is propagated across multiple RF hops (to synchronize nodes that do not both receive the same physical-layer broadcast messages). This enables the acoustic-ranging system to provide acoustic ranging between nodes, independent of the local communication topology between those nodes. The local error extrapolation to this multihop synchronization is straightforward if the assumption is made that the error between each broadcast message is uncorrelated and Gaussian (as it has been observed to be [8, 9]). Given the distribution of Figure 18.8 for a 10-hop system, assuming hop RMS errors of [1,1,1,1,2,2,2,2,3,4], μs yields a 10-hop RMS error of approximately 7 μs. This describes only the random error within the system (given the above assumptions) and does not account for systematic error, as the measured RMS is only a measurement of the distribution of the received broadcast signal's timestamps. However, this does demonstrate that the CPU-to-CPU synchronization error, even between nodes over multiple communication hops, is expected to be well below the 20.8 μs sample size of the SHM acoustic CODECs.

The range and angle between pairs of nodes are determined within the SHM system via acoustic time-of-flight measurements used collaboratively by nodes to determine their relative locations. The distributions in range and angle measurements for a set of FLW tests are shown in Figures 18.9 and 18.10. Figure 18.9 also includes a comparison of the results with preprocessed data, which back-calculated the temperature for acoustic propagation, resulting in reduced errors since, during FLW testing, a constant acoustic-propagation velocity was used for all tests.

Once ranges and angles are shared, on each node a multilateration process calculates all other nodes at x and y coordinates relative to itself [10]. Figure 18.11

provides an indication of the error seen by a node for its neighbors' calculated positions. Figure 18.11 shows as circles the locations of each node in node 235's neighbor-location table for multiple runs at FLW over a course of days. Also shown in this figure, as triangles, are the ground-truth locations of those nodes measured by hand. The data shown in Figure 18.11 represents snapshots taken off of node 235 at 10-minute intervals during multiple runs. This figure displays a number of interesting characteristics of the autonomous position location with SHM. First, there are a few outlier node locations (outside the expected 270m × 35m box in which nodes were deployed). These are predominantly a result of initial error in node placement and are corrected as the nodes continued to range. Second, Figure 18.11 shows the tendency of the multilateration algorithm to twist the node positions symmetrically around the reporting node 235 (at the origin). Third, for nodes further from 235 (the origin), the reduction in accuracy of the positions where node 235 thinks its neighbors are is clearly evident.

18.7 Conclusion

The SHM system is a defense-system application built around a WSN and incorporating state-of-the-art:

1. embedded-processing;
2. energy-constrained-communication;
3. distributed-software components.

The SHM system is one of the first examples of an applied wireless-networked embedded system, independent of an external infrastructure and whose application space requires significant distributed-computing and communication capabilities.

References

[1]   Estrin, D., *Embedded Everywhere: A Research Agenda for Networked Embedded Computers*, Washington, D.C.: National Academy Press, 2001.

[2]   Akyildiz, I. F., et al., "Wireless Sensor Networks: A Survey," *Computer Networks*, Vol. 38, No. 4, 2002, pp. 393–422.

[3]   Pottie, G. J., and Kaiser W. J., "Wireless Integrated Network Sensors," *Communications of ACM*, Vol. 43, No. 5, 2000.

[4]   Altshuler, T. A., "DARPATech 2002 Address on the Self Healing Minefield," at http://www.darpa.mil/DARPATech2002/presentations/ato_pdf/speeches/ALTSCHUL.pdf.

[5]   Merrill, W. M., et al., "Dynamic Networking and Smart Sensing Enable Next-Generation Landmines," *IEEE Pervasive Computing*, Vol. 3, No. 4, 2004, pp. 84–90.

[6]   Sohrabi, K., et al., "Scalable Self-assembly for Ad Hoc Wireless Sensor Networks," *Proc. IEEE CAS Workshop on Wireless Communications and Networking*, Pasadena, CA, September 5–6, 2002.

[7]   Sohrabi, K., et al., "Methods for Scalable Self-assembly of Ad Hoc Wireless Sensor Networks," *IEEE Transactions on Mobile Computing*, Vol. 3, No. 4, 2004, pp. 317–331.

[8]   Elson, Girod J., L., and Estrin D., "Fine-Grained Network Time Synchronization Using Reference Broadcasts," *Proc. OSDI 2002*, Boston, MA, December 2002.

[9]   Elson, J., and Römer K., "Wireless Sensor Networks: A New Regime for Time Synchroniza-
      tion," *Proc. ACM HotNets-I*, Princeton, New Jersey, NJ, October 2002.

[10] Merrill, W. M., et al., "Autonomous Position Location in Distributed, Embedded, Wireless
      Systems," *Proc. IEEE CAS Workshop on Wireless Communications and Networking*, Pas-
      adena, CA, September 5–6, 2002.

[11] Merrill, W. M., et al., "Collaborative Networking Requirements for Unattended Ground
      Sensor Systems," *Proc. 2003 IEEE Aerospace Conference*, Vol. 5, Big Sky, MT, March
      8–15, 2003, pp. 2153–2166.

[12] Raman, S., and McCanne S., "A Model, Analysis, and Protocol Framework for Soft
      State-based Communications," *Proc. ACM SIGCOMM*, Cambridge, MA, September
      1999, pp. 15–25.

[13] See CircleMUD, at http://www.circlemud.org/~jelson/software/fusd.

CHAPTER 19

Workplace Applications for Sensor Networks

W. Steven Conner, John Heidemann, Lakshman Krishnamurthy, Xi Wang, and Mark Yarvis

19.1 Introduction

The current generation of interactive devices and networks fosters a wide class of interactive, ubiquitous computing applications [1]. The recent trend to integrate wireless networking into interactive devices such as PDAs, cellular phones, and portable computers has led to the availability of information, such as news and stock quotes, as well as services such as e-mail, appointment tracking, and multimedia content, from any location at any time. These applications have significantly improved workplace productivity, despite the fact that human participation is often required in the compute loop. These applications have traditionally interacted with *virtual* content such as e-mail, financial records, and text documents.

Today, millions of sensors are scattered throughout workplaces in both industrial and nonindustrial office environments. These sensors include HVAC-monitoring devices such as thermometers, barometers, and moisture gauges, safety monitors such as carbon monoxide and smoke detectors, security monitors such as motion and glass-break detectors, and access-control devices such as RFID badge readers. In most cases, sensors are deployed for a specific application, and access to sensor output is only available locally; a person typically must walk up to a sensor to obtain its current reading. In some cases, sensors may be wired to a nearby closed-loop monitoring station, but such monitoring stations are generally application specific. While these sensors serve useful purposes for the individuals who deploy them, in practice each sensor is typically used only for a single, specific monitoring application.

By networking these devices to provide ubiquitous access to remote information and actuation capabilities, many new applications emerge. The advent of inexpensive, low-power wireless sensors and self-configuring network technologies allows sensors to be easily deployed in a ubiquitous, ad hoc manner. These deployments interface with the physical world and promise to make everyday tasks easier, enhancing our ability to examine and optimize the environments in which we live and work. Recent advances in sensor hardware make it feasible to deploy small wireless sensors in office environments, but many challenges remain. This chapter looks at two case studies in detail to explore those challenges: an application to assist workers in finding conference rooms and another that guides visitors around an office environment.

The characteristics of workplace applications are in many ways similar to other classes of sensing applications, such as industrial and scientific sensing. Nodes must be sufficiently small to integrate seamlessly into their environment. Networks can be very large in terms of number of nodes (hundreds of nodes) as well as their physical space (thousands square feet). The RF propagation environment can include interference from other RF devices and electronic equipment as well as concrete and steel obstructions. Finally, network lifetime is a key factor in demonstrating value. Thus, workplace applications must address all of the same issues of scale, self-configuration (despite a lack of mobility), and energy conservation that must be faced to support a real deployment of any WSN.

In addition to illustrating the challenges in developing and evaluating prototypes of real applications, these applications illustrate problems paramount to the office environment. Workplace applications differ from other classes of sensing in that they tend to include far more direct user interaction with sensors and actuators. As we will discuss, the conference-room application must integrate with the existing networking and sensor infrastructure and interact with users in a useful manner; the visitor-guidance application must consider human-movement constraints and be easy to deploy and maintain.

In addition, both conference-room and visitor-guidance applications require self-configuring wireless networks and low-power operation (as do many other applications in sensor networks). These requirements might be surprising for in-building applications where power and networking are both comparatively plentiful. However, it is not always feasible to locate sensors near power or network outlets. Additional wiring would quickly exceed the cost-benefit ratio of these ad hoc applications. Even in new construction, each wired network port and outlet has a cost that must be justified. Thus, we see low-power operation, energy harvesting, and wireless communication as necessities even in relatively wired environments. However, there is also an opportunity to leverage these sparsely available infrastructural resources for the benefit of the entire network.

We briefly review hardware that can be used to deploy workplace sensor-network applications, followed by a detailed description of two applications: conference-room monitoring (Section 19.3) and visitor guidance (Section 19.4). We briefly discuss several other representative workplace applications in Section 19.5. We conclude by summarizing our experiences and identifying reusable components in these examples.

19.2   Hardware for Workplace Sensor-Network Deployment

Four types of hardware platforms with heterogeneous capabilities are commonly used in the deployment of workplace sensor-network applications: sensor nodes, display nodes, gateway nodes, and handheld nodes. These hardware platforms are tailored for sensing, human interaction with the sensor network, and interfacing the sensor network with workplace networks. Thus, they provide a mix of processing power and I/O capabilities. Each of the hardware building blocks described in this section should be viewed as representative of a class of devices. Table 19.1 provides a comparative description of these devices.

19.2.1   Sensor Nodes

A mote is a generic sensor-node platform that integrates sensing, computation, and communication. Motes are typically low-cost, small, battery-powered devices designed to allow large-scale deployment of sensors in an environment. An example of the Berkeley mote [2] that is commonly used in sensor-network research and applications is the Mica2 (Figure 19.1). The Mica2 mote is constructed using off-the-shelf components and includes an I/O connector to provide a stackable platform for effective integration with sensors and alternative communication boards for experimentation. The Mica2 optimizes power consumption, cost, and size; it is designed primarily to handle limited amounts of data from simple sensors and is not suitable for many sensor-network applications that require collection of high-bandwidth data, such as vibration, sound, or vision. The Intel Imote increases processing capacity to provide an example of a device that can be used to sense more bandwidth intensive data and perform robust in-network communication. Many of the workplace applications described in this chapter use one or more types of motes.

Table 19.1   Comparison of Various Hardware Platforms for Use in WSNs

Node Type	Sample Name and Size	Typical Application Sensors	Radio Bandwidth	MIPS Flash RAM	Typical Active Energy(mW)	Typical Sleep Energy(μW)	Typical Duty Cycle (%)
Generic sensing platform	Mote 1–10 cm^3	General-purpose sensing and communications relay	< 100 Kbps	< 10 < 0.5 Mb < 10 Kb	3V*10–15 mA	3V*10 μA	1–2%
High bandwidth sensing	Imote 1–10 cm^3	Rich sensing (video, acoustic, and vibration)	~ 500 Kbps	< 50 < 10 Mb < 128 Kb	3V*60 mA	3V*100 μA	5–10%
Gateway	Stargate > 10 cm^3	High-bandwidth sensing and communications aggregation; gateway node	> 500 Kbps –10 Mbps	> 100	3V*200 mA	3V*10 μA	> 50%

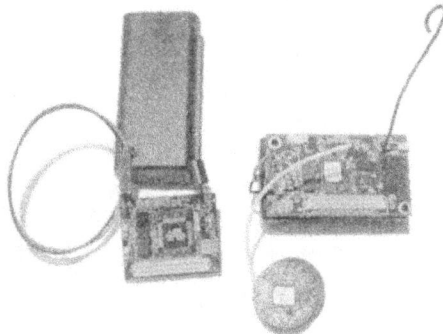

Figure 19.1   Mica2, Mica2Dot, and Rene motes.

19.2.2 Display Nodes

Many workplace applications, including several described in this chapter, require simple user interactions at various points within the network; thus, we must augment a basic sensor with simple, human-oriented I/O capabilities. The *button box* node (Figure 19.2) includes a Mica2 mote and is powered by two AAA batteries. It provides a simple interface that includes two buttons for input and three LEDs and a buzzer for output.

While the button box is useful in many applications, a richer interface is sometimes required. The LCD display node (Figure 19.3) is a small, low-power, wrist-watch form-factor node designed to enable limited human interaction with a sensor network. This device consists of a Mica2 mote integrated with four control buttons and an LCD capable of showing text and simple graphics. These buttons may be used to trigger the node to wake up from deep sleep and also to allow user

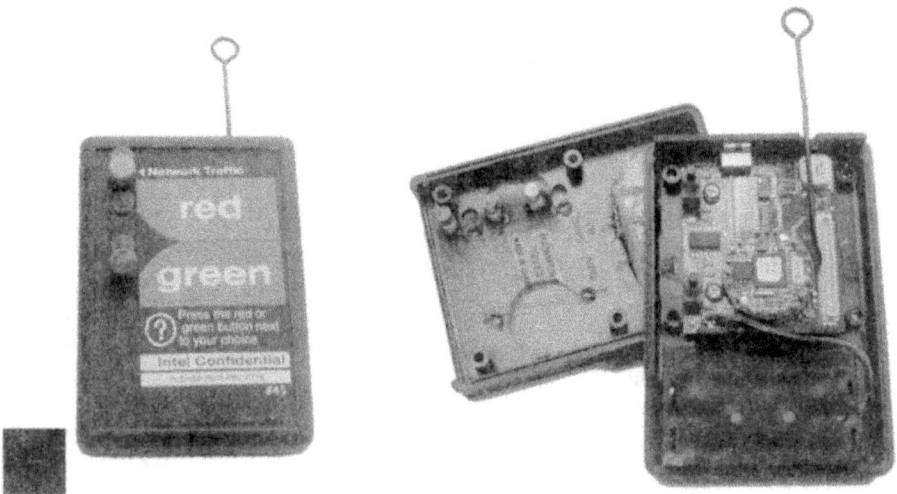

Figure 19.2 An external and internal view of the button box node.

Figure 19.3 LCD display node.

text input. These devices provide an easy and inexpensive method of allowing ubiquitous display of and user interaction with information in the workplace.

19.2.3   Handheld Nodes

Suitable for limited human interaction, handheld computing devices such as PDAs and laptops can also provide sophisticated user interfaces and data-analysis tools. The Canby (Figure 19.4) is a compact flash card form-factor Mica2 mote that allows handhelds and laptops to interact easily with the sensor network. These devices may be used as part of a field tool. An example field tool, TASK [3], allows the handheld to be used to query devices in the proximity of the human by sending out ping messages to which nearby nodes respond. In some workplace applications, a GUI on handheld nodes has been used to provide information from sensors in the human's proximity.

19.2.4   Gateway Nodes

Previously described sensor nodes minimize cost and size by eliminating support for traditional networks such as Ethernet or 802.11. Gateway nodes are devices capable of bridging communication between sensor nodes and higher-end wireless or

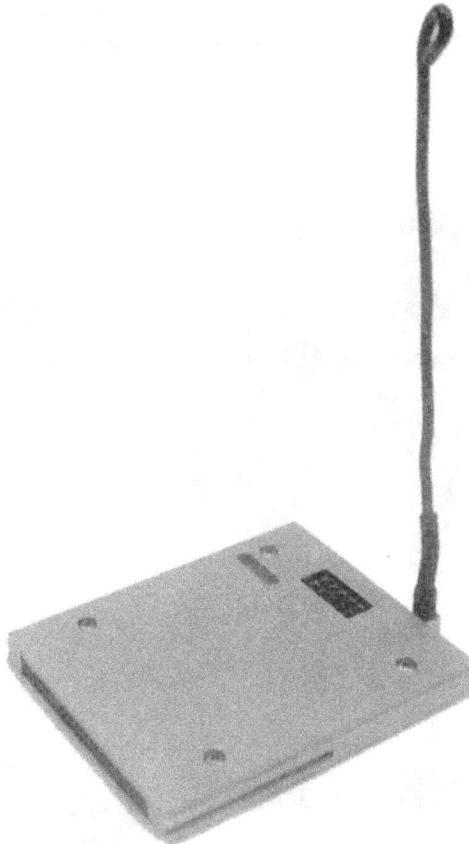

Figure 19.4   Canby compact Flash mote.

wired networks. Gateway nodes often have more computing capabilities than sensor nodes, as well as access to line power. An example of a gateway node is the Stargate platform (Figure 19.5), which includes a 400-MHz Intel XScale™-architecture-based processor, tens of megabytes of RAM, and up to gigabytes of persistent storage. It is capable of interfacing directly with either a Mica2 or an Imote device and can bridge the data from the low-power sensor network to traditional networks, including 802.11, Ethernet, and wide-area networks. In addition, the processing and memory provisions on the Stargate node allow it to act as a Web interface to a sensor network. Sensor readings can be stored in a local database and queried over the Web. The same Web interface can also be used to actuate or manage the sensor network. As described later in this chapter, Stargates may also be used to create hierarchical networks that provide performance enhancements to reduce sensor-network energy consumption and extend the lifetime of battery-powered nodes.

19.3   Conference-Room Application

In many modern office complexes, closed-wall offices have been replaced with high-density cubicles to inspire an atmosphere of open collaboration and accessibility among employees. However, the lack of private offices makes it difficult for employees to hold impromptu meetings. Discussions in or near cubicles often disturb employees who are trying to work nearby. Cafes and other common areas tend to be noisy and distracting and are not good candidates for important or private discussions.

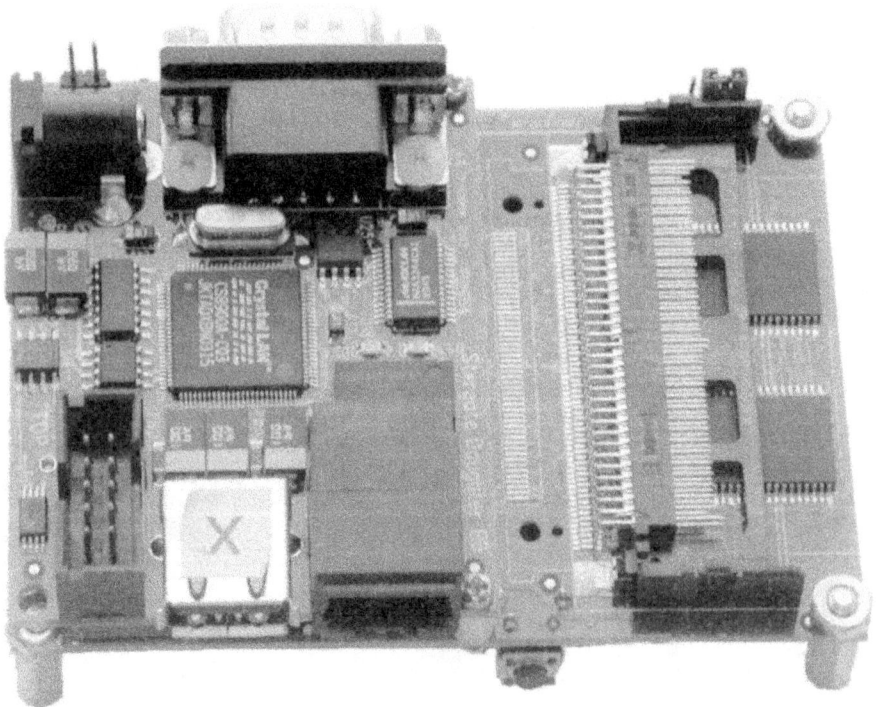

Figure 19.5   An Intel Xscale™-architecture-based gateway node.

Modern buildings have conference rooms for meetings, but these rooms may be reserved days or weeks in advance, and it is often not realistic to reserve a room with little or no notice. However, it is common for meetings to be shorter than the entire reservation time or to be cancelled entirely without canceling the room reservation. Thus, it is often possible for employees to find an empty room for an impromptu meeting through exhaustive search.

Many conference rooms are equipped with motion detectors that are used to turn off the lights when the room is not in use to save electricity. In most cases, these motion sensors are hard-wired to the light switch in a given room and are not accessible from outside the room. As a driving application behind the sensor-network research project, we have networked these motion detectors in one building at Intel using multihop network protocols implemented on a collection of motes, giving employees access to room usage status and allowing them to find empty rooms from their handheld, mobile, and desktop computing devices from anywhere and at any time

19.3.1 Architecture and Operation

The system consists of a network of sensors deployed in and around conference rooms. In-room sensors are connected to motion detectors (Figure 19.6), which monitor room occupancy status. Considering an office building that includes 10 to 15 conference rooms on each of four floors (as is the case at Intel), a complete network deployment consists of 40 to 60 pairs of sensor and display nodes. In a small campus, the network could consist of several hundred nodes.

A gateway node receives the sensor data, which is aggregated and stored to provide status information to desktop users over the Web. Figure 19.7 shows a screen-shot from a Web application that provides live occupancy information for

Figure 19.6 Conference-room motion-sensor node.

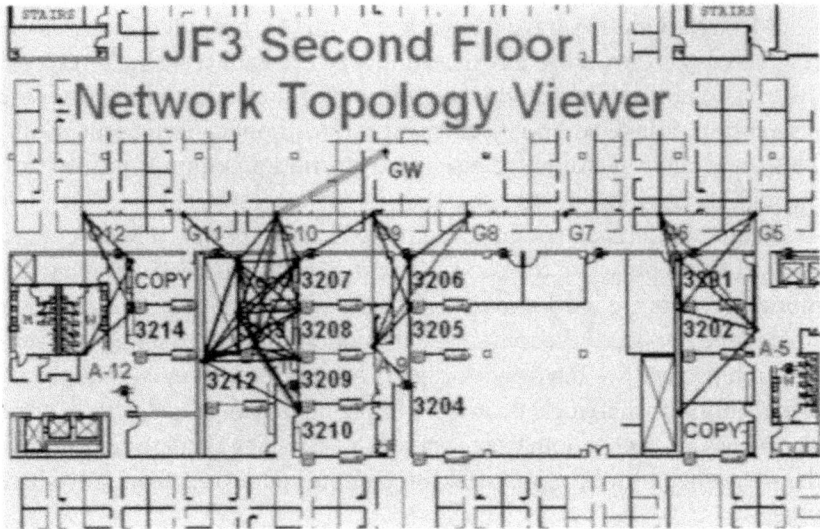

Figure 19.7 A Web page showing live conference-room occupancy and network topology in the building.

rooms on a given building floor. Users of this application can avoid searching for a conference room and walk directly to an empty room. We have also connected PDAs to the sensor network, allowing mobile users to obtain the status of nearby conference rooms directly (Figure 19.8). Occupancy information is also available via status nodes at the end of the aisles that indicate the presence of an empty room in that aisle.

In addition to providing live occupancy data, motion-detector data may also be compiled over time for future analysis. Figure 19.9 illustrates an application that compares gathered room-usage statistics with data from the on-line reservation system. Such data allows the automatic identification of individuals who consistently reschedule meetings without canceling room reservations and allows facility analysts to analyze building-usage patterns. Typical conference-room usage patterns can aid the design of new buildings. These usage scenarios require real-world

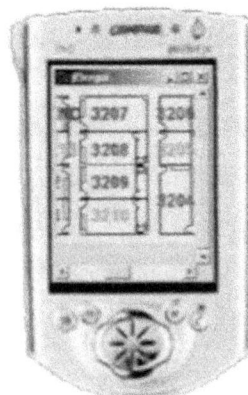

Figure 19.8 A PDA-based application showing the status of rooms in the vicinity of the mobile user.

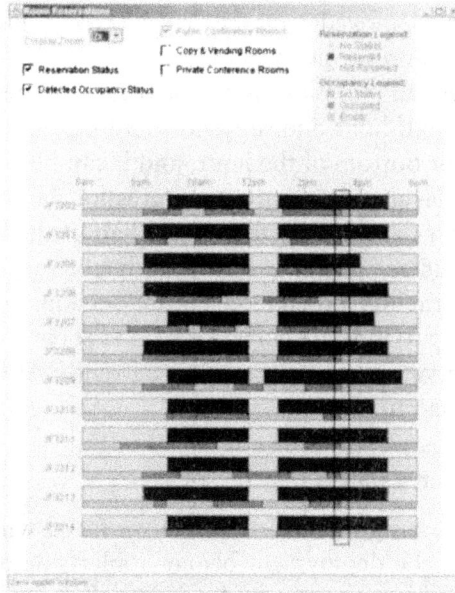

Figure 19.9   An occupancy-history application comparing actual usage data to room reservations.

information to be gathered, processed, stored, and made available in a ubiquitous manner. In addition to these applications, the infrastructure has been reused to gather temperature and battery voltage and current usage at each node.

This application also makes room reservation status, normally available through an Outlook® reservation service, available at each room. Status nodes at the entrance to each conference room indicate current and future reservation status (Figure 19.10). This status information is pushed from the gateway to individual nodes, using the reverse path generated by the data-collection tree. To save battery power, these nodes can turn on when a user presses the status button, prolonging the life of the node.

Figure 19.10   Conference-room reservation-status indicator.

In the currently deployed network, one of three LEDs is lit to specify whether the room is reserved or unreserved, or if the status is not known (typically only if the network is down). One of two buttons can be pressed to determine the reservation status in the current half-hour and the next half-hour. Because users typically use the device at the top or bottom of the hour, and it can be difficult for the user to know the exact time, a flashing light indicates status for the first half of the hour, while a solid light indicates the status in the second half of the hour.

Clearly, the current interface hardware is not very rich, resulting in an overly complex user-interaction procedure. Future versions of this node will include an inexpensive display to indicate current time and room ownership. This service helps resolve room-reservation conflicts that are currently settled by phone calls to conference-room administrators.

19.3.2 Application Challenges

Power was the most significant challenge we faced when installing these applications. To simplify the deployment of our application and to avoid installation of new wires, we used battery-powered devices, allowing an ad hoc network deployment at minimal cost. The power issue is not limited to the conference-room application. In our analysis of building- and factory-monitoring applications, eliminating both power and network wiring at the sensors results in a significant cost savings and improved return on investment.

These deployment and maintenance challenges motivate the design of our protocols. While we leverage power outlets available throughout the building, many of the sensors and devices are battery powered. One of our primary objectives is to take advantage of wall-powered nodes to yield energy savings for battery-powered nodes. However, all battery-powered nodes will not be within one hop of wall-powered nodes. A topology-control protocol is required to leverage heterogeneity, and a synchronization protocol is required to allow battery-powered nodes to sleep yet still communicate [4]. Such battery-conservation techniques can extend network lifetime to meet practical requirements. Building maintenance practices already require replacement of other consumables, such as light bulbs, at intervals of six months or a year.

19.3.3 Communication Protocols

Network protocols for the conference-room application can be divided into two parts: sensing and actuation. The goal of the sensing task is to deliver the occupancy status of each conference room to the Web server. We use a single-destination version of the DSDV protocol [5] to create a many-to-one data-delivery tree to a sink node connected to the Web server. An end-to-end reliability metric, which sums the normalized log of the link-success rates, is used to select paths with the greatest chance of delivering data to the sink, as described in [6]. Each sensor periodically sends a packet to the sink, containing the node number (one byte), the room number (two bytes), and the room's occupancy status (one bit). Nodes along the delivery path append their own node number and occupancy status. For each packet received, the Web server obtains the room-number-to-node-number mapping for the originating node and the occupancy status for several rooms.

Actuation in this application provides room reservation information to nodes outside of each conference room. Since reservations change infrequently, an entire day's worth of reservations for each room are pushed every few hours. To enable one-to-many communication, the server uses the list of forwarding nodes (essentially a trace route) in each incoming occupancy-status packet to track the topology of the data-delivery tree. The server is an ideal candidate for this function as it is not memory constrained. Reservation-status packets generated by the server can then be source-routed by reversing the path from the data-delivery tree. Current reservation packets consist of a reservation-status bitmap and a timestamp with half-hour granularity that corresponds to the start of the bitmap. In a future version of the application, the reservation-status packets will also include the name of reservation owners to help users resolve room conflicts. At the start of each half-hour, a packet containing the current timestamp is flooded into the network to provide time synchronization. Latency of a few seconds or less provides sufficient synchronization for the reservation-status user interface.

While the conference-room application could utilize a flat multihop network, Ethernet and 802.11 connections are common in conference rooms and can be used to create an overlay network. By tunneling sensor-network packets across the IP-based infrastructure backbone, the sensor network can utilize a highly reliable, high-bandwidth communication channel. This architecture also allows the application to scale both physically and in the number of conference rooms supported. Since conference rooms can be spread physically through a building, the hierarchical architecture allows full connectivity of the network without requiring a uniform node distribution. In addition, with additional exit points from the low-bandwidth WSN, bandwidth does not limit the number of conference rooms the network can support.

To create the backbone overlay, we deployed Stargate nodes in several conference rooms on each floor of the building. Each Stargate is attached to a mote and includes software to receive sensor-network packets from the mote and tunnel them to the sink node. Because route update packets from the sink node flow across both the sensor and overlay networks, the Stargate-enabled motes will be able to advertise a favorable routing metric (indicating a highly reliable path is present), causing nodes to form clusters around each Stargate. Data from nodes in these clusters flows across the sensor network to the Stargate node and then across the backbone network to the sink. Use of the overlay network reduces the depth of the data-delivery tree, thereby increasing network reliability and decreasing the amount of energy that nodes must spend forwarding packets.

19.4 Follow-Me Application

Navigating an unknown place can be difficult. While signs may guide the way and computer kiosks may provide room numbers and maps, neither provides active assistance to visitors as they move through a building.

The Follow-Me application is an active visitor-guidance system designed to address this problem. Sensor nodes are deployed around a building on walls, one at each office doorway. Nodes blink their lights to indicate a path, guiding a visitor with a "breadcrumb trail" to their destination.

Although we describe this problem in the context of an office application, Follow-Me represents a class of applications where sensors are deployed to assist navigation. Other examples include marking paths in buildings damaged by earthquake or fire and in underground exploration. The key innovation demonstrated in Follow-Me is the *deployment-order* approach to logical topology configuration.

19.4.1 Hardware

Sensor nodes in this application are button box display nodes (described in Section 19.2.2). Figure 19.11 shows a possible deployment scheme, with 85 button boxes deployed on a single floor of an office building. A multifloor deployment could consist of hundreds of nodes. Node deployment is based on two general guidelines: There should be one node at each office doorway, and the distance between two adjacent nodes should not be too large. The later means we need to place additional nodes along hallways with few doors, such that visitor can follow lights easily. A touch-screen display at the entrance allows visitors to select a destination.

We are in the process of completing deployment at The Information Science Institute (ISI). As of April 2004, our current deployment is smaller, with eight button boxes covering one long hallway at half the desired density and with two button boxes with labeled buttons substituting for the touch-screen display.

19.4.2 Protocols

The Follow-Me application must guide visitors along appropriate paths. While network routing algorithms specialize in path finding, they are not directly applicable to guiding humans, who are constrained by physical walls, and prefer to follow adjacent nodes. Traditional routing algorithms select the shortest path based on radio connectivity, choosing paths through physical walls and skipping physically intermediate nodes when possible. Even a strictly geographic routing algorithm will cut corners and pass through walls if it shortens the physical path.

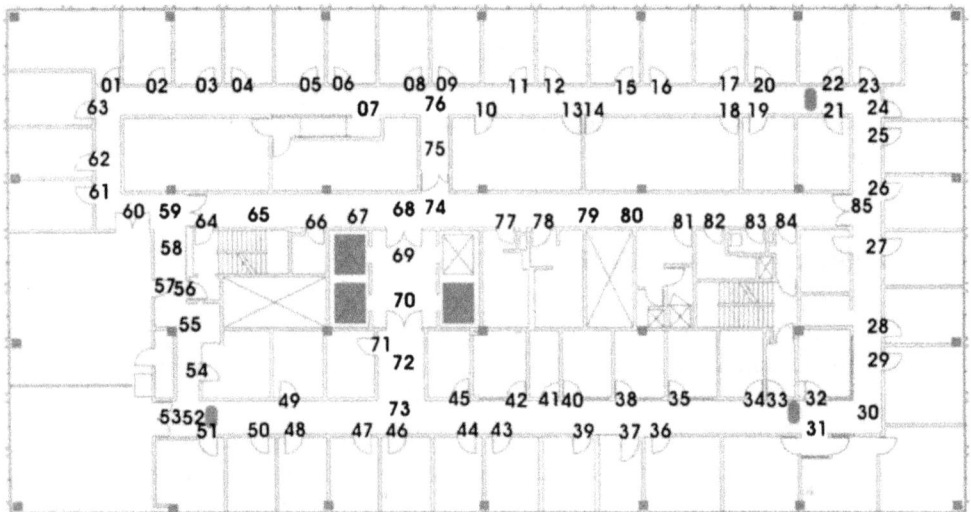

Figure 19.11   Follow-Me deployment example.

Thus, the main technical challenge in Follow-Me is determining the *logical topology* that connects nodes as a human would walk, as opposed to the radio or physical topologies. Figures 19.12 and 19.13 provide a comparison between radio and logical topologies for the same Follow-Me deployment shown in Figure 19.11.

It is not easy to capture logical topology because it is defined by human constraints such as walls and doors that are not visible to sensor nodes. In Section 19.4.2.1, we describe *deployment order*, our algorithm that captures logical topology. It is present when a network is first configured, allowing construction of complex topologies with minimal human interaction. We then consider how we can build Follow-Me by layering a simple routing algorithm over this logical topology.

19.4.2.1   Deployment Order

As described earlier, each node in the Follow-Me application needs to be configured properly with its logical location, which is a set of physical neighbors. We would like nodes to configure themselves automatically. With localization hardware, it is possible to derive logical locations from physical coordinates. However, we

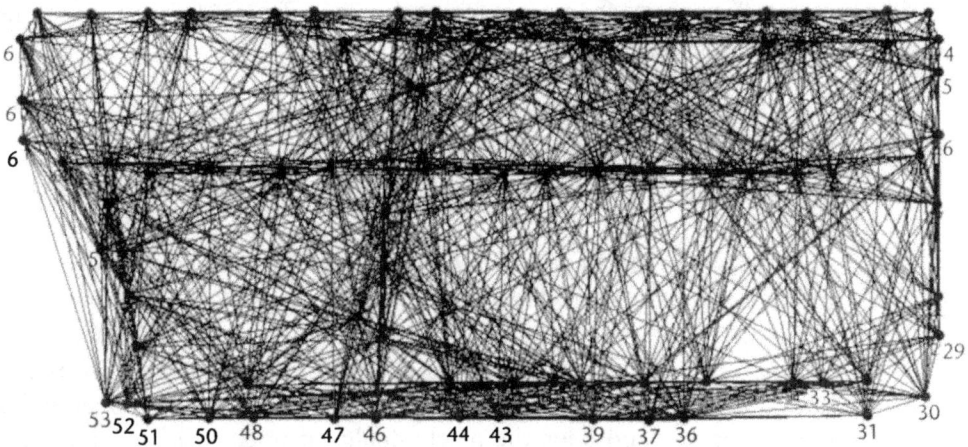

Figure 19.12   A radio connectivity graph for the deployment shown in Figure 19.11.

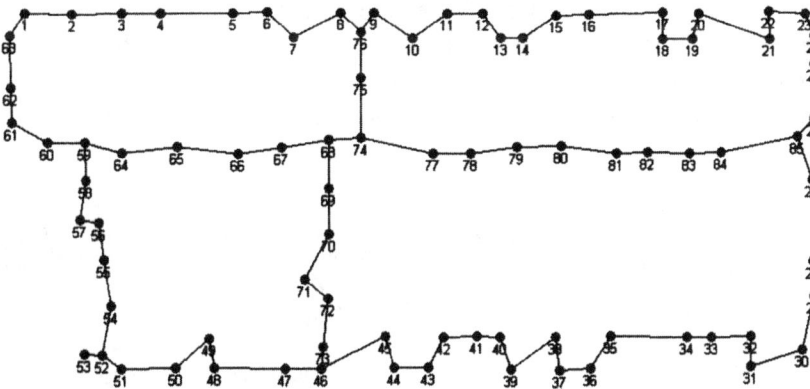

Figure 19.13   A logical topology for the deployment shown in Figure 19.11.

developed a new method for two reasons. First, the system must be easy to deploy and have low cost. This constraint requires methods that work without specialized localization infrastructure or specialized hardware. Second, the system must work well with buildinglike topologies: long, linear segments, parallel hallways, and moderate density.

We developed the deployment-order method for logical location configuration. For many instances, sensor nodes are deployed sequentially. If two nodes are deployed (switched on) one after another within a short time, we can assume they are the closest neighbors to each other. Links between these closest neighbors can create a linear path (not necessarily straight). If nodes detect and remember this path, it can be used later to guide visitors. Some other mechanisms are needed to handle "intersections." One method is to interact with sensor nodes manually to add and remove links. The algorithm on each node can be described by simple state machines. We will discuss both linear paths and intersections below.

19.4.2.2 Linear Paths

To create a linear path, a newly deployed node communicates with previously deployed nodes to determine which one was deployed immediately prior to its own deployment. We can use the state machine shown in Figure 19.14 on each node. The three states are described below:

- *Active.* This is the state after a node is switched on. Nodes in this state send out connection-request packets to look for neighbors.
- *Receptive.* Nodes in this state will reply to connection-request packets and establish links.
- *Passive.* Nodes in this state will not be involved in link operations. This is the state for normal operation.

A node is in the active state right after its deployment (switched on), and it sends out connection-request packets. Every node in the receptive state replies with a connection-reply packet and adds the newly deployed node to its neighbor list. The newly deployed node also adds all replying nodes to its neighbor list as it receives their connection-reply packets.

For example, after the first node is switched on, it won't find any neighbors and will go into the receptive state. When the second node is switched on, it will find the first node. The first and the second node will link to each other. The first node will go into the passive state, and the second node will go into the receptive state. Similarly, the third node will link with the second node, and so on.

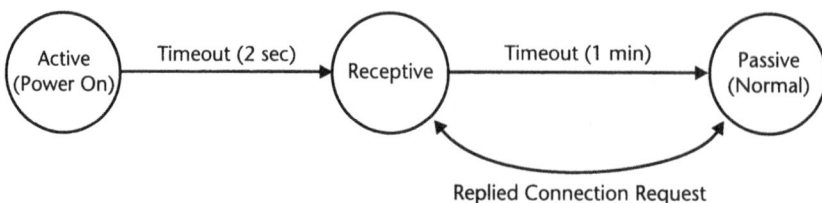

Figure 19.14 Initial state diagram of the deployment-order method.

In other words, only nodes in the receptive state will accept link-request packets and link to newly deployed nodes. A node will leave the receptive state after accepting a link-request packet. This procedure operates on the concept of *receptiveness*.

19.4.2.3   Intersection Handling

We can see that linear paths can be automatically configured while nodes are being deployed. Beyond linear paths, we need to handle cases where nodes have more than two neighbors, which we call *intersections*. Assuming the majority of links belong to linear paths, we can still use the process for linear paths most of the time, plus additional steps for intersections.

In our implementation, we use a button on each sensor node to toggle node states. When the node is in the passive state, pressing the button will bring the node into the active state. When the node is in the receptive state, pressing the button will bring the node into the passive state. The updated state diagram is shown in Figure 19.15.

With the ability to change state, we can add arbitrary connections by making one node active and another node receptive. An example of an intersection is shown in Figure 19.16. In this figure, solid circles indicate nodes in the passive state; empty circles indicate nodes in the receptive state; and shadow filled circles indicate nodes in the active state.

19.4.2.4   Reconfiguration and Maintenance

The ability to tolerate failed nodes is an important feature for almost all sensor-network applications. For the deployment-order method, one-hop node failures can be fixed by the following process.

When a node detects a failed neighbor, it will try to skip this neighbor and link to the neighbor's neighbors directly. This is done by broadcasting a link-fix packet containing the ID of the failed neighbor. Only neighbors of the failed node will respond with their own IDs. New links between the sending node and responding nodes will be established, and old links to the failed node will be removed.

During normal operation, inadvertent configuration changes can occur if someone accidentally presses buttons on button boxes. To prevent this situation, buttons need to be locked after the node-configuration process is complete. At the same time, network managers may still want to change configurations periodically for

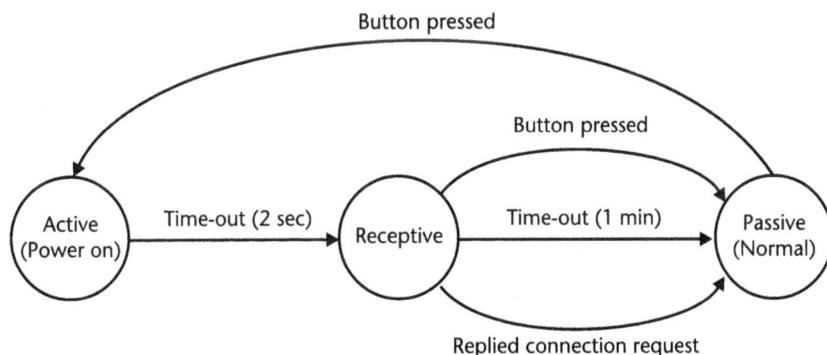

Figure 19.15 Updated state diagram of the order-deployment method.

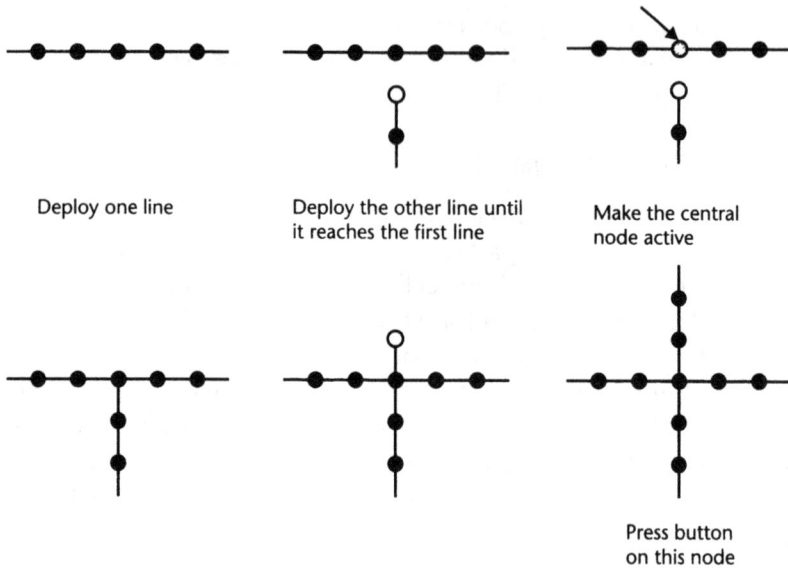

Figure 19.16   Intersection handling of the deployment-order method.

network maintenance. We designed a lock mechanism that utilizes a "key" node to fulfill these goals. More details about the full system can be found elsewhere [7].

19.4.2.5   Routing

Given the logical topology, it is relatively easy to build the Follow-Me application. We use a simple minimum-distance routing algorithm over the logical topology to determine the best path to guide a visitor between two points.

Our current implementation uses flooding to find forward paths and gradient-style routing for reverse paths. This routing combination is very similar to Directed Diffusion [8]. We could also employ DSDV-style routing algorithms as described in Section 19.3.3, provided they operate on the logical topology.

When a visitor arrives at the lobby and selects a destination from a touch screen, the network finds the path as described above, flooding and establishing previous-hop gradients. The destination node gathers routes and selects the best one based on the desired metric. Unlike Directed Diffusion or DSDV, where latency or energy is the metric of choice, we use physical distance traveled as the metric. (Our current implementation assumes all nodes are equidistant, and so, hop count is equivalent to physical distance; we are in the process of relaxing this approximation.)

While we use routing in the logical topology for Follow-Me, a general routing service is applicable to many tasks. For example, we monitor our network from a central point using this same routing algorithm. In this case, our routing algorithm uses radio connectivity rather than logical topology.

19.4.3   User Interaction

A common and effective approach to designing sensor-network applications is to keep sensor nodes simple and to rely on the collaborative behavior of the whole network to achieve complex functions. Unlike systems with a keyboard and a screen, simple devices such as LEDs and buttons are more frequently used, and sensor nodes are spatially distributed in the target environment. The user-interface part of the Follow-Me application shares the same idea.

As described earlier, a touch screen driven by a gateway node can be used for visitors to choose their destinations, and synchronized blinking patterns across the network are used to show paths to visitors. These blinking patterns should create a visual effect of moving light dots or lines, communicating both path and direction information to visitors in an intuitive way.

An interesting user-interaction problem is how to guide multiple visitors at the same time. A possible solution is to show several paths simultaneously using different colors or blinking patterns. To reduce chances of mixing paths among visitors, we can limit blinking to nodes within visible ranges of visitors, provided there is a method to sense visitors' locations.

19.5   Other Applications

In addition to these two applications, several other in-building applications are being considered at Intel, ISI, and elsewhere.

The "voting app" was developed to provide audience feedback to a speaker, without requiring installation of a wired input device at each seat in an auditorium. Audience members are given a button box (described in Section 19.2.2). Using the buttons, the audience can respond to a question from the speaker or suggest that the speaker speed up or slow down. The LEDs are used to indicate the user's current vote. In the future, other types of sensors may be employed to detect voting-box motion and orientation, allowing richer audience participation. Votes are delivered to the speaker's laptop using the same single-destination version of DSDV used in the conference-room application, which forms a data-delivery tree. While data collection could include vote aggregation at each branch of the tree, we chose to deliver each individual vote, which reduces the impact of packet loss on the outcome of a vote. At regular intervals, each node sends a packet containing the originating node's identity and vote, and each forwarding node appends its identity and vote. This application was deployed at several Intel Chief Technology Officer (CTO) keynote presentations, primarily as a demonstration of sensor-networking technology. As such, network topology was also presented, using the route taken by each packet to identify the data-delivery tree dynamically. Like the conference-room application, this application can also utilize a hierarchical topology to increase the scale and reliability of the network.

At ISI we are exploring a security application that exploits multiple classes of sensors to balance privacy and security [9]. We place a video camera in our building lobby, a public area with a security guard where visitors have little expectation of privacy. We augment this sensor with motion detectors throughout the building in hallways and at office doors. Motion detectors can sense and timestamp the presence of an individual, but it cannot tell who the individual is or capture his or her photograph; thus, there is no direct way to observe office occupants. If there has been a theft or security violation, we map the path of an individual from the site of the violation (perhaps a particular office) through a time-related series of motion-sensor detections, back to the lobby, and ideally, to a photograph of the thief. This application seeks to balance privacy and security, allowing investigation of problems, but avoids pervasive cameras and explicit search to extract information.

Labscape [10] was developed by the University of Washington and Intel Research's Seattle lab. Labscape is a smart environment that combines sensing and traditional ubiquitous computing to improve workflow in a cell-biology laboratory. Labscape's focus on creating a real-world application for its users has provided insights into design approaches, evaluation methods, and implementation challenges [10]. Labscape provides workflow automation in two phases: experiment preparation and execution. In the preparation phase, Labscape allows researchers to plan the experiment using graphical flowcharts. These flowcharts guide the execution of the experiment and are placeholders to document results. During the execution of an experiment, users may log each step and annotate it with experimental results. Data logging during the execution phase is simplified using bar code scanners and RFIDs to identify physical objects. The system also allows users to link pictures, diagrams, and hand-written notes. Unlike traditional lab work that requires separate document and record logging, Labscape seamlessly integrates the documentation step into the experiment's execution phase. From the Labscape project, researchers learned that building systems that integrate user interaction and sensing is extremely difficult; integrating sensing is easy but presenting the data to the user is much harder.

The applications that have been explored in this chapter are those that use workplace sensing to improve day-to-day activities. Additionally, WSNs can play a significant role in improving traditional building automation, control, and maintenance. Large buildings are subject to microclimates, which are dealt with today on a manual basis based on complaints from users of the building. Wireless sensing provides the opportunity for fine-grained temperature and humidity monitoring and control. Monitoring vibration from heaters, coolers, pumps, and motors, using wireless sensors, enables proactive maintenance to predict and prevent failures. Likewise, sensing temperatures inside switch boxes allows early detection of shorts and prevention of fires and failures. The combination of traditional and new applications enables a world of revolutionary uses of wireless sensors.

19.6   Reusable Tools and Techniques

The applications described in the preceding sections suggest the applicability of sensor networks to office environments. More important than these specific applications are the tools and techniques that are reusable in similar applications. Three such areas stand out: routing, leveraging existing infrastructure, and exploiting simple external interactions.

All of the applications described require multihop wireless routing. The details of the protocols vary, with DSDV or a simple diffusionlike protocol and shortest path, minimal loss, or logical topology as the primary routing metric. But despite the rich connectivity from widely available 802.11 and Ethernet in the office environment, the power and flexibility advantages of a lightweight, sensornet-specific, multihop routing protocol are important.

This observation does not imply that we should ignore existing infrastructure. In fact, our second observation is that there is a great advantage to exploiting that infrastructure where possible. Examples of this include using overlay routing to improve reliability in the conference-room application, exploiting a Web interface or handheld GUI in the conference-room application, and employing a PC-based touch screen in the Follow-Me application.

Finally, complementing this use of GUIs, we observe that very simple interactions can be used successfully. The conference-room application illustrates this observation in the contexts of sensing and user interaction: direct sensing of the flashing LED on an existing motion detector greatly simplifies the problem of detecting if a room is occupied, while the simple "busy" or "free" status of the room can be indicated by red and green lights on the LED box. Three examples in Follow-Me are the use of deployment-order information to collect logical information, the use of a simple state machine to patch together complex topologies, and the use of flashing lights plus the node's physical position to guide visitors to their destination.

19.7 Conclusion

The workplace provides a rich and challenging environment for applications of sensor networks. This chapter has described two such applications, conference-room scheduling and visitor guidance, in detail and additional applications in less detail. From these applications, common hardware and software themes are apparent. Unlike outdoor applications, hardware and software designed for workplace applications emphasize user interaction and the use of existing infrastructure.

While the specific workplace applications described above are of interest to specific users and industries, the techniques developed to implement and deploy these applications are generally useful to enable a broad class of workplace applications. Routing algorithms, overlay networks, and easy configuration are specific technologies that all make sensor networks applicable to in-building applications. A key challenge to the sensor-network community is to demonstrate that workplace applications can provide a return on investment through ease of use, ease of management, and workplace productivity.

References

[1]   Weiser, M. "The Computer for the 21st Century," *Scientific American*, Vol. 265, No. 3, September 1991, pp. 94–104.

[2]   Hill, J., et al., "System Architecture Directions for Networked Sensors," *Proc. ACM ASPLOS 2000*, Cambridge, MA, November 2000, pp. 93–104.

[3]   TASK TinyOS Toolkit, at http://berkeley.intel-research.net/task.

[4]   Conner, W. S., et al., "Experimental Evaluation of Synchronization and Topology Control for In-Building Sensor Network Applications," *Proc. Second ACM International Workshop on Wireless Sensor Networks and Applications (WSNA '03)*, San Diego, CA, September 2003, pp. 38–49.

[5]   Perkins, E. C., and Bhagwat P., "Highly Dynamic Destination-Sequenced Distance-Vector Routing (DSDV) for Mobile Computers," *Proc. Conference on Communications Architectures, Protocols and Applications*, London, U.K., August 1994, pp. 234–244.

[6] Yarvis, M. D., et al., "Real-World Experiences with an Interactive Ad Hoc Sensor Network," *Proc. International Workshop on Ad Hoc Networking (IWAHN 2002)*, Vancouver, British Columbia, Canada, August 2002, pp. 143–151.

[7] Wang, X., Silva F., and Heidemann J., "Infrastructureless Location Aware Configuration for Sensor Networks," *Proc. Sixth IEEE Workshop on Mobile Computing Systems and Applications, (WMCSA 2004)*, English Lake District, U.K., December 2004.

[8] Intanagonwiwat, C., Govindan R., and Estrin D., "Directed Diffusion: A Scalable and Robust Communication Paradigm for Sensor Networks," *Proc. Sixth Annual International Conference on Mobile Computing and Networking (MobiCOM '00)*, Boston, MA, August 2000, pp. 56–67.

[9] Rajgarhia, A., Stann F., and Heidemann J., "Privacy-Sensitive Monitoring with a Mix of IR Sensors and Cameras," Technical Report ISI-TR-582, USC, Information Sciences Institute, November 2003.

[10] Arnstein, L., et al., "Labscape: A Smart Environment for the Cell Biology Laboratory," *IEEE Pervasive Computing Magazine*, Vol. 1, No. 3, July–September 2002, pp. 13–21.

About the Editors

Nirupama Bulusu is an assistant professor of computer science at the Maseeh College of Engineering and Computer Science at Portland State University. An author of numerous highly cited publications on sensor networks, she received her Ph.D. in computer science in 2002 from the University of California at Los Angeles.

Sanjay Jha is an associate professor at the School of Computer Science and Engineering at the University of New South Wales. He is also a project leader for the Smart Internet CRC and an associate of the Australian National Information and Communication Technology Center (NICTA). Dr. Jha is the coauthor of *Engineering Internet QoS* (Artech House, 2002).

Index